Landscape Series

Volume 32

Benjamin Kofi Nyarko, Dept of Geography & Regional Planning, University of Cape Coast, Cape Coast, Ghana

Henrique Pereira, German Centre for Integrative Biodiversity Research (iDiv), Martin Luther University Halle-Wittenberg, Leipzig, Sachsen, Germany

Alexander Prishchepov, Dept of Geosci & Natural Resource Mgmt, University of Copenhagen, Copenhagen, Denmark

Robert M. Scheller, North Carolina State University, Raleigh, NC, USA

Kalev Sepp, Inst. Agricultural & Environ. Sciences, Estonian University of Life Sciences, Tartu, Estonia

Anton Shkaruba, Inst. of Agricultural & Environ Sciences, Estonian University of Life Sciences, Tartu, Estonia

Janet Silbernagel Balster, Silvernail Studio for Geodesign, LLC, Black Earth, WI, USA

Ileana Stupariu, Dept. Regional Geography and Environment, University of Bucharest, Bucharest, Romania

Raymond Tutu, College of Humanities, Delaware State University, DOVER, DE, USA

Teiji Watanabe, Fac of Environmental Earth Sci, A-301, Hokkaido University, Sapporo, Hokkaido, Japan

Wei-Ning Xiang, University of North Carolina at Charlotte, Charlotte, NC, USA

Qing Zhao, Institute of Applied Ecology, Chinese Academy of Sciences, Shenyang, China

Springer's innovative Landscape Series is committed to publishing high quality manuscripts that approach the concept of landscape and land systems from a broad range of perspectives and disciplines. Encouraging contributions that are scientifically-grounded, solutions-oriented and introduce innovative concepts, the series attracts outstanding research from the natural and social sciences, and from the humanities and the arts. It also provides a leading forum for publications from interdisciplinary and transdisciplinary teams across the globe.

The Landscape Series particularly welcomes contributions around several globally significant areas for landscape research, which are anyhow non-exclusive:

- Climate and global change impacts on landscapes and ecosystems including mitigation and adaptation strategies
- Human Dimensions of Global Change

- Biodiversity and ecosystem processes linked to ecosystems, landscapes and regions
- Biogeography
- Ecosystem and landscape services including mapping, assessment and modelling
- Land System Science
- Regional ecology (including bioregional theory & application)
- Human-Environment Interactions and Social-Ecological Systems & Frameworks (SESF) - including theories, practice and modelling

Volumes in the series can be authored or edited works, cohesively connected around these and other related topics and tied to global or regional initiatives. Ultimately, the Series aims to facilitate the application of landscape research and land system science to practice in a changing world, and to advance the contributions of landscape theory and research and land system science to the broader scholarly community.

Thomas Wohlgemuth • Anke Jentsch
Rupert Seidl
Editors

Disturbance Ecology

Editors
Thomas Wohlgemuth (iD)
WSL Swiss Federal Research Institute
Birmensdorf, Zürich, Switzerland

Anke Jentsch (iD)
Universität Bayreuth
Bayreuth, Bayern, Germany

Rupert Seidl (iD)
Technical University of Munich
Freising, Bayern, Germany

ISSN 1572-7742   ISSN 1875-1210   (electronic)
Landscape Series
ISBN 978-3-030-98758-9   ISBN 978-3-030-98756-5   (eBook)
https://doi.org/10.1007/978-3-030-98756-5

This book is a translation of the original the German language edition "Störungsökologie" by Thomas Wohlgemuth et al., published by Haupt Verlag AG in © Haupt Bern 2019. All Rights Reserved.

This Springer imprint is published by the registered company Springer Nature Switzerland AG
The registered company address is: Gewerbestrasse 11, 6330 Cham, Switzerland

Contents

About the Editors

Thomas Wohlgemuth is head of the research unit Forest Dynamics at the Swiss Federal Institute for Forest, Snow and Landscape Research WSL in Birmensdorf (Switzerland). He investigates reforestation after windthrow, forest fire, and drought-induced mortality, as well as early growth of tree species under experimental conditions. His focus is on the interaction between disturbance and climate change, and the effects of this interaction on forest communities.

Anke Jentsch is Professor of Disturbance Ecology and Vegetation Dynamics at the University of Bayreuth (Germany). Her research covers, in particular, disturbance and biodiversity, natural hazards, and climate change. Her scientific interests focus on the understanding of ecosystem dynamics and community resilience. Her work includes extensive field experiments in Central Europe on the effects of weather extremes on ecosystem functions.

Rupert Seidl is Professor of Ecosystem Dynamics and Forest Management at the Technical University of Munich (TUM) in Germany. He works towards a better understanding of forest change and harnesses these insights to improve ecosystem management. Methodologically, he combines dynamic simulation models with remote sensing data and field observations to quantify past and potential future changes in forest structure, composition, and function.

List of Textboxes

Chapter 1
Disturbance Ecology. A Guideline

Thomas Wohlgemuth ⓘ, Anke Jentsch ⓘ, and Rupert Seidl ⓘ

Abstract The guideline contextualizes the field of disturbance ecology over time and space. It addresses fundamental elements of disturbance ecology, such as drivers of disturbances, adaptations of plants to disturbance, disturbance effects on biodiversity, disturbance resilience, disturbances in the context of ecosystem management, and climate change impacts on disturbance regimes. The chapter introduces the structure of the book and highlights the audience for which the book was written.

Keywords Biodiversity · Climate change · Conservation · Disturbance regimes · Ecosystems · Land use · Landscape ecology · Management

T. Wohlgemuth (✉)
Forest Dynamics Research Unit, Swiss Federal Institute for Forest, Snow and Landscape Research WSL, Birmensdorf, Switzerland
e-mail: thomas.wohlgemuth@wsl.ch

A. Jentsch
Bayreuth Center of Ecology and Environmental Research (BayCEER), University of Bayreuth, Bayreuth, Germany

R. Seidl
Ecosystem Dynamics and Forest Management Group, School of Life Sciences, Technical University of Munich, Freising, Germany

Berchtesgaden National Park, Berchtesgaden, Germany

© The Author(s), under exclusive license to Springer Nature Switzerland AG 2022
T. Wohlgemuth et al. (eds.), *Disturbance Ecology*, Landscape Series 32, https://doi.org/10.1007/978-3-030-98756-5_1

1.1   Disturbances in Community Ecology
and Ecosystem Dynamics

Disturbance ecology originated from vegetation science and quantitative landscape ecology, both fields that have a rich history. Vegetation science and its broader relative, geobotany (broadly, the biogeographical study of plants),[1] have dealt with the systematic sequence or succession of plant communities (Kratochwil and Schwabe 2001), creating overviews and characterizing vegetation units (e.g. Braun-Blanquet 1964; Ellenberg 1996). For a long time, the focus of research was on the ecological equilibrium of individual plant communities in given locations, especially in Central Europe. Today, many disciplines build on this past work in order to investigate the reactions of ecosystems and especially of plant communities to environmental change. The comprehensive study of ecosystems – from individuals to plant communities and vegetation landscapes – has generated new insights: that plant communities are dynamic systems, that they vary greatly in space (Sousa 1984), and that recurring patterns of temporal sequences occur during succession (Clements 1916; Watt 1947; Gurevich et al. 2006; Walker and Wardle 2014). Signs of previous interruptions of vegetation succession, such as fire scars or standing deadwood, indicate transient resource changes (Davis et al. 2000) and the alteration of undisturbed successional patterns. Thus, these signs provide evidence of a deviation from the ecological equilibrium – that is, they indicate a perturbation/disturbance (Odum et al. 1979).

   In North America, the classical and static view of vegetation types was abandoned in the 1970s and 1980s, with a dynamic perspective of plant communities in the landscape – including disturbances – taking hold (Chap. 2). Synthesis work by Levin and Paine (1974), White (1979), and White and Pickett (1985) on the importance of disturbances for vegetation dynamics set the foundation for this new, dynamic perspective of vegetation. Also in the 1970s, groundbreaking studies on the interaction of species richness and disturbance were published (Grime 1973; Connell 1978). In the following decades, disturbance was a major focus of ecological research in the English-speaking world and was increasingly reflected also among vegetation ecologists across all continents (Fig. 1.1).

   How did the field of 'disturbance ecology' develop in Central Europe? Impacts of disturbance are already implied in phytosociology (e.g. Braun-Blanquet 1964; Ellenberg 1996), a school of thought that was dominant in Europe for much of the second half of the twentieth century. In this approach, traces of disturbance are detectable in units and specific terms, for example, the interruption of succession leading to 'secondary succession', and the 'permanent communities' that result as a consequence of repeated disturbances. Nonetheless, these aspects have rarely been

---

[1] In the past, there has been some difference in the terms used between the European literature and the Anglo-American literature. In Europe, floristic geobotany was the term often used for the branch of biogeography concerned with the distribution of plant species. The equivalent term in the Anglo-American literature was phytogeography (Mueller-Dombois and Ellenberg 1974).

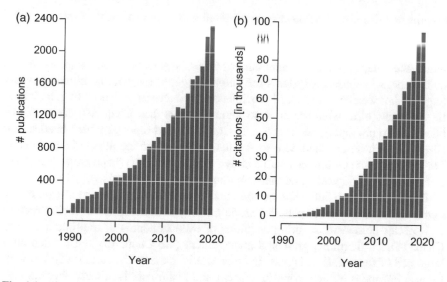

Fig. 1.1 Number of scientific publications per year containing the search terms 'disturb*' and 'ecol*', from 1990 to 2020 (**a**), and the annual number of citations to these publications (**b**). (Data retrieved from the Web of Science (WOS) on 17.03.2021)

the subject of in-depth studies in Europe as the main interest of research has been focused on plant communities that are 'stable' in their species composition and are thus in an equilibrium. In Europe, the broad scientific discussion of disturbances began much later than in the 'Anglo-American' world. This may be because in the 1970s phytosociology was still a dominant paradigm in many places, and its findings were increasingly applied in the field of nature conservation. The terms *perturbation* and *disturbance* (White 1979) were not yet established in Europe in the 1990s, and the German-language equivalent *Störung*, which often has negative connotations, was hardly used, for example, in forestry. As any kind of management intervention can be understood as an abrupt and spatially discrete event of tree mortality (see Chap. 14), the concept behind the word even generated strong opposition in some communities, which persist to this day. However, the term *disturbance* has increasingly been established in ecosystem research also in Central Europe (White and Jentsch 2001). Its significance has increased considerably over the past decades, both because of an increased relevance of disturbance processes and also because of substantial advances in research. This is also reflected in a widespread change in conservation paradigms towards protecting natural processes rather than individual species or communities (e.g. Scherzinger 1996; Müller 2015).

Disturbances occur in all plant communities (see Chap. 2) and contribute significantly to the spatial and temporal heterogeneity of ecosystems that enables many species to coexist (see Chap. 4). For several decades research has been conducted on the stability of ecosystems, which led to the concept of ecological resilience, that is, the ability of disturbed ecosystems to retain their functions and structures in the face of disturbance (see Chap. 5). In response to regularly occurring disturbances, organisms have also developed physiological and morphological adaptations (see Chap. 6).

Forests in Europe are exposed to abiotic disturbances such as fire (see Chap. 7), wind (see Chap. 8), and, in mountainous areas, avalanches (see Chap. 9). Biotic disturbances in these ecosystems are often triggered by extreme climatic events (e.g. dry-hot weather) or succeed previous disturbances (interaction) (e.g. bark beetle outbreaks after windthrow). Also plant diseases (see Chap. 10) cause disturbance, as do phytophagous insects (see Chap. 11) and especially bark beetles (see Chap. 12). These all contribute to the heterogeneity of the affected ecosystems. Large herbivores (i.e. animals with a bodyweight of ≥ 50 kg) shape the vegetation to different extents depending on their population size (see Chap. 13).

By far the largest influence on the current vegetation in Europe is exerted by humans. Practically all forests in Europe have been transformed in their structure and species composition during centuries or even millennia of multiple uses (see Chap. 14). Furthermore, grassland management (see Chap. 15) is exercised on a large part of the unforested areas. In other words, most of the current vegetation in Central Europe is a direct result of current and historical disturbance regimes (see Chap. 3), made up of both natural and anthropogenic disturbance agents.

1.2 Modern Disturbance Research

Abrupt changes in environmental conditions have developed into a key object of investigation in ecology since the 1980s (Johnson and Miyanishi 2007). An important approach in dynamic vegetation modelling uses disturbance-caused gaps (Churchill and Hansen 1958; Forcier 1975; Glenn-Lewin and van der Maarel 1992) as a starting point for forest development (Kienast and Kuhn 1989; Bugmann 1996). Newer model types increasingly integrate disturbances as an important element of system dynamics (Seidl et al. 2011; see Box 17.1, Chap. 17). Biodiversity research has identified numerous findings on the influence of disturbances on species communities (e.g. Thom and Seidl 2016; Thorn et al. 2018; see Chap. 4). A broad and important topic is the role of disturbances in the long-term maintenance of ecosystem heterogeneity, and thus the importance of disturbance for the conservation of biodiversity (Kulakowski et al. 2017). In turn, high levels of biodiversity increase the resistance to disturbances, especially to extreme climatic events such as prolonged drought as well as early and late frost (Isbell et al. 2015; see Chap. 6). This connects to topics of functional diversity (e.g. Hector et al. 2010; Ratcliffe et al. 2017), which provide the basis for experimental disturbance ecology, investigating, for example, the influence of weather extremes and land-use changes on ecosystem functions (Jentsch et al. 2007).

1.3 Disturbance Regimes Are Changing

The biomes of the world are characterized by typical disturbance regimes, shaped by both the prevailing climate and land uses (see Chap. 2). These regimes are currently undergoing significant change. Climate change is expected to shift disturbance regimes and extreme events (IPCC 2021), for example, with changes in the seasonality of precipitation, an increase of heavy rainfall events, and higher frequencies of strong winds expected for the future. Also, higher temperatures generally lead to a greater risk of wildfires and an increasing insect activity (see Chap. 16). Along with the changing climate, land use is also changing in manifold ways as societies adapt to the new environmental conditions (Shukla et al. 2020). Furthermore, global markets and political decisions trigger land-use changes and thus changes in the disturbance regimes, for example, in Europe (e.g. the bioeconomy strategy of the EU; European Commission 2017).

Changing disturbances in space and time also pose risks for ecosystem management (see Chap. 17). Compared with the situation decades or centuries ago, the current demands for ecosystem services by society are quite diverse, which is mirrored by the growing public awareness since the Rio Convention on Biological Diversity (UN 1992). Quantifying these ecosystems services (Hassan et al. 2005; see Chap. 18) and assessing how changing disturbance regimes affect or constrain these services is an important task for research.

1.4 Who Is This Book For?

The 18 chapters of this textbook provide a broad overview of disturbance-related research results from various disciplines, with a focus on Central Europe. A total of 31 experts from universities and research institutions in predominantly German-speaking countries contributed to this textbook and ensure that the broad and diverse aspects of disturbance ecology are addressed. A previous version of this book was published in German in 2019 and was the first textbook on the subject of disturbance ecology in German. Numerous requests for an English translation demonstrated the broad interest in the book from colleagues in other parts of Europe and the world. This English version is addressed to students, researchers, and practitioners from around the world who are interested in a Central European perspective on disturbance ecology. The current book is a direct translation of the German version and not a second edition; nevertheless, some references have been updated and minor corrections were made in the course of the translation. We hope that this book is not only of interest to colleagues working on issues of disturbance ecology, but that it is also able to excite colleagues that are new to the field of disturbance ecology. Ultimately, we want to share our fascination for disturbance dynamics and hope to inspire readers to dive into the mesmerizing world of disturbances in nature.

References

Braun-Blanquet J (1964) Pflanzensoziologie. Grundzüge der Vegetationskunde, 3rd edn, Wien/ New York, 865 p

Bugmann H (1996) A simplified forest model to study species composition along climate gradients. Ecology 77:2055–2074

Churchill ED, Hansen HC (1958) The concept of climax in arctic and alpine vegetation. Bot Rev 24:127–191

Clements FE (1916) Plant succession: an analysis of the development of vegetation. Carnegie Institution of Washington, Washington, DC, 515 p

Connell JH (1978) Diversity in tropical rain forests and coral reefs: high diversity of trees and corals is maintained only in a non-equilibrium state. Science 199:1302–1310

Davis MA, Grime JP, Thompson K (2000) Fluctuating resources in plant communities: a general theory of invasibility. J Ecol 88:528–534

Ellenberg H (1996) Vegetation Mitteleuropas mit den Alpen in ökologischer, dynamischer und historischer Sicht (5). Ulmer, Stuttgart, 1095 p

European Commission (2017) Review of the 2012 European bioeconomy strategy. European Commission, Directorate-General for Research and Innovation: Directorate F Bioeconomy, Brussels, 84 p

Forcier LK (1975) Reproductive strategies and co-occurrences of climax tree species. Science 189:808–810

Glenn-Lewin DC, van der Maarel E (1992) Pattern and processes of vegetation dynamics. In: Glenn-Lewin DC, Peet RK, Veblen TT (eds) Plant succession. Chapman and Hall, London, pp 11–59

Grime JP (1973) Competitive exclusion in herbaceous vegetation. Nature 242:344–347

Gurevich J, Scheiner SM, Fox GA (2006) The ecology of plants. Sinauer, Sunderland, 574 p

Hassan R, Scholes R, Ash N (2005) Millennium ecosystem assessment. In: Ecosystems and human well-being: current state and trends, vol 1. Island Press, Washington, DC, 917 p

Hector A, Hautier Y, Saner P, Wacker L, Bagchi R, Joshi J, Scherer-Lorenzen M, Spehn EM, Bazeley-White E, Weilenmann M, Caldeira MC, Dimitrakopoulos PG, Finn JA, Huss-Danell K, Jumpponen A, Mulder CPH, Palmborg C, Pereira JS, Siamantziouras ASD, Terry AC, Troumbis AY, Schmid B, Loreau M (2010) General stabilizing effects of plant diversity on grassland productivity through population asynchrony and overyielding. Ecology 91:2213–2220

IPCC (2021) Climate Change 2021: the physical science basis. Contribution of Working Group I to the Sixth Assessment Report of the Intergovernmental Panel on Climate Change [Masson-Delmotte V, Zhai P, Pirani A, Connors SL, Péan C, Berger S, Caud N, Chen Y, Goldfarb L, Gomis MI, Huang M, Leitzell K, Lonnoy E, Matthews JBR, Maycock TK, Waterfield T, Yelekçi O, Yu R, Zhou B (eds)]. Cambridge University Press, Cambridge, UK (in press)

Isbell F, Craven D, Connolly J, Loreau M, Schmid B, Beierkuhnlein C, Bezemer TM, Bonin C, Bruelheide H, de Luca E, Ebeling A, Griffin JN, Guo Q, Hautier Y, Hector A, Jentsch A, Kreyling J, Lanta V, Manning P, Meyer ST, Mori AS, Naeem S, Niklaus PA, Polley HW, Reich PB, Roscher C, Seabloom EW, Smith MD, Thakur MP, Tilman D, Tracy BF, van der Putten WH, van Ruijven J, Weigelt A, Weisser WW, Wilsey B, Eisenhauer N (2015) Biodiversity increases the resistance of ecosystem productivity to climate extremes. Nature 526:574–577

Jentsch A, Kreyling J, Beierkuhnlein C (2007) A new generation of climate-change experiments: events, not trends. Front Ecol Environ 5:365–374

Johnson EA, Miyanishi K (2007) Plant disturbance ecology: the process and the response. Elsevier, Amsterdam, 698 p

Kienast F, Kuhn N (1989) Simulating forest succession along ecological gradients in southern Central Europe. Vegetatio 79:7–20

Kratochwil A, Schwabe A (2001) Ökologie der Lebensgemeinschaften: Biozönologie. Eugen Ulmer, Stuttgart, 756 p

Kulakowski D, Seidl R, Holeksa J, Kuuluvainen T, Nagel TA, Panayotov M, Svoboda M, Thorn S, ~~Kuuluvainen O, Whitlock C, Wohlgemuth T, Bebi P~~ (2017) ~~A walk on the wild side: disturbance dynamics and the conservation and management of European mountain forest ecosystems.~~ For Ecol Manag 388:120–131

Levin SA, Paine RT (1974) Disturbance, patch formation and community structure. Proc Natl Acad Sci USA 71:2744–2747

Mueller-Dombois D, Ellenberg H (1974) Aims and methods of vegetation ecology. Wiley, New York, 547 p

Müller J (2015) Prozessschutz und Biodiversität: Überraschungen und Lehren aus dem Bayerischen Wald. Natur Landsch 90:421–425

Odum EP, Finn JT, Franz EH (1979) Perturbation theory and the subsidy-stress gradient. Bioscience 29:349–352

Ratcliffe S, Wirth C, Jucker T, van der Plas F, Scherer-Lorenzen M, Verheyen K et al (2017) Biodiversity and ecosystem functioning relations in European forests depend on environmental context. Ecol Lett 20:1414–1426

Scherzinger W (1996) Naturschutz im Wald: Qualitätsziele einer dynamischen Waldentwicklung. Eugen Ulmer, Stuttgart, 448 p

Seidl R, Fernandes PM, Fonseca TF, Gillet F, Jönsson AM, Merganicová K, Netherer S, Arpaci A, Bontemps JD, Bugmann H, González-Olabarria JR, Lasch P, Meredieu C, Moreira F, Schelhaas MJ, Mohren F (2011) Modelling natural disturbances in forest ecosystems: a review. Ecol Model 222:903–924

Shukla PR, Skea J, Calvo Buendia E, Masson-Delmotte V, Pörtner H-O, Roberts DC, Zhai P, Slade R, Connors S, van Diemen R, Ferrat M, Haughey E, Luz S, Neogi S, Pathak M, Petzold J, Portugal Pereira J, Vyas P, Huntley E, Kissick K, Belkacemi M, Malley J (2020) Climate Change and Land: an IPCC special report on climate change, desertification, land degradation, sustainable land management, food security, and greenhouse gas fluxes in terrestrial ecosystems – summary for policymakers (revised version). International Panel on Climate Change IPCC, 36 p

Sousa WP (1984) The role of disturbance in natural communities. Annu Rev Ecol Syst 15:353–391

Thom D, Seidl R (2016) Natural disturbance impacts on ecosystem services and biodiversity in temperate and boreal forests. Biol Rev 91:760–781

Thorn S, Bässler C, Brandl R, Burton PJ, Cahall R, Campbell JL, Castro J, Choi CY, Cobb T, Donato DC, Durska E, Fontaine JB, Gautier S, Hebert C, Hothorn T, Hutto RL, Lee EJ, Leverkus A, Lindenmayer D, Obrist MK, Rost J, Seibold S, Seidl R, Thom D, Waldron K, Wermelinger B, Winter B, Zmihorski M, Müller J (2018) Impacts of salvage logging on biodiversity: a meta-analysis. J Appl Ecol 55:279–289

UN (1992) Convention on biological diversity. Rio de Janeiro, 30 p

Walker LR, Wardle DA (2014) Plant succession as an integrator of contrasting ecological time scales. Trends Ecol Evol 29:504–510

Watt AS (1947) Pattern and process in the plant community. J Ecol 35:1–22

White PS (1979) Pattern, process, and natural disturbance in vegetation. Bot Rev 45:229–299

White PS, Jentsch A (2001) The search for generality in studies of disturbance and ecosystem dynamics. Prog Bot 62:399–449

White PS, Pickett STA (1985) Natural disturbance and patch dynamics: an introduction. In: Pickett STA, White PS (eds) The ecology of natural disturbance and patch dynamics. Academic, New York, pp 3–13

Part I
Definitions and Assessments

Chapter 2
Disturbances and Disturbance Regimes

Anke Jentsch ⓘ, **Rupert Seidl** ⓘ, **and Thomas Wohlgemuth** ⓘ

Abstract Disturbance ecology is a field of ecology, which includes vegetation ecology, ecosystem dynamics, and biogeochemistry of nutrient cycles. Here, contents and topics of disturbance ecology are presented, as well as definitions of disturbance events and disturbance regimes, descriptors of disturbance regimes, and methods for the quantitative characterization of disturbance species in landscape elements of Central Europe. Elements of ecosystem dynamics such as disturbance interactions, disturbance cycles, and disturbance cascades are introduced. Important scales of disturbance ecology such as frequency and magnitude of disturbance events are discussed, successional processes depending on disturbances are classified, and the model of dynamic equilibrium in landscape ecology is explained. The theories of niche differentiation, the Intermediate Disturbance Hypothesis, and the role of disturbances for biodiversity and productivity are briefly introduced in this chapter and discussed in more detail in the following chapters of this book.

Keywords Disturbance event · Disturbance regime · Patch scale · Landscape · Succession · Equilibrium

A. Jentsch (✉)
Bayreuth Center of Ecology and Environmental Research (BayCEER), University of Bayreuth, Bayreuth, Germany
e-mail: anke.jentsch@uni-bayreuth.de

R. Seidl
Ecosystem Dynamics and Forest Management Group, School of Life Sciences, Technical University of Munich, Freising, Germany

Berchtesgaden National Park, Berchtesgaden, Germany

T. Wohlgemuth
Forest Dynamics Research Unit, Swiss Federal Institute for Forest, Snow and Landscape Research WSL, Birmensdorf, Switzerland

T. Wohlgemuth et al. (eds.), *Disturbance Ecology*, Landscape Series 32,
https://doi.org/10.1007/978-3-030-98756-5_2

2.1 Contents and Subjects of Disturbance Ecology

Disturbance ecology deals with events in space and time, their cycles, and their ecological effects (White and Jentsch 2001; Jentsch and White 2019). Disturbances in ecosystems are often accompanied by the loss or transformation of living biomass. However, the stability and diversity of ecosystems and their functions are also among the major topics of disturbance ecology. Accordingly, one of the most important findings in disturbance ecology is that disturbances promote biodiversity and a dynamic equilibrium (see Chap. 4). All ecosystems are shaped by their disturbance regimes and are preserved by them. Therefore, disturbance ecology is concerned with resilience and functional stability (see Chap. 5), as well as with extreme events and abrupt changes in landscapes (Turner et al. 2020).

Disturbances are mostly analysed on the scale of ecosystems and communities, and the analysis is often based on the vegetation and the contribution of the disturbances to biodiversity, ecosystem dynamics, and ecosystem services (see Chap. 18). All over the world, ecosystems develop differently according to local site characteristics, which are determined by abiotic factors (e.g. water, heat, light, nutrients, and salts), mechanical factors (e.g. wind, fire, snow, herbivory, and mowing), and biotic factors (e.g. species pool, biotic interactions, and trophic networks) (see Chap. 2). The consistency of these factors leads to characteristic species compositions and communities, which are best adapted to the local conditions. These can include closed primary forests, which are not influenced by anthropogenic use (Fig. 2.1), as well as semi-dry grasslands, which are created by grazing with livestock.

Fig. 2.1 Disturbance as a discrete event: windthrow in one of the last remaining primary forests in Europe near Uppsala, Sweden. There are a few primary forests left in Europe; most have been replaced by commercial forests or near-natural forest protection areas with different landscape conservation and management objectives. (Photo: A. Jentsch)

Over longer periods of time, the vegetation in turn also influences the static and dynamic site characteristics, especially soil formation processes and microclimate, but also the disturbance regimes. For example, a vegetation cover that is low all year round reduces erosion processes or reduces the infiltration of rainwater into the soil. The vegetation structure influences the weather and on larger spatial scales also the climate of entire regions. Similarly, the invasion of individual plant and animal species, which act as ecosystem engineers, can lead to the change of a disturbance type or even a disturbance regime (see Box 18.1 in Chap. 18). Vegetation, climate, and disturbances are thus in a dynamic exchange with each other and influence each other.

In Europe, typical natural disturbances include windthrow, fire, flooding (Fig. 2.2), outbreaks of insect pests, snow break, late frost, heavy rain, droughts, and periods of hot weather. Typical anthropogenic disturbance regimes of cultural landscapes include the various forms of open land use such as mowing and grazing, logging in forests, or the control of floodplain dynamics.

How can disturbance events be defined? How can the temporal dynamics of landscapes be quantified when these events rapidly and fundamentally change ecosystems, communities, and populations in the short term, but at the same time shape and maintain the character of the landscape in the long term?

Fig. 2.2 Flooding regimes in floodplain landscapes are accompanied by different precipitation rhythms and magnitudes. The 'flood of the century' on the Elbe in 2003 changed the normal floodplain dynamics, the landscape patterns, and the possibilities of infrastructure and grassland use. (Photo: UFZ)

2.2   Definition of Disturbance Events
and Disturbance Regimes

Disturbances are defined both relatively and absolutely (White and Jentsch 2001). A disturbance is described in *relative* terms as an event that represents a deviation from the normal dynamics of an ecosystem. In classical ecological literature, in addition to the term 'deviation' (Odum et al. 1979), the term 'perturbation' (Pickett and White 1985) has also been used. In this sense, disturbances are often understood as those events that change the characteristic processes of the ecosystem, for example, the prevention of fires in the grasslands and forests of the Mediterranean area or the absence of grazing in semi-dry grasslands. Other events that are characteristic of the ecosystem, such as fires in boreal forests, snow pressure in alpine megaphorbs during winter, or the falling of a tree in an old forest, are not considered as disturbances according to the relative definition.

However, the relative definition of disturbances is problematic for three reasons. First, it is based on the assumption that we know the 'normal' dynamics of ecosystems and can therefore distinguish between stability (i.e. the persistence of dynamic patterns) and discontinuity (i.e. fluctuation within defined limits). Second, even if the first assumption is true, the statistical frequency of disturbances in space and time with respect to a defined reference period would need to be known. Third, a static system must be assumed in which the environmental factors remain the same – temporarily or even in the long term. However, in view of the rapid global change with land-use changes, nitrogen deposition, climate change, and invasion processes, this perspective, which still prevailed in the second half of the twentieth century, has become as invalid as the concept of static, potential natural vegetation (PNV; Tüxen 1956), according to which ecosystems without disturbances ideally develop into characteristic climax communities (Chiarucci et al. 2010). However, recent studies suggest that plant communities behave stochastically after disturbances such as weather extremes and that very different successional patterns with different species compositions can occur (Kreyling et al. 2011). If disturbances are defined relatively, a comparison of disturbance and regeneration dynamics between ecosystems is no longer possible.

In *absolute* terms, a disturbance is a measurable, abrupt change of ecosystem state variables, no matter whether these changes occur periodically or only episodically, and whether they are predictable or happen unexpectedly (Jentsch and White 2019). All fire events in Mediterranean forests are therefore disturbances, regardless of their intensity or frequency. Here, the direct consequence of a disturbance is the loss of biomass (Grime 1979) or the rapidly changing availability of resources (White and Pickett 1985; Davis et al. 2000; Jentsch and White 2019). In the absolute definition, disturbance is defined by a common currency, for example, biomass reduction (Grime 1979), resource input (e.g. Sousa 1984; Tilman 1985), or species dominance reduction (e.g. Wohlgemuth et al. 2002). The effects of the disturbance can be measured, for example, on the vegetation structure. This absolute, mechanistic definition of disturbance requires no further specification of attributes, such as

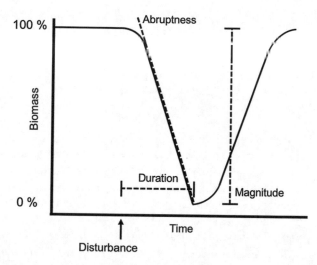

Fig. 2.3 Three criteria for the definition of a disturbance event: (1) discrete beginning and duration (abruptness), (2) short duration relative to the lifespan of the dominant organisms or ecosystems, and (3) strength/magnitude as a proportionate change in a measurement variable, such as biomass. (White and Jentsch 2001, redrawn)

the distinction between anthropogenic and natural causes. Accordingly, a disturbance is a discrete event in time that causes quantitative and qualitative changes in a community and influences resource availability or the physical environment (Pickett and White 1985). This is usually accompanied by the loss of biomass.

For this textbook 'Disturbance Ecology' we suggest the following definitions:

2.2.1 Disturbance Event

Disturbances are temporally and spatially discrete events that lead to the loss of living biomass and change the availability of resources in biotic communities.

The characteristics of a disturbance event are an abrupt beginning, duration, and intensity (Fig. 2.3). The duration of a discrete disturbance is short relative to the lifespan, reproduction rate, growth rate, or succession rate of a species or an ecosystem. The abruptness of a disturbance event is defined by its intensity and duration. Intensity can be measured as the proportionate change of a parameter, for example, the loss of biomass. Processes that act continuously and without direct influence on the ecosystem structure are referred to as 'stresses' rather than disturbances. Processes that act over long periods of time, have no event character, and/or do not result in biomass loss are referred to as 'environmental variability' rather than disturbances (Box 2.1).

Box 2.1: Quantification of Disturbance Regimes in Dynamic Landscapes

Anke Jentsch ⓘ
Bayreuth Center of Ecology and Environmental Research (BayCEER),
University of Bayreuth, Bayreuth, Germany
e-mail: anke.jentsch@uni-bayreuth.de

In practice, disturbance ecologists face a variety of challenges with regard to understanding the diversity of disturbance types and environmental factors. In addition to *abiotic factors* such as climate (e.g. precipitation, temperature, radiation, seasonality, microclimate, extreme events), relief (e.g. slope direction and inclination, relief position), soil (soil type, and availability of nutrients and water) and biotic factors (competition, herbivory, mycorrhization, soil fauna, and microbiome), and the interactions between the abiotic and biotic factors, there are various *natural* and *anthropogenic disturbances* (e.g. flooding, fire, mowing, grazing, soil compaction) and the interactions between different disturbances that jointly affect species composition, structure, and population dynamics of ecosystems. Together, all disturbance types determine the local disturbance regime.

In order to assess the influence of disturbance regime on plant community composition of a given patch, site, or larger landscape unit, information about the prevailing land-use regime is necessary. Other factors that also play an important role in determining plant species occurrences, in addition to the type of disturbance/management intervention (e.g. mowing), include the frequency (e.g. the number of cuts per year), intensity (e.g. mowing height), and timing (e.g. spring or late summer) of the disturbance/intervention.

Likewise, in addition to anthropogenic land use, natural disturbances (e.g. erosion, flooding, extreme drought) have an effect on community composition and need to be quantified. Accordingly, a determination key for landscape dynamics and disturbance regimes can be applied (Box Table 1; see Buhk et al. 2007), which records both natural and anthropogenic disturbances and categorizes them in quantitative and qualitative terms. The disturbance regime of a landscape is the totality of all occurring disturbance types together with their spatial and temporal characteristics (Box Fig. 1).

(continued)

Box 2.1 (continued)

Box Table 1 Classification of disturbance agents in landscape elements of Central Europe

A. Characteristic (natural and anthropogenic) landscape elements in Europe

(1) Coniferous forest, (2) Broadleaved forest, (3) Mixed forest, (4) Meadow, (5) Pasture, (6) Field, (7) Fallow land, (8) Running water, (9) Standing water body, (10) Track/road, (11) Settlement, (12) Hedge, (13) Stonewall/scree, (14) Spring

Such landscape elements can be, for instance, characterized in more detail in the frame of nationwide habitat classifications

B. Characteristic (natural and anthropogenic) disturbance agents in Europe

(1) Mowing, (2) Grazing, (3) Trampling, (4) Logging, (5) Windthrow, (6) Tillage/bioturbation (7) Flooding, (8) Pond drainage, (9) Soil compaction, (10) Fire, (11) Use of pesticides, (12) Nutrient input

C. Classification of the disturbance regime according to landscape characteristics (compare Buhk et al. 2007)

Temporal attributes		Spatial attributes	
Frequency	Centuries	Size	Point, linear
	Decades		25% of the area
	Annual		50% of the area
	2× per year		75% of the area
	3× per year		Full area
	>3× per year	Form	Linear
	Steady in time		Laminar
	Permanently intensive		Point
		Distribution	Homogeneous
			Heterogeneous
Seasonality	1. quarter	Selectivity	None
	1. and 4. quarter		Age
	1.–3. quarter		Species
	2.–3. quarter		Location
	2.–4. quarter		Corridors and borders of land parcels (e.g. fields/stands/forests)
	3. quarter		
	3.–4. quarter		
	4. quarter		
Duration	<1 day		
	<1 week		
	<1 month		
	<1 year		
	≥1 year		

(continued)

Box 2.1 (continued)

Box Fig. 1 Anthropogenic disturbance regimes represent the dominant land use in European cultural landscapes: e.g. the coexistence of forest use, arable farming, viticulture, and pasture use in Italy. (Photo: A. Jentsch)

Reference

Buhk C, Retzer V, Beierkuhnlein C, Jentsch A (2007) Predicting plant species richness and vegetation patterns in cultural landscapes using disturbance parameters. Agr Ecosyst Environ 122:446–452

2.2.2 Disturbance Regime

The disturbance regime describes the temporal and spatial dynamics of all disturbances affecting a landscape and their interactions over a longer period of time.

In contrast to single disturbance events, the disturbance regime of a landscape describes the temporal and spatial dynamics of all disturbances as well as their interactions over a longer period of observation (Turner 2010; Burton et al. 2020). The elements of a disturbance regime are the type of disturbances, spatial and temporal characteristics, magnitude (intensity), specificity, and interactions with other disturbances (Sousa 1984; White and Pickett 1985; Moloney and Levin 1996; White and Jentsch 2004; Burton et al. 2020). Spatial characteristics include the area, shape, and spatial distribution of disturbances. Temporal characteristics include the duration, frequency, seasonality, and return interval of disturbances. The magnitude of a disturbance is determined by the intensity of the event and the extent of the impact on the ecosystem. Specificity describes the sensitivity of disturbances with respect to species, size classes, parcels, or successional stage. Synergisms include the interactions between different types of disturbances. Disturbance interactions include loops, that is, the alteration of one disturbance in terms of extent or severity by a previous or co-occurring disturbance, and cascades, that is, a sequence of more than two disturbances, each triggered by the preceding one (see Box 2.2). The most important parameters for the characterization of disturbance regimes are summarized in Table 2.1.

Box 2.2: Disturbance Interactions, Disturbance Cycles, and Disturbance Cascades

Anke Jentsch [iD]
Bayreuth Center of Ecology and Environmental Research (BayCEER),
University of Bayreuth, Bayreuth, Germany
e-mail: anke.jentsch@uni-bayreuth.de

Phil J. Burton
University of Northern British Columbia, Prince George, BC, Canada

Lawrence R. Walker
Department of Life Sciences, University of Nevada, Las Vegas, NV, USA

After a disturbance event has happened, there are different possibilities of ecosystem dynamics: (1) the event may have no impact on further disturbance events; (2) the event may result in further events of the same type; (3) the event may prevent events of the same type; or (4) the event may trigger new types of disturbances.

1. Recovery takes place after a disturbance without further disturbances, which can be accompanied by different dynamics and successional processes depending on the location.
2. If disturbance events of the same type are repeated in close succession, recovery is difficult. In the deflected state, a quasi-stable pioneer stage is formed, such as in avalanche channels on slopes or in fields with regular soil upheaval. Thus, there is positive, reinforcing feedback between disturbances.
3. If disturbance events consume resources, such as flammable biomass during a fire, the temporary lack of resources prevents a repetition of the same event for a certain time. Hence, there is a negative, dampening feedback between disturbances (Wilson and Agnew 1992).
4. If disturbance events trigger different disturbance events, disturbance regimes with disturbance interactions are created. These interactions

(continued)

Box 2.2 (continued)

(linked disturbances and compound disturbances) can take place in relatively predictable iterative or circular cycles (disturbance loops), for example, bark beetle outbreaks after windthrows. However, they can also result in cascades that are difficult to predict and that trigger further disturbance events in a domino effect.

Disturbance interactions of all types can be characterized more precisely by the following parameters:

1. Hierarchical links between successive disturbance events.
2. Trajectory pattern, for example, disturbance loop or disturbance cascade.
3. Degree of deterministic predictability of the disturbance interactions, for example, regarding the type and sequence of disturbance events: high or low.
4. Recurrence interval of the pattern of disturbance interactions, for example, annually, every few years, every few decades, every few centuries.

The proportion of predictable and probable interactions between disturbance events can be regarded as a measure of the complexity of the interaction network. Often the predictability of a further disturbance event is lowest directly after an extreme disturbance because an extreme event causes the ecosystem to deviate as much as possible from the reference dynamics. This is often accompanied by a longer time span before the next disturbance event occurs.

Disturbance cycles often show ecosystem typical return intervals, which depend on regeneration periods and characterize the corresponding biome. In contrast, *disturbance cascades* seldom reoccur in a similar way since they often depend on spatial configurations that are unpredictable and rarely occur repeatedly. Disturbance interactions, which include disturbance cascades with chain reactions, can also be characterized by temporal parameters such as the length of intervals between events, spatial parameters such as local to supra-regional effects, or trophic and organizational dimensions such as organismic to landscape or biome spanning effects. The temporal intervals between single events of disturbance interaction networks can be shortened, lengthened, or show no pattern at all. In practice, it is particularly important to assess the probability of serial disturbance events and improve the predictability of individual events by cooperating with scientists. Typical examples of disturbance interactions in circular disturbance cycles are windthrow, insect outbreaks, and fire in boreal forests, but also the phenomenon of forest dieback in Central, Northern, and Eastern Europe (Box Fig. 2). Typical examples of disturbance interactions in cascades with the domino effect are heavy rain, landslides, flooding, erosion, and destruction of infrastructure in temperate cultural landscapes.

(continued)

Box 2.2 (continued)

Box Fig. 2 Forest dieback – a damaged high-elevation stand at the Nusshardt in the Fichtelgebirge (Germany). For decades, pollutant emissions led to 'acid rain', which acidified the lime-poor soil, causing physiological damage to various tree species. In extremely dry summers, many of the already weakened trees died, which triggered further disturbance events. After the introduction of catalytic converters for passenger cars and the reduction of sulphur emissions in heavy industrial plants, the forest dieback has been considered to be over since about 2003. (Photo: A. Jentsch)

Those disturbance interactions whose course depends on the starting or boundary conditions can be changed or averted by specific management interventions. In Mediterranean forests, for example, the combustible biomass is currently being reduced in order to change the course of fires (see Chap. 7). Firebreaks prevent the spread of flames to neighbouring stands and thus the development of large fires. The targeted burning of small areas (prescribed burning) should prevent the outbreak of larger fires. The removal of deadwood after windthrow (Box Fig. 3) also serves to remove breeding material from bark beetles. Such feedback can prolong the time between disturbance events.

Both interactions and feedback can be classified based on the disturbance history of a landscape (see Box 2.1). However, often the data for many biotic and abiotic parameters and records of historical events are missing.

(continued)

Box 2.2 (continued)

Box Fig. 3 Windthrow is a disturbance event that produces large amounts of dead biomass over a short time and thus often leads to further disturbances, such as bark beetle outbreaks, fungal infestation, and fire. The photograph shows a stand following bark beetle outbreak, which in turn followed a windthrow event that affected the neighbouring stands. (Photo: A. Bürgi, WSL)

Reference

Wilson JB, Agnew ADQ (1992) Positive-feedback switches in plant communities. Adv Ecol Res 23:263–336

If the frequency of a disturbance is quantified relative to the lifespan and the growth rate of dominant organisms, then generally valid processes can be identified, allowing a comparison between different ecosystems. Relatively speaking, the spatial and temporal patterns of disturbance and regeneration are similar at different scales and in different landscapes.

Non-spatial factors such as frequency and magnitude (intensity) of a disturbance determine the immediate impact of the event on resources and members of a community. They determine, for example, how strongly the species composition or nutrient balance is changed by a disturbance and which successional stages take place during the regeneration phase until the original situation is re-established. The intensity of disturbances is measured differently depending on the type of disturbance. For example, fire severity can be divided into four levels (Keeley 2009): *scorched* (leaf fall due to heat), *light* (litter on the ground is charred, trees and humus

Table 2.1 Components of a disturbance regime (see Box 2.1) following Moloney and Levin (1996), Paine et al. (1998), and Turner (2010)

Indicator	Description
Space	
Extent, distribution	Size, form, and distribution of disturbed areas: e.g. size of patches in a forest stand disturbed by wind
Form	Form of the disturbed areas: e.g. fractal dimension in wind-disturbed forests; linear extent of flooding
Time	
Frequency	Mean number of events in a defined period: e.g. one windthrow event within 100 years results in a disturbance frequency of 0.01
Return interval	Mean number of years between two disturbance events: e.g. 100 years
Magnitude	
Intensity	Physical force of a disturbance event per area unit and time: e.g. maximum wind gust speed during a storm, in m/s
Severity	Ecological impact on a disturbed patch: e.g. single windthrown trees vs. patches of many downed trees, % of dead biomass
Spatio-temporal interaction (see Box 2.2)	
Linked disturbance	Causal relation between two disturbances: e.g. increasing risk of bark beetle gradation after wind disturbance
Compound disturbance (cascades)	Amplifying effects between two con-temporal disturbance events resulting in unexpected reactions: e.g. Föhn storm 'Uschi' in the year 2002 and the following heatwave of 2003 produced a bark beetle outbreak up to the subalpine elevation belt

Note: Spatial indicators are described in terms of size and form; temporal indicators are described in terms of frequency and return interval; indicators of magnitude are described in terms of intensity and severity. The spatio-temporal interactions include the linked disturbances and the cascade disturbances. More details on interacting disturbances can be found in Burton et al. (2020)

layer are intact), *moderate* to *severe* (parts of the organic soil, as well as herbs and shrubs, are charred, individual trees are dead), and *deep burning/crown fire* (extensive areas of trees are dead). A five-category scale is common for hurricanes, and the magnitude of earthquakes is measured on the Richter scale, which does not have an upper limit. Spatial factors such as the size, distribution, and form of disturbances are of particular importance for the mechanisms of vegetation recovery after a disturbance and for the long-term dynamics of the ecosystem. For instance, the size and shape of a disturbance area determine the role of matrix or edge vegetation in re-colonization. This role is greatest in the case of linear disturbances such as flooding. Furthermore, disturbances in space and time often do not occur independently of each other. The size, shape, and spatial interaction of the individual disturbances determine the rate at which disturbed areas are re-colonized and finally the structure of the landscape mosaic. Thus, the spatio-temporal dependencies between disturbances are an important aspect in the accurate description of disturbance regimes (see Box 2.1).

2.3 Relevant Scales of Disturbance Ecology

Various concepts and hypotheses have been developed with respect to the significance of disturbances for the dynamics of landscapes and for the preservation of biodiversity and ecosystem functions. These concepts and hypotheses can be organized along the relevant scales of disturbance ecology (White and Jentsch 2001; Jentsch and White 2019): spatially, the categories 'patch' and 'landscape' are of interest; and temporally, the categories 'event' and 'regime' are of interest.

The effects of a disturbance within the disturbed area are of particular interest regarding the change in resource availability, the dominance patterns, and the organic residues. These influence the mechanisms and the speed of recovery after a disturbance. At the landscape scale, the effects of disturbances on biodiversity, ecosystem functions, and dynamic equilibrium are examined. Interactions between disturbances are also of interest. At all scales, the functional properties of species are of interest; these represent the adaptations to the overall spatio-temporal dynamics (Box 2.3).

Box 2.3: The Role of Remote Sensing in Disturbance Ecology

Cornelius Senf ⓘ
Ecosystem Dynamics and Forest Management Group, Technical University of Munich, Munich, Germany

Remote sensing is a group of methods used to obtain information about the Earth's surface by contactless measurements of reflected or emitted electromagnetic radiation, typically using aeroplanes or satellites, also known as airborne or satellite-based remote sensing. Passive remote sensing measures the solar radiation reflected by the Earth's surface using an optical sensor mounted on an aircraft or satellite, which is sensitive to the visible wavelength ranges (0.4–0.7 µm), the near-infrared (0.7–3 µm), or the long-wave thermal radiation emitted by the Earth's surface (8–14 µm). In contrast to passive remote sensing system, active remote sensing systems emit radiation themselves and therefore are independent of incoming solar radiation. Active systems include the use of lasers to measure the travel time between the emission and detection of a reflected laser pulse (LiDAR; *Light Detection And Ranging*) and radio waves (2–30 cm) (RADAR; *Radio Detection And Ranging*) to scan objects.

The radiation measured by remote sensing sensors can provide insights into the states of vegetation on Earth and their changes over time. For example, the reflectance characteristics of vegetation as measured by optical sensors are highly sensitive to the chlorophyll and water content of vegetation (Box Fig. 4). The state of vegetation can therefore be described by spectral

(continued)

Box 2.3 (continued)

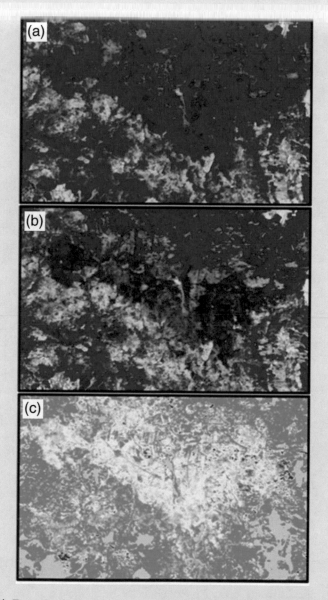

Box Fig. 4 Example of a remote sensing detection of bark beetle infestation in the Bavarian Forest National Park. The sequence shows a Landsat image (in Tasseled Cap Transformation) from 1990 (**a**) and 2000 (**b**) with a clear change from the coniferous forest (blue) to open ground with dead vegetation (orange/brown). (**c**) A corresponding change in needle water content, estimated from the six spectral channels of the Landsat sensor (green = small change, violet = strong change)

(continued)

Box 2.3 (continued)

indices such as the Normalized Difference Vegetation Index[1] (NDVI), the Normalized Burn Ratio[2] (NBR), or the Tasseled Cap Transformation.[3] Active systems (especially LiDAR) are sensitive to the vertical stratification of vegetation. Remote sensing therefore allows for a spatially explicit characterization of physiological and structural vegetation characteristics and their changes over time.

By far the most important role of remote sensing in the context of disturbance ecology is the detection and mapping of disturbances. This is often achieved by manually digitizing disturbances in airborne aerial photographs. This method provides very accurate results, but is often not feasible over large areas or long periods of time. Following the increasing availability of free optical satellite data (e.g. Landsat or Sentinel-2), methods for the automated detection of spectral changes caused by disturbances have been developed (Kennedy et al. 2009). These methods allow accurate mapping of disturbances over large areas and at annual timesteps (Hermosilla et al. 2016). The possibility to distinguish between different types of disturbances (e.g. abiotic and biotic disturbances or clear-cutting), and to detect disturbances of low intensity (e.g. defoliation), has increased significantly with the increasing quality of sensors and with improved image processing capabilities (Senf et al. 2015). Furthermore, the use of satellite-based remote sensing opens a historical view on disturbances (since 1972 with the launch of the first Landsat satellite, or even earlier using spy satellites), which in turn provides important insights into post-disturbance succession (Pflugmacher et al. 2012). Moreover, satellite-based remote sensing allows for a consistent view on disturbance dynamics, independent of political borders, accessibility, and historical data coverage (Hansen et al. 2013). Remote sensing is thus an important tool in disturbance ecology; this is reflected in the increasing number of publications using remote sensing as the basis for data collection.

---

[1] The Normalized Difference Vegetation Index, based on the difference between the near-infrared and red reflectance measured from passive remote sensing sensors, is a measure of green vegetation activity.

[2] The Normalized Burn Ratio, based on the difference between the short wavelength and near-infrared reflectance measured from passive remote sensing sensors, is a measure used to detect changes in vegetation caused by fire and serves as a proxy of burn severity.

[3] The Tasseled Cap transformation is a method to transform the information contained in multi-spectral passive satellite data into a few independent components of vegetation brightness, greenness, and wetness.

(continued)

Box 2.3 (continued)

References

Hansen MC, Potapov PV, Moore R, Hancher M, Turubanova SA, Tyukavina A, Thau D, Stehman SV, Goetz SJ, Loveland TR (2013) High-resolution global maps of 21st-century forest cover change. Science 342:850–853

Hermosilla T, Wulder MA, White JC, Coops NC, Hobart GW, Campbell LB (2016) Mass data processing of time series Landsat imagery: pixels to data products for forest monitoring. Int J Digit Earth 9:1035–1054

Kennedy RE, Townsend PA, Gross JE, Cohen WB, Bolstad P, Wang YQ, Adams P (2009) Remote sensing change detection tools for natural resource managers: understanding concepts and tradeoffs in the design of landscape monitoring projects. Remote Sens Environ 113:1382–1396

Pflugmacher D, Cohen WB, Kennedy RE (2012) Using Landsat-derived disturbance history (1972–2010) to predict current forest structure. Remote Sens Environ 122:146–165

Senf C, Pflugmacher D, Wulder MA, Hostert P (2015) Characterizing spectral-temporal patterns of defoliator and bark beetle disturbances using Landsat time series. Remote Sens Environ 170:166–177

2.3.1   Disturbance Frequency and Magnitude

Frequency and extent of disturbances are often inversely correlated (Fig. 2.4); this has also been described as the frequency–area power-law with respect to the occurrence of forest fires (Malamud et al. 2005). Numerous observations suggest that events of a small scale occur frequently (e.g. bioturbation or cryoturbation), while events of a large scale occur more rarely (e.g. fire and windthrow). This is especially true for disturbance regimes where the state of the ecosystem (e.g. the presence of combustible biomass) is a crucial precondition for the occurrence of the next event. Fire regimes, forestry use, or mowing rhythms in open landscapes typically show such an ecosystem-specific feedback (Box 2.4).

The suppression hypothesis states that disturbance events become stronger the more and more often their occurrence is suppressed. In the case of fire regimes, the combustible biomass accumulated by fire suppression can lead to catastrophic large fires (referred to as the 'fire paradox'; see Chap. 7). Some invasive species can also change formative disturbance regimes as ecosystem engineers (e.g. increased fire frequencies) (see Box 18.1 in Chap. 18).

Small, frequent disturbances are predictable in their probability of occurrence and therefore enable a species composition specially adapted to these disturbance regimes, which often consists of competitively weak but disturbance-adapted species (Fig. 2.5). Large, rare disturbances, on the other hand, occur only occasionally.

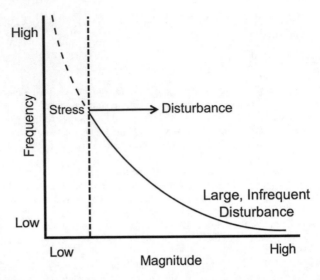

Fig. 2.4 Intensity and frequency of disturbance events often behave in an inversely proportional manner. Strong disturbance events are usually rare because it takes time for regeneration until enough resources are available for the next disturbance event. Events of extremely low strength (left of the dotted line) are usually not called disturbance but stress. (White and Jentsch 2001)

Box 2.4: Disturbance Regime of an Alpine River Landscape: The Tagliamento

Andreas von Heßberg
Bayreuth Center of Ecology and Environmental Research (BayCEER),
Department of Disturbance Ecology and Vegetation Dynamics,
University of Bayreuth, Bayreuth, Germany
e-mail: andreas.hessberg@uni-bayreuth.de

Alpine rivers generate a high-flow dynamic with large scree and sediment loads (detrital); these loads include a high amount of deadwood. Floods from snowmelt and heavy rainfall constantly reshape the riverbed. As a result, pioneer zones, islands, and littoral zones of varying stability emerge. The typical landscape pattern of an alpine river consists of the wide furcation zone of the middle course. In this zone, the main river channel branches out into many different channels which then merge together again and again; this results in the formation of many islands at different successional stages. In such a translocation zone, which differs conceptually and biogeographically from willow and poplar low open forests (German: *Weichholzaue*) and floodplain riparian forests (German: *Hartholzaue*) (Mucina et al. 2016), the dragged detrital material is deposited. Because of the spatial differences in topography, the flow velocity, and consequently the turbulent shear stress, of the water varies

(continued)

Box 2.4 (continued)

considerably. This results in a heterogeneous and highly dynamic landscape pattern. Especially at higher water levels, alpine rivers regularly fill their own riverbeds because of the large sediment load. Therefore, the main current continuously changes its direction of flow, which causes the river channels to branch out more often (Kretschmer 1996; Tockner et al. 2005).

Heavy rainfall events affect the run-off dynamics of alpine rivers, and often this can happen over very short periods. The run-off from such an event can cause the river to rise by up to 2 m within a few hours and ebbs away to the previous level a day later. Pioneer plants, which germinate and establish on sediments still wet from the receding flood, can be washed away again by the next flood or become overgrown with sand and gravel. The dynamics of the river result in a high-frequency disturbance regime and a high proportion of pioneer areas (Lippert et al. 1995).

Alpine rivers offer habitats to an unusually high number of animal and plant species because of the highly dynamic ecosystem, and usually a high number of specialized non-competitive pioneer species. In the recent translocation zone of the Tagliamento (Box Fig. 5), the largest free alpine river in

Box Fig. 5 The alpine river Tagliamento – the reference river for the NATURA 2000 network and for the European Water Framework Directive – flows for 170 km from the eastern Dolomites in northeastern Italy to the northern Adriatic Sea. The annual precipitation in the Tagliamento catchment area (2900 km²) reaches up to 3500 mm. The picture shows the furcation zone of this typical alpine river with some washed-up tree trunks

(continued)

Box 2.4 (continued)

Europe, species such as the German tamarisk (*Myricaria germanica* (L.) Desv.), the riverside wolf spider (*Arctosa cinerea* Fabricius), and the Eurasian stone-curlew (*Burhinus oedicnemus* L.) occur. This conspicuous species diversity is maintained by the river's disturbance regime, the physical conditions of detrital transport and deposition, the structural richness as well as the biodiversity of the adjacent environment, and the high energy input into this river system (Kuhn 1995, 2005).

Deadwood, and in particular large tree trunks with root plates, carried by alpine rivers is an important element of the disturbance regime. As the flood recedes, trunks get stuck in the riverbed, turn with the river current, and align themselves with the main current with the root plates pointing upstream. While a flood is receding, such logs form a barrier to the current. Bow waves form higher flow velocities and cause erosion, eddies, and calmed zones with sedimentation. Deposits of fine sediment store rainwater particularly well during the dry seasons, ensuring the survival of the pioneer species. Within one vegetation period, a green island can emerge from the alluvial deadwood. After further flooding, more abiotic and biotic material accumulates. In this way, the island grows for a few years until patches of softwood (*Salix* spp., *Alnus* spp., *Populus* spp.) establish in the middle of the gravel bed. On such islands, mostly hybrid poplars may shoot up, with a dense thicket of shrubs in the understorey (Gurnell et al. 2000), often also colonized by invasive neophytes. The more islands there are in the riverbed, the smaller is the river cross section of the flood. These forces of the current increase shear stress and erosion and laterally erode the established islands and re-mobilize the substrate. The continuous supply of detrital material and deadwood causes the formation of new islands downstream (Kollmann et al. 1999). In this way, alpine rivers form highly dynamic ecosystems with a high degree of structural diversity and biodiversity.

References

Gurnell AM, Petts GE, Hannah DM, Smith BPG, Edwards PJ, Kollmann J, Ward JV, Tockner K (2000) Wood storage within the active zone of a large European gravel-bed river. Geomorphology 34:55–72

Kollmann J, Viele M, Edwards PJ, Tockner K, Ward JV (1999) Interactions between vegetation development and island formation in the Alpine river Tagliamento. Appl Veg Sci 2:25–36

Kretschmer W (1996) Hydrobiologische Untersuchungen am Tagliamento (Friaul, Italien). Jahrb Ver Schutz Bergwelt 61:123–144

(continued)

Box 2.4 (continued)

Kuhn K (1995) Beobachtungen zu einigen Tiergruppen am Tagliamento. Jahrb Ver Schutz Bergwelt 60:71–86

Kuhn K (2005) Die Kiesbänke des Tagliamento (Friaul, Italien) – ein Lebensraum für Spezialisten im Tierreich. Jahrb Ver Schutz Bergwelt 70:37–44

Lippert W, Müller N, Rossel S, Schauer T, Vetter G (1995) Der Tagliamento – Flusmorphologie und Auenvegetation der größten Wildflußlandschaft der Alpen. Jahrb Ver Schutz Bergwelt 60:11–70

Mucina L, Bültmann H, Dierßen K, Theurillat JP, Raus T, Čarni A, Šumberová K, Willner W, Dengler J, García RG, Chytrý M, Hájek M, Di Pietro R, Iakushenko D, Daniëls FJA, Bergmeier E, Santos Guerra A, Ermakov N, Valachovič M, Schaminée JHJ, Lysenko T, Didukh YP, Pignatti S, Rodwell JS, Capelo J, Weber HE, Solomeshch A, Dimopoulos P, Aguiar C, Hennekens SM, Tichý L (2016) Vegetation of Europe: hierarchical floristic classification system of vascular plant, bryophyte, lichen, and algal communities. Appl Veg Sci 19:3–264

Tockner K, Surian N, Toniutti N (2005) Geomorphologie, Ökologie und nachhaltiges Management einer Wildflusslandschaft am Beispiel des Fiume Tagliamento (Friaul, Italien) – ein Modellökosystem für den Alpenraum und ein Testfall für die EU-Wasserrahmenrichtlinie. Jahrb Ver Schutz Bergwelt 70:3–17

Fig. 2.5 The grey hair-grass (*Corynephorus canescens* (L.) P. Beauv.), a pioneer and target species of nature conservation, grows only in open areas created by the activity of ants in sandy habitats, here on former, stabilized, cryptogam-covered inland dunes in southern Germany. (Photo: A. Jentsch)

They often lie beyond the historical variation of the affected ecosystems (Landres et al. 1999), regarding the lifespans and reproduction periods of the species contained therein. The affected species may only be weakly adapted, or not adapted at all, to such disturbances (Box 2.5).

Box 2.5: Pulse Dynamics in Ecology

Anke Jentsch (iD)
Bayreuth Center of Ecology and Environmental Research (BayCEER),
University of Bayreuth, Bayreuth, Germany
e-mail: anke.jentsch@uni-bayreuth.de

Peter S. White
Biology Department, University of North Carolina, Chapel Hill, NC, USA

Pulse events and the dynamics they trigger are a widespread phenomenon in ecology. These include both biotic and abiotic processes. Examples of biotic processes that trigger pulse events include episodic reproductive events in rodents and insects. Examples of abiotic events include the El Niño–Southern Oscillation (ENSO) climatic phenomenon (irregular), the annual water circulation of freshwater lakes in spring and autumn, or fire regimes that periodically reduce plant biomass. Two different temporal and spatial scales are relevant for the description of pulse dynamics: (1) individual pulse events and (2) recurrent pulse events. These pulse events include disturbance regimes in the different vegetation zones of the Earth (White and Pickett 1985; White and Jentsch 2001; Holt 2008; Yang and Naeem 2008; Jentsch and White 2019). Disturbances are important pulses in ecosystems, but there are also many other pulsed processes that influence ecosystem dynamics.

In the ecological literature five different types of pulse dynamics have been distinguished (Yang and Naeem 2008):

1. Abrupt changes in environmental conditions, for example, heatwaves
2. Abrupt abiotic changes in the supply of resources, for example, the occasional upwelling of deep marine waters
3. Abrupt biotic changes in resource supply owing to abrupt demographic changes caused by mass mortality and predator–prey relationships
4. Abrupt changes of resources by transformation from living to dead structures, for example, disturbances leading to loss of living biomass
5. Spatial input or output of resources through relocation, for example, during mass movements (avalanches, landslides) or flooding in floodplains

Since pulse type 5 involves forces that move nutrients, organic material, soil, or geological substrate via wind, water, and gravity, this type could additionally be classified as pulse type 2 (abiotic changes in resource supply) or pulse type 4 (biomass-altering disturbances).

(continued)

Box 2.5 (continued)

   An individual pulse event can be defined as an abrupt change in ecosystem parameters and processes, where the degree of abruptness can be determined as the strength of the pulse divided by its duration. Pulse events can vary greatly in strength and duration (Box Fig. 6a). While strength and duration of pulse events are continuous variables, ecological system responses, such as resistance to pulse forces and niche width limits of species, are usually subject to discontinuities or interruptions of growth, which may lead to rapid changes of state or threshold dynamics after pulse events. On larger, landscape scales, pulse dynamics are described by their spatial extent (size and distribution) and temporal attributes (frequency of occurrence and predictability) (Box Fig. 6b, c). The original pulses can trigger a series of further pulses

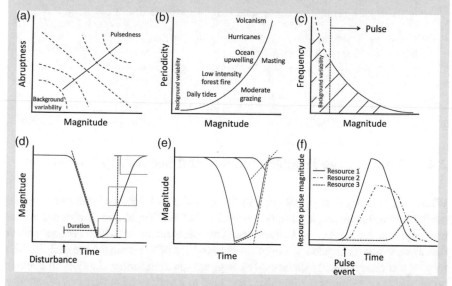

Box Fig. 6   The characteristics that define pulse dynamics. (**a**) Pulse events vary by magnitude of change and abruptness. (**b**) Pulse regimes vary in periodicity, with the degree of variation in periodicity determining predictability. (**c**) The inverse relationship between magnitude and frequency of pulse events; the area below the curve is shaded because low-magnitude pulses can occur at any frequency, but high-magnitude pulses are generally constrained to low frequency. (Modified from White and Jentsch 2001). (**d**) The rate of change after pulse events (illustrated here by a high-magnitude biomass-altering disturbance) is initially limited by rate of colonization and organism response (lower box), and is finally limited by diminishing resources or space (upper box), with a maximum recovery rate at intermediate time since pulse initiation (middle box). (**e**) The initial rate of change (dashed lines) varies with pulse magnitude. (**f**) Pulse events initiate secondary pulses that can lead to synergisms such as feedback loops or cascades. (From Jentsch and White 2019, redrawn)

(continued)

Box 2.5 (continued)

(Box Fig. 6d–f), which are characterized by biotic and abiotic processes and can in turn convert resources from organic to inorganic pools, for example, nitrogen release by fire. With respect to spatial and temporal scales, the resource-oriented concept of pulse dynamics (Yang and Naeem 2008; Jentsch and White 2019) and the pattern-oriented concept of patch dynamics (White and Pickett 1985) are similar.

Jentsch and White (2019) have developed a new concept of pulse dynamics in ecology, which integrates different perspectives from landscape ecology, biogeochemistry, disturbance ecology, biodiversity, and biogeography.

References

Holt RD (2008) Theoretical perspectives on resource pulses. Ecology 89:671–681

Jentsch A, White PS (2019) A theory of pulse dynamics and disturbance in ecology. Ecology 100:e02734

White PS, Jentsch A (2001) The search for generality in studies of disturbance and ecosystem dynamics. Prog Bot 62:399–449

White PS, Pickett STA (1985) Natural disturbance and patch dynamics: an introduction. In: Pickett STA, White PS (eds) The ecology of natural disturbance and patch dynamics. Academic, New York, pp 3–13

Yang LH, Naeem S (2008) The ecology of resource pulses. Ecology 89:619–620

2.4 Disturbances Trigger Primary and Secondary Succession

When disturbances destroy all organic life of a site and fundamentally change the availability of resources, a primary succession follows. Typical examples are volcanic eruptions, landslides, and mudslides. The mechanisms by which recovery from disturbance occurs depend strongly on the severity of the disturbance and the history of the ecosystem. If disturbances trigger only temporary changes in species composition or resource availability, they lead to a variety of patterns of secondary succession, for example, 'gap dynamics', 'patch dynamics', or 'cyclic recovery'. Often, the severity of a disturbance event determines which type of dynamics will occur (Fig. 2.6).

The organic residues after a disturbance (biological legacy) include, for example, the surviving organisms, but also dormant seeds in the soil (soil seed bank), microorganisms, fungi, insects, and dead organic material such as humus, root fragments, and structural residues such as tree stumps. The amount and distribution of organic material, and the presence and functional properties of living organisms and soil properties have a major impact on the starting points, the type, and the speed of recovery after a disturbance (Figs. 2.3 and 2.6). In terms of recovery after a

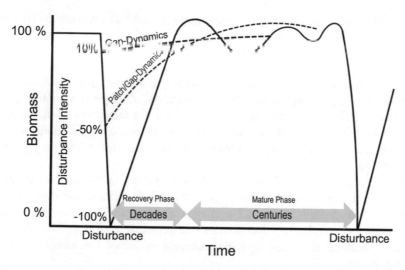

Fig. 2.6 Different patterns of vegetation dynamics are classified according to the strength of the previous disturbance event. (Based on van der Maarel 1993). Here, the amount of destroyed biomass relative to the existing biomass before the disturbance event is indicated

disturbance, the importance of organic residues after a disturbance increases as the severity of the disturbance event decreases, that is, the amount of organic residues is smaller after a severe disturbance and therefore the recovery is slower (White and Jentsch 2001, 2004).

Disturbances can lead to both an input and a loss of resources. If resources are released by disturbances, these are newly available for some organisms, but other organisms may find that they suddenly lack one or more resources. In terms of succession, the early colonizers absorb the released resources, which can then be used by later-successional species (White and Jentsch 2004). The recovery time after disturbances is initially limited by local colonization rates and site-specific growth rates. Later, the recovery rate is limited by the availability of resources. The maximum recovery rate is at medium periods after disturbance events (White and Jentsch 2001; Jentsch and White 2019).

2.5 Disturbances in Landscape Ecology: The Dynamic Equilibrium

Landscape ecology can be used to find an appropriate dimension for the recording of disturbance regimes. This is because a dynamic equilibrium in landscapes (Turner et al. 1993) consists of two ratios: (1) the ratio of the size of the disturbed area to the area size of the entire landscape or ecosystem and (2) the ratio of the frequency of disturbance events to the duration of the required recovery time until complete recovery. The smaller the disturbance relative to the size of the landscape (e.g. in the case of small-scale bioturbation by rodent populations) and the lower the frequency of disturbance relative to the recovery time (e.g. in the case of sparse grazing by migratory shepherds), the more probable a dynamic equilibrium is (Fig. 2.7). A disequilibrium can be caused by very large disturbances like massive floods, a strong windthrow, a large fire, or a volcanic eruption. There is a dynamic equilibrium as long as there are no local extinction events, although the variations in the area share of different successional stages and species within the whole landscape area can be large (Turner et al. 1993; Jentsch 2004). In a dynamic equilibrium, the entire diversity with all its elements remains in the system and passes through stable and variable phases.

A steady state is relatively rare in the different vegetation zones of the Earth (see Chap. 3), but it occurs in disturbance regimes with frequent, small events such as bioturbation by rodents in steppe landscapes. The state of dynamic equilibrium in landscapes characterized by disturbances reaches a critical threshold of instability when disturbances are extremely large relative to landscape size and recovery time and occur rarely.

While some disturbances in an unfragmented landscape spread quickly and over large areas, the same disturbances in a fragmented landscape have strong limits in space and time. Prominent examples are fire or insect calamities in natural

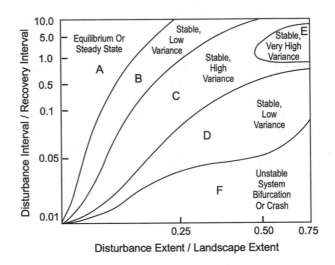

Fig. 2.7 Depending on the extent and interval of disturbances, landscapes are in a dynamic equilibrium (**a–d**) or are subjected to an irreversible change (**f**). Landscapes show stable phases and low variability in species composition when disturbances are relatively small and frequent (**a, b**, e.g. bioturbation in grasslands) or relatively large and rare (**d**, e.g. late frosts in temperate forests). Landscapes can also exhibit high variability if disturbances are relatively large and irregular (**c**, e.g. fire in boreal forests). Extreme sites like tidal zones are characterized by a high variability of environmental conditions and at the same time a high stability in species composition (**e**). A threshold value (line between **d** and **f**) marks dynamics that lead to irreversible system changes and regime shift (**f**), such as overgrazing in combination with perennial droughts or volcanic eruptions. (Based on Turner et al. 1993)

landscapes compared to cultural landscapes. If an ecosystem is homogenized (e.g. shortly after a disturbance or because of long undisturbed conditions), it will ini̇tally lose species, and be less resistant to future displacements. Similarly, interruptions of natural disturbance cycles can lead to changes in the abundance of many species.

A landscape in dynamic equilibrium includes disturbance regimes that allow the coexistence of many age stages and a wide range of species.

2.6 Disturbances and Niche Differentiation in Plant Communities

Disturbances generate spatial heterogeneity, initiate successions, and promote the emergence of biodiversity and diverse biotic communities (see Chap. 4). Disturbances generate selection pressure and evolutionary adaptations, which lead to niche differentiation in communities (Jentsch and White 2019). Such functional adaptations are subject to two mechanisms: (1) niche complementarity and (2) niche redundancy. On the one hand, after a disturbance, the species spectrum and the dominance pattern within the species community change according to the change in resources. On the other hand, disturbance often affects dominant species, which results in an increase of less dominant species. The less dominant species often have a higher tolerance to disturbances (complementarity), even if their functional properties and contributions to the functional structure of ecosystems are similar to those of formerly dominant species (niche redundancy). Dominant and less dominant species change in abundance under changing environmental conditions and thus contribute to functional stability. Thus, functional redundancy plays a major role in maintaining the continuity of ecosystem functions during changing environmental conditions and in ensuring resistance or functional resilience (see Chap. 5) against disturbances. Both mechanisms, complementarity and redundancy, contribute to ecosystem stability.

2.7 Intermediate Disturbance Hypothesis

The Intermediate Disturbance Hypothesis (IDH) states that the local species diversity varies along gradients of disturbance frequency, disturbance recovery time, and disturbance area. The species diversity is greatest under intermediate disturbance (size and frequency), while systems with frequent and/or severe disturbances as well as those with rare and/or low severity disturbances show reduced diversity (Connell 1978). This pattern is based on two assumptions: (1) the existence of a hierarchical competition between species, and (2) the evolutionary development of the trade-off between competitive ability and disturbance tolerance. Here, disturbance-tolerant species are usually less competitive regarding the effective uptake of limiting resources. Thus, niche partitioning is performed along the disturbance gradient, that is, there is an arrangement of the species with their different characteristics in different spatio-temporal niches (niche partitioning). Pioneer species, which appear shortly after a disturbance, form the first part of the species pool, and other more competitive (with respect to uptake of limiting resources) species, which later become more abundant, form the second part of the pool. Under intermediate conditions, species from both species pools can coexist and thus contribute to maximizing diversity. A further discussion of the IDH is given in Chap. 4.

2.8 Disturbances, Biodiversity, and Productivity

The effects of disturbances on biodiversity (see Chap. 4) are significantly modified by the productivity of the site. Based on the IDH (Connell 1978), Huston (1994) specified that it is only for sites of medium productivity that the highest species diversity occurs at intermediate disturbance rates. For sites of high productivity, the highest species diversity occurs at high disturbance rates, and for sites of low productivity, the highest species diversity occurs at low disturbance rates (Fig. 2.8).

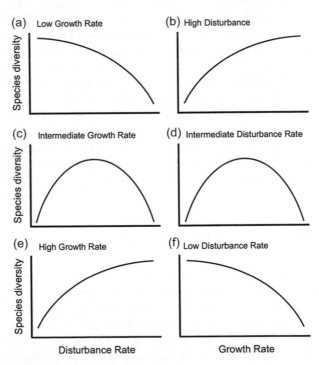

Fig. 2.8 The impact of the disturbance on biodiversity depends on the frequency of the disturbances and the productivity or growth rate of the site. Occasional disturbances lead to maximum biodiversity only at medium productivity. (Based on Huston 1994)

References

Burton PJ, Jentsch A, Walker LR (2020) The ecology of disturbance interactions: characterization, prediction and the potential of cascading effects. Bioscience 70:854–870

Chiarucci A, Araujo MB, Decocq G, Beierkuhnlein C, Fernandez-Palacios JM (2010) The concept of potential natural vegetation: an epitaph? J Veg Sci 21:1172–1178

Connell JH (1978) Diversity in tropical rain forests and coral reefs: high diversity of trees and corals is maintained only in a non-equilibrium state. Science 199:1302–1310

Davis MA, Grime JP, Thompson K (2000) Fluctuating resources in plant communities: a general theory of invasibility. J Ecol 88:528–534

Grime JP (1979) Plant strategies and vegetation processes. Wiley, Chichester, 222 p

Huston MA (1994) Biological diversity: the coexistence of species on changing landscapes. Cambridge University Press, Cambridge, 685 p

Jentsch A (2004) Disturbance driven vegetation dynamics: concepts from biogeography to community ecology and experimental evidence from dry acidic grasslands in central Europe. Diss Bot 384:1–218

Jentsch A, White PS (2019) A theory of pulse dynamics and disturbance in ecology. Ecology 100:e02734

Keeley JE (2009) Fire intensity, fire severity and burn severity: a brief review and suggested usage. Int J Wildland Fire 18:116–126

Kreyling J, Jentsch A, Beierkuhnlein C (2011) Stochastic trajectories of succession initiated by extreme climatic events. Ecol Lett 14:758–764

Landres PB, Morgan P, Swanson FJ (1999) Overview of the use of natural variability concepts in managing ecological systems. Ecol Appl 9:1179–1188

Malamud BD, Millington JD, Perry GL (2005) Characterizing wildfire regimes in the United States. Proc Natl Acad Sci USA 102:4694–4699

Moloney KA, Levin SA (1996) The effects of disturbance architecture on landscape-level population dynamics. Ecology 77:375–394

Odum EP, Finn JT, Franz EH (1979) Perturbation theory and the subsidy-stress gradient. BioScience 29:349–352

Paine RT, Tegner MJ, Johnson EA (1998) Compounded perturbations yield ecological surprises. Ecosystems 1:535–545

Pickett STA, White PS (1985) The ecology of natural disturbance and patch dynamics. Academic, Orlando, 472 p

Sousa WP (1984) The role of disturbance in natural communities. Ann Rev Ecol Syst 15:353–391

Tilman D (1985) The resource-ratio hypothesis of plant succession. Am Nat 125:827–852

Turner MG (2010) Disturbance and landscape dynamics in a changing world. Ecology 91:2833–2849

Turner MG, Romme WH, Gardner RH, Neill RVO, Kratz TK, O'Neill RV (1993) A revised concept of landscape equilibrium: disturbance and stability on scaled landscapes. Landsc Ecol 8:213–227

Turner MG, Calder WJ, Cumming GS, Hughes TP, Jentsch A, LaDeau SL, Lenton TM, Shuman BN, Turetsky MR, Ratajczak Z, Williams JWA, Williams AP, Carpenter SR (2020) Climate change, ecosystems, and abrupt change: science priorities. Philos Trans B 375:20190105

Tüxen R (1956) Die heutige potentielle natürliche Vegetation als Gegenstand der Vegetationskartierung. Angew Pflanzensoz 13:5–42

van der Maarel E (1993) Some remarks on disturbance and its relation to diversity and stability. J Veg Sci 4:733–736

White PS, Jentsch A (2001) The search for generality in studies of disturbance and ecosystem dynamics. Prog Bot 62:399–449

White PS, Jentsch A (2004) Disturbance, succession, and community assembly in terrestrial plant communities. In: Temperton VM, Hobbs RJ, Nuttle T, Halle S (eds) Assembly rules and restoration ecology: bridging the gap between theory and practice. Island Press, Washington DC, pp 342–366

White PS, Pickett STA (1985) Natural disturbance and patch dynamics: an introduction. In: Pickett STA, White PS (eds) The ecology of natural disturbance and patch dynamics. Academic, New York, pp 3–13

Wohlgemuth T, Bürgi M, Scheidegger C, Schütz M (2002) Dominance reduction of species through disturbance – a proposed management principle for central European forests. For Ecol Manag 166:1–15

Chapter 3
Disturbance Regimes and Climate Extremes of the Earth's Vegetation Zones

Anke Jentsch (ORCID) **and Andreas von Heßberg**

Abstract The vegetation zones of the Earth are characterized by different climatic influences as well as natural and anthropogenic disturbance regimes. Low temperatures, short vegetation periods, climatically induced disturbances (such as cryoturbation and fire), as well as disturbances caused by insect outbreaks, large herbivores, and logging characterize the polar to boreal zones. From the temperate zone to the Mediterranean zone, and from the winter humid zone to the subtropical/tropical arid zone, thousands of years of anthropogenic land use have significantly changed the vegetation. While winter storms, frosts, and heavy rainfall are climatically decisive factors in the temperate zone, fire, grazing, and drought play a major role in the frost-free, warmer zones. In the semi-humid subtropics and tropics, heavy rain and storm events are the predominant disturbances.

Keywords Ecozones · Climate · Event regime · Vegetation · Limiting factor

3.1 Introduction

Natural and anthropogenic disturbances are omnipresent in the different vegetation zones of the Earth. They shape ecosystems, create and maintain the typical character of the ecosystems with spatial heterogeneity and temporal dynamics, create free space again and again, turn back the clock of succession, provide free substrate, provide nutrients, modify biotic interactions, and are an important evolutionary force for the emergence and conservation of biodiversity (Walker 1999; White and Jentsch 2001; Jentsch and Beierkuhnlein 2011; Jentsch and White 2019). Disturbances account for both dynamic equilibrium in ecosystems and abrupt regime shift (Turner et al. 2020). Typical disturbances in different areas of the Earth

A. Jentsch (✉) · A. von Heßberg
Bayreuth Center of Ecology and Environmental Research (BayCEER),
University of Bayreuth, Bayreuth, Germany
e-mail: anke.jentsch@uni-bayreuth.de

© The Author(s), under exclusive license to Springer Nature Switzerland AG 2022
T. Wohlgemuth et al. (eds.), *Disturbance Ecology*, Landscape Series 32,
https://doi.org/10.1007/978-3-030-98756-5_3

include fire, storms, insect outbreaks, avalanches, landslides, floods, droughts, mega-herbivores, soil burrowers, but also the whole spectrum of anthropogenic land use in cultural landscapes, such as logging in forests and harvesting on agricultural land (Richter 2001), and the interactions between the disturbances (Burton et al. 2020). The occurrence of disturbances in all ecosystems, their occurrence across different temporal and spatial scales, and their evolutionary-selective role give them outstanding relevance with respect to the dynamics of all landscape areas on Earth (White and Jentsch 2001; Jentsch and White 2019).

This chapter provides an overview of the most important disturbance regimes of all major biomes of the Earth with their typical plant formations (Fig. 3.1). From the poles to the Equator we cover (1) the polar and subpolar zone; (2) the boreal zone; (3) the temperate zone; (4) the Mediterranean zone; (5) the subtropical and tropical arid zones; (6) the humid subtropics; (7) the summer humid tropics; and (8) the ever-humid tropics. In addition, we address disturbances in high mountains, coastal areas, and volcanic islands (see Boxes 3.1 and 3.2 'Permafrost' and 'Disturbances in Marine Ecosystems').

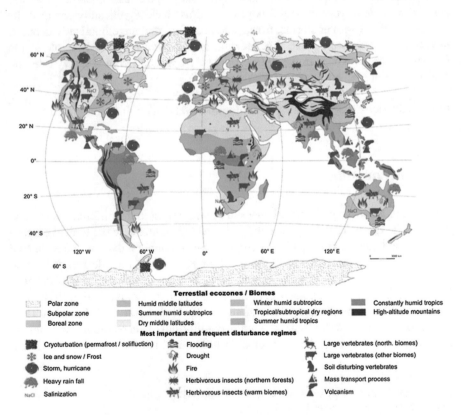

Fig. 3.1 The vegetation zones of the Earth with the large mountain ranges (black) and the disturbance regimes that characterize them (grey pictograms). (Vegetation zone areas according to Richter 2001 and vegetation zone categories according to Schultz 2016)

Box 3.1: Permafrost

Max Schuchardt
Bayreuth Center of Ecology and Environmental Research (BayCEER),
University of Bayreuth, Bayreuth, Germany
e-mail: max.schuchardt@uni-bayreuth.de

Permafrost areas cover 22% of the land surface of the northern hemisphere in polar, subpolar, and alpine regions. Because of the vast extent of the permafrost, it has an important influence on climate and ecosystem dynamics, but melting of permafrost as a consequence of climate change also poses a threat to human infrastructure.

Global Climate Dynamics

Permafrost soils have accumulated and conserved organic carbon from dead plants and animals over thousands of years (Box Fig. 1). Because of accelerated climate warming at high latitudes, this sink function will likely change in the future (IPCC 2013). Higher temperatures do not only promote the thawing of deeper layers, whereby more organic-fixed carbon is mineralized, but they also favour microbial decomposition at higher temperatures during longer vegetation periods. It is assumed that 1330–1580 Pg (1 Pg = 1 gigaton = 1×10^{15} g) of terrestrial carbon is stored in permafrost (Schuur et al. 2015; Celis et al. 2017; Box Fig. 2). This is almost twice as much carbon as is currently present in the Earth's atmosphere. The large-scale thawing of permafrost is likely to release as much carbon as global deforestation has already done. However, the effect on global warming could be 2.5 times stronger because of the contribution of methane (Schuur and Abbott 2011). Longer growing seasons, milder winters, and changes in the winter precipitation regime are likely to favour plant growth, especially of shrubs in the polar and subpolar zone (Elmendorf et al. 2012). Stronger plant growth will counteract soil CO_2 emissions by storing more carbon (Sweet et al. 2015). However, the subpolar tundra areas are considered to have become a net carbon source (Celis et al. 2017).

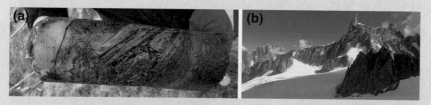

Box Fig. 1 (**a**) Permafrost core showing ice wedges and preserved organic material from Abisko, Northern Sweden. (**b**) Steep mountain peaks interspersed with permafrost in the Mont Blanc massif. (Photos: M. & J. Schuchardt)

(continued)

Box 3.1 (continued)

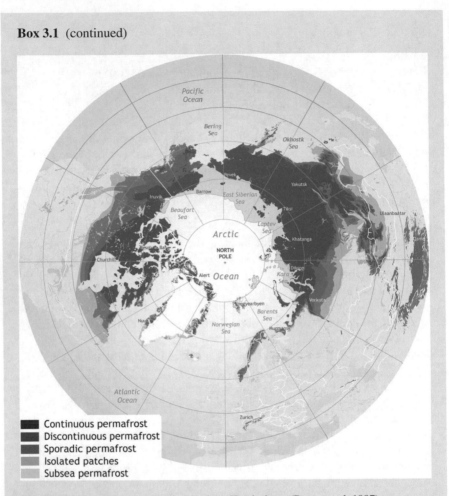

- ███ Continuous permafrost
- ███ Discontinuous permafrost
- ███ Sporadic permafrost
- ▒▒▒ Isolated patches
- ░░░ Subsea permafrost

Box Fig. 2 Permafrost zones in the Northern Hemisphere. (Brown et al. 1997)

Risk to People and Infrastructure

In alpine regions, permafrost acts like a kind of cement on both sediment and bedrock. Porewater and ice wedges in cracks and crevices in the rock glue it together and ensure slope stability, especially in vegetation-free zones with a slope inclination of more than 37° (Gruber and Haeberli 2007). Global warming changes the thermal regime of the frozen rock. Model calculations have shown that the annual mean temperature at the summit of the Zugspitze (2962 m, border of Germany and Austria) rose by 2.7 °C between 1915 and 2015. The increase in ambient temperature has caused the permafrost to retreat 7 m on the south flank and 2 m on the north flank into the mountain core (Gallemann et al. 2017). The thawing of permafrost layers can destabilize large rock massifs and lead to huge rockfalls and landslides. Such mass

(continued)

Box 3.1 (continued)

movements of over 1 million m³ have been observed, for example, in the Swiss Alps at the Dents du Midi and Dents Blanches in 2006 (Gruber and Haeberli 2007).

On a smaller scale, lower snow cover can have an effect on frost weathering. Less snow means less insulation and therefore a smaller buffer layer, which means that the rock is exposed to more freeze–thaw cycles over a year. Rockfalls are thus likely to be more frequent (Kellerer-Pirklbauer 2017). In the Swiss Alps, where average temperatures in the hot summer of 2003 were 3.0 °C higher than the long-term average, the number of rockfalls increased significantly, which was largely attributed to the thawing processes in the permanently frozen rock (Schiermeier 2003; Gruber et al. 2004).

In addition to the increased risk potential in alpine regions, permafrost thawing poses a risk to human infrastructure, such as roads, railroads, power lines, and oil- and gas pipelines. A prominent example of an important infrastructure under threat is the Svalbard Global Seed Vault on Svalbard, in which seeds from all over the world are stored deep-frozen at a depth of up to 120 m in the permafrost rock massif (US Arctic Research Commission 2003).

References

Brown J, Ferrians OJ, Heginbottom JA, Melnikov E (1997) Circum-arctic map of permafrost and ground-ice conditions. U.S. Geological Survey in Cooperation with the Circum-Pacific Council for Energy and Mineral Resources. Circum-Pacific Map Series CP-45, scale 1:10,000,000, 1 sheet

Celis G, Mauritz M, Bracho R, Salmon VG, Webb EE, Hutchings J, Natali SM, Schädel C, Crummer KG, Schuur EAG (2017) Tundra is a consistent source of CO_2 at a site with progressive permafrost thaw during six years of chamber and eddy covariance measurements. J Geophys Res-Biogeo 122:1471–1485

Elmendorf SC, Henry GH, Hollister RD, Björk RG, Boulanger-Lapointe N, Cooper EJ, Cornelissen JH, Day TA, Dorrepaal E, Elumeeva TG (2012) Plot-scale evidence of tundra vegetation change and links to recent summer warming. Nat Clim Change 2:453

Gallemann T, Haas U, Teipel U, Von Poschinger A, Wagner B, Mahr M, Bäse F (2017) Permafrost-Messstation am Zugspitzgipfel: Ergebnisse und Modellberechnungen. Geol Bavarica 115:1–77

Gruber S, Haeberli W (2007) Permafrost in steep bedrock slopes and its temperature-related destabilization following climate change. J Geophys Res-Earth 112:F02S18

Gruber S, Hoelzle M, Haeberli W (2004) Permafrost thaw and destabilization of Alpine rock walls in the hot summer of 2003. Geophys Res Lett 31:L13504

IPCC (2013) Summary for policymakers. In: Stocker TF, Qin D, Tignor M, Allen SK, Boschung J, Nauels A, Xia Y, Bex V, Midgley PM (eds) Climate change 2013: the physical science basis. Contribution of Working Group I to the Fifth Assessment Report of the Intergovernmental Panel on Climate Change. Cambridge University Press, Cambridge.   https://www.ipcc.ch/site/assets/uploads/2018/03/WG1AR5_SummaryVolume_FINAL.pdf

Kellerer-Pirklbauer A (2017) Potential weathering by freeze-thaw action in alpine rocks in the European Alps during a nine year monitoring period. Geomorphology 296:113–131

(continued)

Box 3.1 (continued)

Schiermeier Q (2003) Alpine thaw breaks ice over permafrost's role. Nature 424: 712–712

Schuur EAG, Abbott B (2011) Climate change: high risk of permafrost thaw. Nature 480:32

Schuur EAG, McGuire AD, Schädel C, Grosse G, Harden JW, Hayes DJ, Hugelius G, Koven CD, Kuhry P, Lawrence DM (2015) Climate change and the permafrost carbon feedback. Nature 520:171–179

Sweet SK, Griffin KL, Steltzer H, Gough L, Boelman NT (2015) Greater deciduous shrub abundance extends tundra peak season and increases modeled net CO_2 uptake. Glob Change Biol 21:2394–2409

US Arctic Research Commission (2003) Climate change, permafrost, and impacts on civil infrastructure. Special Report 01-03. U.S. Arctic Research Commission Permafrost Task Force, Arlington, Virginia, USA, 62 p

Box 3.2: Disturbances in Marine Ecosystems

Julian Gutt
Alfred Wegener Institute Helmholtz Centre for Polar and Marine Research
Bremerhaven Germany
e-mail: julian.gutt@awi.de

In the oceans, as on land, both anthropogenic and natural disturbances occur (Kaiser et al. 2011). Anthropogenic disturbances are dominant in European waters whilst natural disturbances are particularly apparent in the tropics and polar regions. Often it is difficult to separate discrete disturbances from continuously developing stresses and specific environmental conditions in independent habitats (IPCC 2019; United Nations 2017).

Natural Disturbances

Natural disturbances have an impact on the seas and coasts in Europe only at a regional scale, where most native organisms are environmentally tolerant and communities have a high self-repair capacity (resilience). However, the natural environmental drivers cannot always be clearly separated from anthropogenic factors.

Ice winters impact the mudflats of the North Sea when temperatures are unusually low and, as a consequence, the sediment as well as the sea surface locally freeze. In such winters, sediment dwellers freeze to death and ice floes 'shave' mussel beds. Such disturbances are more frequent in the Arctic Ocean, for example, along the rocky shores of Svalbard.

Submarine landslides (e.g. on the Norwegian continental shelf) are rare, but they can cover several tens of thousands of square kilometres and trigger tsunamis. Thus, they contribute to habitat fragmentation on the seafloor rather than to the short-term dynamics of seafloor communities.

(continued)

Box 3.2 (continued)

In *cold vents*, water with dissolved or gaseous methane and hydrogen sulphide leaks from the sediment. Chemotrophic microbes form the basis for unique communities similar to those found around hydrothermal vents. Common multicellular organisms in such habitats are tube worms and mussels, which attract fish and other mobile organisms. If carbonate crusts develop, then sea anemones colonize an otherwise rare hard substratum. In the waters of Great Britain, there are over 173,000 such seeps, which are formed in the short term and disappear again in the long term.

In the case of marine *volcanic eruptions*, as on land, all life is primarily destroyed. Later, the lava serves as a special substratum for recolonization.

Harmful algal blooms develop when there is rapid growth of unicellular algae; the algae produce toxins that become dangerous for the 'seafood' consumed by humans and directly for humans.

An example of *diseases* among marine endotherms is the spread of distemper virus in seals in northwestern Europe in 1988; this virus killed 60% of the harbour seals (*Phoca vitulina* L.) in the North Sea population. Another outbreak of the virus occurred in 2002.

Anthropogenic Disturbances

Fisheries significantly change natural food webs because key ecological species (food fishes, predators) are selectively removed from the ecosystem. This applies to pelagic species (e.g. herring and tuna) and bottom-dwelling fish (e.g. cod and flatfish). In addition, bottom trawling (towing the trawl net along the sea floor) destroys seabed habitats (e.g. cold coral reefs) (Box Fig. 3). Among other species, dolphins and albatrosses die worldwide in kilometre-long free-floating drift nets as by-catch.

In European waters, most *non-native species* have been introduced through the discharge of ballast water from ships and via the Suez Canal. Of approximately 1000 such species, only a few cause ecological damage. For example, the Pacific red king crab (*Paralithodes camtschaticus* Tilesius), originally from the Pacific Ocean and intentionally released in the 1960s into the Barents Sea (off the northern coasts of Norway and Russia), now dominates regionally the bottom fauna down to the Lofoten Islands. The introduction of the Indo-Pacific killer alga, *Caulerpa taxifolia* (M. Vahl) C. Agardh, from an aquarium into the Mediterranean Sea drastically damaged species-rich seagrass meadows because this species is not grazed as it contains toxins and therefore it outcompetes native species.

Disturbances are overlaid by *additional stressors* (Pörtner et al. 2021), including man-made climate change together with ocean acidification; pollution, including over-fertilization and noise; waste, also in the form of microplastic and persistent organic pollutants (POPs); destruction of habitats by dredging of sediments; and creation of new habitats and artificial substrata

(continued)

Box 3.2 (continued)

Box Fig. 3 Off the Irish coast at about 700 m water depth, a cold coral reef has been destroyed by fishing nets. The inset shows an intact colony of the coral species *Desmophyllum pertusum* L. (Photos: IFREMER)

(e.g. in the form of pipelines or the underwater foundations of offshore wind turbines).

Disturbances in seas and oceans outside Europe also have significant ecological effects: the El Niño climate phenomenon in the southeast Pacific; iceberg scouring in the Antarctic; hurricanes that devastate coral reefs; a variety of factors that cause the bleaching of corals (especially sea surface warming, turbidity, salinity decrease, overexposure to sunlight and exposure to air due to extreme low tides); storms that destroy kelp forests; explosive population growth of predators, such as the crown-of-thorns starfish (*Acanthaster planci* L.), which preys intensively on the coral polyps and, thus, damages reefs across large areas.

References

IPCC (2019) Summary for policymakers. In: Pörtner H-O, Roberts DC, Masson-Delmotte V, Zhai P, Tignor M, Poloczanska E, Mintenbeck K, Alegría A, Nicolai M, Okem A, Petzold J, Rama B, Weyer NM (eds) IPCC special report on the ocean and cryosphere in a changing climate. Intergovernmental Panel on Climate Change, Geneva, 35 p

Kaiser MJ, Attrill MJ, Jennings S, Thomas DN, Barnes DKA, Brierly AS, Graham NAJ, Hiddink JG, Howell K, Kaartokallio H (2011) Marine ecology: processes, systems, and impacts, 3rd edn. Oxford University Press, Oxford, 608 p

(continued)

Box 3.2 (continued)

Pörtner HO, Scholes RJ, Agard J, Archer E, Arneth A, Bai X, Barnes D, Burrows M, Chan L, Cheung WL, Diamond S, Donatti C, Duarte C, Eisenhauer N, Foden W, Gasalla MA, Handa C, Hickler T, Hoegh-Guldberg O, Ichii K, Jacob U, Insarov G, Kiessling W, Leadley P, Leemans R, Levin L, Lim M, Maharaj S, Managi S, Marquet PA, McElwee P, Midgley G, Oberdorff T, Obura D, Osman E, Pandit R, Pascual U, Pires APF, Popp A, Reyes-García V, Sankaran M, Settele J, Shin YJ., Sintayehu DW, Smith P, Steiner N, Strassburg B, Sukumar R, Trisos C, Val AL, Wu J, Aldrian E, Parmesan C, Pichs-Madruga R, Roberts DC, Rogers AD, Díaz S, Fischer M, Hashimoto S, Lavorel S, Wu N, Ngo HAT (2021) IPBES-IPCC co-sponsored workshop report on biodiversity and climate change; IPBES and IPCC, 24 p. https://doi.org/10.5281/zenodo.4782538

United Nations (2017) The first global integrated marine assessment. World Ocean Assessment I. Cambridge University Press, Cambridge, 976 p

The Earth's ecozones with their characteristic plant formations and life forms are mainly determined by two factors which vary particularly along longitudinal lines and elevation gradients: (1) the mean annual precipitation and (2) the mean annual biotemperature. These determine the climate zones. In turn, the ratio between precipitation and potential evaporation by radiation determines the climatic humidity provinces from hyper-arid to perhumid, which decisively shape the plant formations of the Earth (Fig. 3.2a, b).

3.2 Polar/Subpolar Zone

3.2.1 *Characterization and Ecological Conditions*

The polar zone includes the Arctic and Antarctic zones. Its current area is 14.8% of the total land area of the Earth; about 73% of this area is covered by ice (i.e. about 10.8% of the total land area of the Earth) (Schultz 2016). The subpolar tundra zone covers 15–24% of the Earth's surface (Bliss 1971; Tarnocai et al. 2009). Polar and subpolar regions have a polar climate, which is characterized by extreme physical environmental conditions with seasonally varying solar radiation. Depending on the latitude, polar regions experience several months with little or no light in winter. However, at midsummer, solar radiation has values of up to 70×10^8 kJ/ha. Therefore, in many areas plant growth is barely, or not at all, possible since even in the short summers the monthly mean temperatures rarely increase above 0 °C (polar frost-debris areas) and the soils only thaw superficially (permafrost). Only a few organisms can survive in this vegetation zone. The minimum temperature for the growth of higher plants is generally ≥5 °C (Richter 2001). Because of the almost non-existent mineralization rate, these soils have accumulated large amounts of raw humus (approximately 1330–1580 gigatons carbon in the northern permafrost zone)

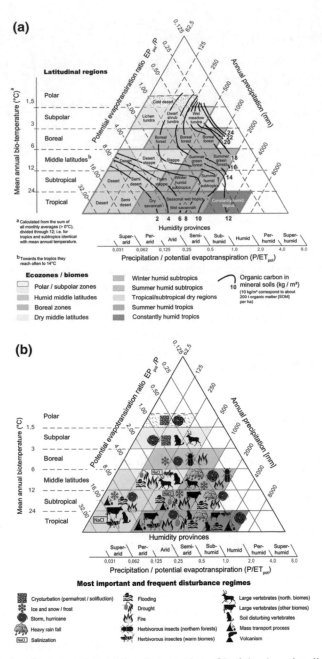

Fig. 3.2 The plant formations (**a**) and disturbance regimes (**b**) of the six major climate zones of the Earth as a function of mean annual temperature, annual precipitation, and evapotranspiration. (Plant formations according to Holdridge 1967)

over thousands of years (Schuur et al. 2015). These carbon sinks could turn into a huge carbon source in the near future (Schuur and Abbott 2011; Ping et al. 2015). Despite low annual precipitation, polar and subpolar regions often have water-saturated soils because permafrost prevents percolation. The annual net primary production of the land is between 0 and 2 (rarely up to 4) tons/ha (Schultz 2016).

3.2.2 Typical Disturbance Regimes and Extreme Climatic Events

Typical disturbance regimes of the polar and subpolar zone include frost, storms, cryoturbation, bioturbation, and reindeer grazing.

Frost is ubiquitous in the polar and subpolar zone. Plants of these regions are adapted to frost during the vegetation period, in contrast to plants of the temperate zone, which cannot tolerate frost in summer. The storage of secondary compounds (glucose, polysaccharides, proteins) for frost stability of the cells consumes a lot of energy (see Chap. 6). Therefore, plants grow very slowly. Many years can pass from germination to the first flowering. Woody dwarf trees and dwarf shrubs grow protected underground and develop leaves and flowers just above the ground (Körner 2003).

Storms mobilize a lot of loose fine material from the mostly vegetation-poor soils, especially in the areas surrounding ice fields where fragmented soil material is exposed. Strong winds carry fine material and snow crystals, causing mechanical damage to plants. Close to large ice fields, seasonal storms occur as so-called cold katabatic winds, called 'Piteraq' in Greenland, (van As et al. 2014). These are strong winds in which high-density air from higher elevations move down slope under gravity. They are more common in winter. The plants react to this by developing, for example, a cushion-like growth with a relatively hard surface and small leaves. The lower the plants grow, the more they are covered by snow in winter (hemicrypto-phytes) and the better they are protected against frost and abrasion (Pfadenhauer and Klötzli 2015).

Permafrost areas account for 23.9% of the total land surface (excluding glaciers and ice sheets) of the Northern Hemisphere (Zhang et al. 2003). *Cryoturbation* causes the mixing of the water-saturated soil by thawing and freezing processes on about 31% of the permafrost area (Hugelius et al. 2014). This is associated with a size sorting of the topsoil material and an integration of organic carbon into deeper soil layers (Krab et al. 2018). Any movement, and thus disturbance of the soil structure, can also result in damage to the plants that grow in these areas (Fig. 3.3). Taproots or lignifying roots can be severely damaged. However, fine roots of grasses and herbaceous plants are better adapted to these restructuring processes (Goulet 1995). At high turbulence intensity, plant growth can be completely inhibited (e.g. on frost pattern soils such as non-sorted circles) (Walker et al. 2008). On flat areas with a slope of $\geq 2°$, cryoturbation results in the formation of 'solifluction tongues' as the result of different downhill flow rates of soil (Troll 1944).

Fig. 3.3 Soil flow at the northernmost tree line in Alaska. (Photo: A. Jentsch)

Bioturbation occurs in the non-water-saturated tundra of the subpolar zone and is caused by a variety of burrowing rodents (lemmings, mice, picas, gophers, marmots). The animals loosen the soil, transferring aboveground biomass to deeper soil horizons and promoting the mineralization of the peaty raw humus layer of the topsoil (Yu et al. 2017). The burrowing activities also cause mechanical disturbances in the surrounding area. Raw soil areas are created where various species can establish themselves.

Seasonally migrating reindeer herds *graze* wide areas of the subpolar tundra. The herds are domesticated to varying degrees and graze the vegetation and thus contribute to nutrient shifting and fertilization in an otherwise highly nitrogen-limited biome. The sharp-edged hooves of a large number of animals in a passing herd result in disturbances of the topsoil and thus create new germination beds. In winter, the reindeer scratch with their hooves for lichens under the snow cover and can cause considerable damage to the vegetation. The grazing influence of reindeer on the vegetation can be large enough to cause a change from lichen-dominated to grass-dominated systems. However, this depends on the grazing intensity. On the other hand, the preferential grazing behaviour of reindeer herds can counteract the expansion of trees and bushes into the tundra as a result of climate change (Bernes et al. 2015).

Global warming in the polar and subpolar zone often occurs as a seasonal heat pulse. This is one of the defining factors of the far north. It intensifies the disturbance dynamics in this highly sensitive biome (see Chap. 16). The high latitudes have warmed up twice as much as the rest of the planet in recent decades (IPCC 2013). Increased winter temperatures and higher amounts of snow change microbial degradation rates, but also promote the survival of frost-sensitive species as the vegetation period extends beyond that which has been experienced in the past (Saccone et al. 2017). An increased spring temperature predicted for large parts of the subpolar zone will change the length of the vegetation period and thus the species composition in the long term (Ernakovich et al. 2014). Glacial retreat and increases in summer temperatures will initiate processes of primary succession. Ice deserts become frost-debris soils with polar plant communities. Initially, these primary areas have many disturbances caused by mass movements of coarse and fine material and aeolian deposits.

3.3 Boreal Zone

3.3.1 Characterization and Ecological Framework

The boreal zone covers 13% of the global land surface (Schultz 2016) and is limited to the Northern Hemisphere. It has a cold temperate climate with long, cold winters and short, moderately warm summers. The vegetation period of the boreal zone is usually 4–5 months. The northern border of the boreal zone corresponds almost exactly with the +10 °C July isotherm. The zone is divided into 'cold-continental' regions with low precipitation and 'cold-oceanic' regions with high precipitation. In the cold-continental regions, winter minimum values of <−60 °C and summer maximum values of +30 °C can be found [e.g. in Sakha (Yakutia), northeast Russia]. The cold-oceanic areas, on the other hand, are located near the coasts of the Atlantic and Pacific Oceans. The vegetation is mainly contiguous boreal coniferous forest (taiga), which is dominated by pines, spruces, and firs in the west, and larches in the east. The forests are interspersed with moors and mountainous tundra. In some northern areas, the coniferous forests even grow on permafrost soils, which, despite the low rainfall, provide an adequate water supply for the vegetation. On sandy, water-permeable soils, the permafrost often begins only at a depth of several metres, and larch forests with lichen-rich ground vegetation grow in these areas. In the more humid areas of the boreal zone, *Pinus* spp., *Picea* spp., *Abies* spp., *Tsuga* spp., and *Pseudotsuga* spp., as well as many Ericaceous dwarf shrubs, dominate the ground vegetation. During the last glacial maximum (~25,000–15,000 years ago), large parts of the boreal zone were covered by an ice sheet, which is why the soils across much of the zone are relatively young. Acidic, nutrient-poor soils predominate, which tend to podzolize because of the incomplete decomposition of the acidic needle and *Vaccinium* litter.

3.3.2 Typical Disturbance Regimes and Extreme Climatic Events

Typical disturbance regimes in the boreal zone include fire, insects, ice and snow breakage, logging, and changes in permafrost dynamics caused by soil warming. Avalanches are also important disturbances in the boreal mountains.

Fires play a dominant role in all boreal forests (see Chap. 7), including on permafrost soils. The occurrence of thunderstorms in combination with the availability of flammable material in the form of needle litter or lichen-dominated, summer-dry ground vegetation means that forest fires can start relatively easily. The frequency of fire events depends on the degree of continentality and ranges from 50 years in highly continental dry areas to 200 years in wetter coastal areas. The dominant coniferous tree species (*Larix* spp., *Pinus* spp.) adapt to the recurrent ground fires by forming a thick bark (see Chap. 6). The thicker bark chars on the outside and protects the inner, more sensitive layer. In a long-burning crown fire, however, the entire tree often burns, and the stand can then collapse over a large area. Because of the cold climate, such stands regenerate very slowly, depending on the water supply of the soil.

Because of the dominance of conifers in boreal forests, *insect feeding* on needles and under the bark contributes to large-scale dynamics within this vegetation zone. Mass outbreaks of single insect species (e.g. different bark beetles; see Chaps. 11 and 12) are relatively common and known from the whole boreal zone (e.g. Heavilin et al. 2007). Often the beetles are attracted by chemical scent signals of injured trees caused by other disturbances like wind and snow breakage, drought, fire, or logging. The affected forest area can be completely infested and destroyed by the outbreaks. Forest regeneration takes place with pioneer forest species such as birch (*Betula* spp.), poplar (*Populus* spp.), willow (*Salix* spp.), as well as with the original species. As a result of climate change, the boreal habitat has become increasingly accessible to 'pests' (Price et al. 2013).

Snow and ice breakage occurs particularly in cold-oceanic areas of the boreal zone, where high humidity, mild winter temperatures, and rapid changes in weather conditions can cause precipitation to freeze on the branches and grow into thick layers of snow and ice. The resulting heavy load of snow and ice can lead to the breaking of branches, tops, and whole trunks. Depending on the species, damaged trees can recover from such disturbances at different rates, but their competitive strength remains reduced over a longer period.

Logging as an anthropogenic disturbance regime (see Chap. 14) characterizes the parts of the boreal zone where humans have established appropriate access. Especially on the northern permafrost soils, large-scale logging can create open areas that thaw superficially because of increased solar radiation and remain unwooded for a long time (Fig. 3.4).

Permafrost dynamics caused by soil warming occurs where forests become patchy as the result of disturbance. The permafrost table sinks because of the warming, and subsidence of the soil as the result of water loss is possible. Thermokarst lakes are formed which, due to the short vegetation period and the cool climate, quickly turn into peat bogs, with very low growth rates (<1 mm increase in height growth per year). In a later phase, the dense moor vegetation shields the soil from solar radiation so that the permafrost table rises, causing the water body to freeze; this leads to the formation of large hummocks ('pingos') across the entire landscape.

Peat exploitation is also practised in boreal peatlands and fens. Boreal peatlands contain a significant amount of terrestrially bonded carbon because of their

Fig. 3.4 Natural floodplain dynamics in the foreground and forest clearing areas in the background after fresh snowfall in British Columbia, Canada. (Photo: A. Jentsch)

enormous area size and low mineralization rates. It is estimated that 500 ± 100 Gt of carbon is stored in northern peatlands; for comparison, the atmosphere stores about 750 Gt of carbon (Yu 2012; Lees et al. 2018). In some countries (e.g. Ireland, Finland), peat has been extracted and burned to provide heat and electricity; however, the industrial use of peat for heat is being phased out. The draining of peatlands by humans, but also the desiccation of peatlands as a consequence of increased temperatures, has led to an increasing release of carbon into the atmosphere (IPCC Physical Science Basis 2013). Agriculture and grazing for subsistence play a subordinate role.

In addition to the changes in permafrost dynamics, climate change will lead to other changes in the carbon cycle in the boreal zone. In particular, the frequency of thunderstorms, and therefore the frequency of fires, is expected to increase (see Chap. 16), which will increase carbon emissions. An increase of large forest fire areas in the boreal zone has been observed since the 1980s (Fauria and Johnson 2008). In the cold-oceanic areas, more frequent cool-humid winters can be expected, which will result in increased ice and snow breakage. Because of a shift in precipitation regimes, large insect outbreaks can quickly spread owing to the occurrence of uniform coniferous forests across large areas (see Chap. 12).

3.4 Temperate Zone

3.4.1 Characterization and Ecological Framework

The temperate zone can be split into humid and dry mid-latitudes. The humid mid-latitudes are between 35° and 60°, with the largest occurrences in the Northern Hemisphere being on the east and west sides of the North American and Eurasian land masses. Only small, but also coastal, areas are found in the Southern Hemisphere in South America (coast of southern Chile), Australia (southeastern Australia and Tasmania), New Zealand (South Island), and on the subantarctic islands. All sub-areas are characterized by a maritime-influenced, temperate climate. The temperate zone covers 9.7% (14.5 million km²) of the Earth's land area (Schultz 2016). Towards the poles, it borders on the boreal zone, and towards the Equator, it borders the winter humid zone on the western sides of the continents, and the ever-humid subtropics on the eastern sides of the continents.

The humid mid-latitudes show a seasonally differentiated annual temperature cycle and the associated remarkable aspect change of the vegetation with deciduous broadleaf and mixed forests, winter leaf fall, and the occurrence of spring geo-phytes, that is, perennial plants that regenerate from underground buds or under-ground storage organs such as bulbs, corms, rhizomes, or tubers. A temperate climate dominates, that is, the winters are cool, and the range of seasonally chang-ing day lengths is less than in the boreal zone, but larger than in the equatorial zones. A relatively rapid decomposition cycle of about 4 years, short mineral cycles with high turnover rates, and an annual net primary production of about 8–13 tons/ha has led to the formation of the typical European cultural landscape with diverse anthro-pogenic use and disturbance regimes (see Chap. 15).

In the Northern Hemisphere, the dry mid-latitudes in continental Eurasia and in the Midwest of America extend polewards to 55°N and in some areas directly bor-der on the tropical/subtropical dry zones. They cover 11.1% (16.5 million km²) of the land area of the Earth (Schultz 2016). Because of the hot summers and cold winters, the vegetation period is thermally limited (especially by temperatures below freezing) in the north and limited by the precipitation regime and humidity in the south.

3.4.2 Typical Disturbance Regimes and Extreme Climatic Events

Typical disturbance regimes of the temperature zone include winter storms, winter heat pulses (often accompanied by late frost in spring and early frost in late sum-mer), heavy rainfall events with floods, summer hot spells with droughts, as well as mowing, grazing, and agricultural and forest management.

Winter storms occur in the temperature zone as extratropical cyclones in the context of low-pressure systems, in the form of blizzards in North America or hur-ricane winds in Europe; these weather systems can travel hundreds of kilometres (see Chap. 8) and can cause major damage (e.g. to managed forests and infrastruc-ture). The maximum wind speeds vary between 150 and 250 km/h and decrease towards the edge of the storm systems. The risk of windbreakage tends to increase with the age of the trees/stocks and ranges from branch breakage, trunk breakage, and creation of gaps in stands, to complete windthrow of forest stands over large areas. Monocultures and homogeneous old stands show a higher risk of wind damage than naturally grown forests (Odenthal-Kahabka 2004; Johnson and Miyanishi 2007). From a global perspective, winter storms in mid-latitudes are the second most cost intensive natural hazard for the global insurance industry after tropical cyclones (Münchner Rück 1999, 2017).

In the Northern Hemisphere, *winter heatwaves* in the coldest season of the temperature zone are caused by macro weather situations of the North Atlantic Oscillation (NAO) with temperatures far above 0 °C. Such heat pulses reduce the winter and frost hardiness of the plants and can lead to premature budding and leaf development. If the warm phase is followed by another cold period or if late frost occurs on phenologically widely developed leaf shoots, the plants can be severely damaged.

Frosts, especially late frosts, which can occur after the onset of leaf sprouting until mid-May in the Northern Hemisphere, reduce ecological fitness and often cause major economic damage in agriculture. Shrubs and trees, especially vines as well as fruit and nut trees, are most affected (Werner-Gnann et al. 2017; Chmielewski et al. 2018). Frost damage to these crop species leads to partial or complete loss of the annual harvest, but not to the death of the plants. Some deciduous tree species are particularly sensitive to late frost, for example, oaks (*Quercus* spp.), sycamore maple (*Acer pseudoplatanus* L.), and beech (*Fagus sylvatica* L.) (Fig. 3.5) (Kreyling et al. 2012), whereas other deciduous species tolerate very cold conditions in winter (see Chap. 6). It should be noted that there is a considerable within-species variation of late frost tolerance for some tree species. Frosts can damage or kill the reproductive organs (flowers), and also lead to the death of meristem tissues, which can delay annual longitudinal growth in affected species, thus reducing their competitiveness in comparison to other, non-damaged tree species, especially during regeneration (e.g. beech vs. oak).

Fig. 3.5 Late frost in May after leaf emergence in a beech forest. (Photo: A. Jentsch)

Floods are caused by heavy rainfall events, which in a relatively short time lead to the water saturation of the soil. Since the plants cannot absorb the water so quickly, the water runs off the surface of the land, resulting in a rapid inflow to receiving streams, which leads to the water exceeding the normal capacity of the smaller streams and consequent inundation of the floodplains. The individual discharge of local floods then combine in the larger rivers and can lead to extreme flooding, causing massive erosion damage and flooding of buildings and infrastructure. Heavy rainfall can occur in different areas at different times of the year. In winter, because of frozen soil, both soil infiltration and plant interception are often reduced. Also, in summer, after a long drought, soils can be too dry to absorb large amounts of precipitation. Wetland habitats are characterized and maintained by regular flooding.

Summer hot spells often last only a few days but are usually accompanied by longer periods of drought. Most of the vegetation in the temperature zone shows little response to heat and drought, but arable crops and meadows of the lowlands and forests of the low mountain ranges can be damaged. Droughts here lead to reduced capacity of photosynthesis and losses in harvest and growth or the death of aboveground biomass. In addition, droughts and hot spells can lead to outbreaks of insect pests.

Mowing, grazing, and agriculture are the formative, anthropogenic disturbance regimes of the temperate cultural landscapes (see Chap. 15). In agriculture, mechanical soil disturbances occur regularly, which can lead to soil erosion in dry periods or during floods owing to a temporary lack of plant cover. Soil compaction, pesticide use, and over-fertilization lead to species-poor agro-ecosystems. Grazing selects certain plant species in the pastures. Most of the aboveground biomass is harvested from meadows in high (intensive) or low (extensive) mowing rhythms. Nutrients are replenished by fertilization both on arable land and partially in permanent grassland. In the steppes of the dry mid-latitudes, fields and grassland are often irrigated.

Climate change in the temperature zone is mainly associated with increasing water deficits in the arid mid-latitudes (Kovats et al. 2014). As a consequence, the boundaries of forest steppes, long grass steppes, short grass steppes, desert steppes, semi-deserts, and deserts may shift (Sala and Maestre 2014). These areas are almost entirely subject to agricultural use: long grass steppes through large-scale arable farming, and short grass and desert steppes through ranching (Schultz 2016). In the humid mid-latitudes, the probability of winter storms as well as winter heatwaves increases (Trenberth et al. 2007). In general, global warming leads to a higher climatic variability, prolongation of the vegetation period, and an earlier onset of leaf emergence. Winter precipitation increasingly falls as rain rather than snow, and the greater variability of precipitation in summer brings droughts followed by heavy rainfall events (IPCC 2013).

3.5 Mediterranean Zone

3.5.1 Characterization and Ecological Framework

The subtropical zone has a broad hygric differentiation from fully arid to summer humid to summer dry and always humid conditions (Richter 2001). The summer-dry subtropics are called the Mediterranean subtropics or Mediterranean zone. Mediterranean ecosystems are found in the Mediterranean Basin, in California, in southwest Australia, in the Cape region of South Africa, and in a small area of Chile. The Mediterranean climate is dominated by a seasonal alternation between cyclone passages with high winter precipitation and anticyclone periods with high-pressure areas during dry summers. Extended periods of low winter temperatures and heavy late frost events are rare.

The dominant tree species are evergreens with mostly scleromorphic leaves, such as oaks (*Quercus ilex* L., *Q. suber* L., *Q. cerris* L., *Q. coccifera* L.), olive (*Olea europaea* L.), laurel (*Laurus* spp.), eucalyptus (*Eucalyptus* spp.), and conifers such as pines (*Pinus* spp.) and sequoias (*Sequoiadendron* spp.). The anthropogenically dominated cultural landscape in the Mediterranean Basin is characterized by dense shrubbery, while succulent species, including prickly pears (*Opuntia* spp.) and agaves (*Agave* spp.), thrive on semi-arid sites. In the Mediterranean zone there is a wide range of different habitats and therefore a very high biodiversity with a high proportion of endemics (Rundel 1999). In areas where the summer drought lasts more than 7 months and the annual rainfall is less than 300 mm, the climate is classified as subtropical/tropical dry towards the south.

3.5.2 Typical Disturbance Regimes and Extreme Climatic Events

Typical disturbance regimes of the summer-dry Mediterranean zone include fire, erosion, grazing, karst formation (badlands), and millennia of agricultural use.

Fires are part of the natural and formative dynamics of ecosystems in all Mediterranean areas. In the Mediterranean zone fires are often caused by summer thunderstorms. Because of the high winter productivity and thus rapid recovery of fuel load, fires in the Mediterranean zone have recurrence intervals of only a few decades (Buhk et al. 2007). Dry litter from broadleaved and coniferous trees is highly flammable because of its high content of essential oils and resins and promotes regular bush fires close to the ground (see Chap. 7). Crown fires occur when there is a long period between ground fires, which allows the accumulation of dead biomass and the build-up of large amounts of fuel. For this reason, active fire management is carried out today in many areas, especially in Australia and California. The vegetation in Mediterranean areas is adapted to fire, heat, and smoke in a variety of ways. Immediately after a bush fire, there is an abundance of light and

Fig. 3.6 Wet laurel forest (here on the island of La Palma, Canary Islands) is the original forest vegetation in the Mediterranean zone. (Photo: A. Jentsch)

mineralized nutrients, and consequently, perennial fire-adapted species can regenerate quickly. For some pyrophytic species with serotinous cones, seeds are only released after a fire (e.g. *Pinus halepensis*). After crown fires, a forest stand may collapse, which may lead to rapid degradation of the summer-dry soils and ultimately soil erosion.

Erosion of topsoils can be caused by heavy summer thunderstorms or prolonged heavy winter rainfall and is most severe after a fire or on areas from which the protective vegetation has been cleared. Because of the anthropogenic overuse of the entire Mediterranean Basin, a large part of the tertiary-formed humus-rich soils has disappeared since antiquity. There are currently very few intact examples of the laurel and holm oak forests that were once widespread around the Mediterranean Basin (Rundel 1999). An example of a relict laurel forest on the Canary Islands is shown in Fig. 3.6.

Grazing with goats and sheep is relatively common across large areas of the Mediterranean Basin, but only to a small extent in the other four Mediterranean regions. During summer drought, overgrazing can cause severe damage and vegetation degradation. This can lead to the permanent loss of trees (scrub encroachment) or the loss of bushes (desertification) and eventually to soil erosion. In some areas of the Mediterranean Basin, transhumance is practised: that is, animals are grazed in the mountains in summer and are moved to the valleys in winter. This type of pastoralism enables the grazing pressure to be distributed across a wider area and reduce the likelihood of damage.

Karst formations can be exposed as the result of anthropogenic action (as well as through natural processes) and occur in Mediterranean areas where there has been heavy overgrazing on carbonate rocks. This process had already partially taken place in antiquity and led to the formation of 'badlands' (i.e. areas which are unsuitable for almost all plant species and further human use).

Land-use change: In almost the entire Mediterranean Basin, demographic change with an accompanying intensification of agriculture has led to the displacement of ~~ ~~~ ~~~~~~~~~~~~ ~~~~ ~~~~~~~~ ~~~~~~ ~~ ~~~ ~~~~~~~~ ~~~~~~~~~~~~ ~~ ~~~~~~~~~~~ (Oteros-Rozas et al. 2013; Sklavou et al. 2017). As a consequence of climate change increased summer temperatures and prolonged dry periods are predicted, which can represent an additional stressor, especially for agriculturally overused regions (IPCC 2013).

3.6 Subtropical and Tropical Arid Zones

3.6.1 Characterization and Ecological Framework

Tropical and subtropical arid zones account for about 21% of the total land area of the Earth (Schultz 2016). They include the deserts and semi-deserts, as well as summer humid xeric shrublands. These zones form transitional areas to the summer humid tropics and the ever-humid subtropics. The winter-moist grass and shrub steppes form transitional areas to the winter-moist subtropics.

The main characteristics of the subtropical and tropical dry zone are above all the aridity and the extremely high annual solar radiation ($700–800 \times 10^8$ kJ/ha); together with the ever-humid tropics, this solar radiation is higher than the other ecological zones. At most 3 months of the year are humid. Deserts receive less than 250 mm of water per year.

3.6.2 Typical Disturbance Regimes and Extreme Climatic Events

Typical disturbance regimes of the subtropical and tropical dry zone include savannah fires, outbreaks of insects, grazing degradation with erosion, droughts, and plagues of rodents with bioturbation.

Fires are regularly recurring disturbances in the subtropical and tropical dry zones, especially in the savannahs and steppes. Because of the high aridity, the grasses, trees, and shrubs that grow here are very dry. Fire is a natural disturbance in this zone that causes rejuvenation of the vegetation and a remineralization of nutrients. Thereby fire removes dead material and creates space and resources for the growing vegetation. Before the areas were inhabited by humans, savannah fires were started by lightning strikes, but currently, most savannah fires are deliberately started by humans. The reasons for starting the fires include pest control, nutrient cycling, and the maintenance of the grassland and savannah ecosystems (van der Werf et al. 2008).

Fig. 3.7 Hardwood forests in Morocco degraded by drought-induced overgrazing. (Photo: A. Jentsch)

Insect outbreaks are among the most devastating natural disturbances for humans, as they periodically invade agro-ecosystems. Locusts, for example, can devour vast swathes of land in just a few days. There are more than 350 species of locusts world-wide, which are usually solitary, but under certain circumstances, the locusts show gregarious behaviour and can form massive swarms that periodically cause distur-bances to agro-ecosystems. An indigenous locust population becomes a plague especially when changes in land use lead to increased resource availability and thus allow an enormous locust population increase. In the case of invasive species, the absence of potential predators also plays a role (Cerritos 2011). Because of their high protein content, locusts are seen as a possible food source that could satisfy the growing demand for food by humans and mitigate the devastating agricultural impacts of locusts.

Pasture degradation with associated erosion is a major threat in the arid subtrop-ics and tropics wherever grazing with cows or goats is practised (Fig. 3.7). If a livestock herd stays in a certain area for a longer period, overgrazing, trampling damage, erosion, and local loss of topsoil are the consequences. The vegetation is often not able to recover from such damage. As a consequence, there is an undesired selection favouring annual grasses over perennial grasses, followed by a significant decrease in biodiversity in the grass and herb layer. Further damage includes reduced fertility and reduced infiltration capacity, as well as reduced water storage capacity and reduced nutrient availability. This also leads to loss of biodiversity and species and may lead to changes in fire regimes (Deng et al. 2014).

Droughts are generally defined as periods of unusually dry weather that last long enough to cause severe hydrological imbalance (IPCC 2012). Natural causes of droughts are reduced precipitation, reduced water vapour in the air, fewer rising air masses, changes in the atmospheric and oceanic weather cycle, and changes in the global weather cycle caused by the El Niño–Southern Oscillation (ENSO). The El Niño–Southern Oscillation (ENSO) is an irregular periodic variation in the winds and sea surface temperatures over the tropical eastern Pacific Ocean. El Niño and La

Niña are the warm and cool phases of ENSO, respectively. ENSO disrupts the normal climate patterns of much of the tropics and subtropics (and beyond), causing intense storms in some places and droughts in other places. Anthropogenic causes of drought damage include changes in land use, agricultural depletion, deforestation, and irrigation with associated soil erosion. The Sahel region, which is located in the subtropical dry zone in Africa between the Sahara to the north and the Sudanian savannah to the south, stretching between the Atlantic Ocean and the Red Sea, has been the location of the most severe droughts of recent decades (Scheffer et al. 2001). In addition to the major droughts in the 1970s and 1980s, the Sahel region was again affected by severe droughts in 2005, 2008, 2010, and 2012, affecting more than 18 million people (Oxfam International 2013). At the end of drought periods, subtropical and tropical dry zones often experience a massive increase in populations of burrowing rodents and the resulting significant bioturbation (Richter and Ise 2005).

The effects of *climate change and land-use change* in the subtropical and tropical arid zones are currently difficult to predict. In general, a spread of fire regimes is expected as well as a higher fire frequency. However, there are models that expect an increase in fire frequency in certain regions but predict a decrease in fire frequency in other regions. Because of climate change, droughts are expected to increase in duration and intensity in many regions, leading to increased grazing degradation and migration of humans away from these areas (IPCC 2012).

3.7 Humid Subtropics

3.7.1 Characterization and Ecological Framework

The subtropical zone has a broad hygric differentiation from fully arid through summer humid to winter humid and ever-humid conditions. In this section, the ever-humid subtropics as well as the summer humid subtropics are addressed.

The ever-humid subtropics account for about 4% of the land area of the Earth and are located on the eastern side of the continents (Schultz 2016). Temperatures are warm temperate with cool winters and hot summers. As a result of the humid maritime trade winds, there is precipitation all year round with most precipitation occurring in summer. These seasonal precipitation events are called monsoons. In winter, mainly cool and dry continental winds blow. Because of this humid subtropical climate, laurel forests and subtropical rainforests occur in this region.

The summer humid subtropics cover about 16.4% of the land area of the Earth (Schultz 2016). The dry and rainy seasons in this zone are caused by variation in the location of the intertropical convergence zone (ITCZ). ITCZ is an area near the Equator where the northeast trade winds and southeast trade winds converge in a permanent low-pressure zone; the heat and moisture result in convection of the air, which then leads to abundant precipitation. The summer humid subtropics zone is

Fig. 3.8 Mudslides following heavy rain and light earthquakes can cause devastating damage, Chiapas, southern Mexico. (Photo: A. Jentsch)

especially characterized by a strong variability in the amount of precipitation. The summer humid subtropics can be divided into wet savannah, dry savannah, xeric shrublands, subtropical dry forest, and monsoon forest. The dry season lasts between 2.5 and 7.5 months, while the rainy season can last between 4.5 and 9.5 months. The farther a region is from the Equator, the lower the daily afternoon rainfall during the wet season and the longer the dry season.

3.7.2 Typical Disturbance Regimes and Extreme Climatic Events

Typical disturbance regimes of the summer humid and ever-humid subtropics include tropical cyclones (i.e. 'hurricanes', 'typhoons'), tornadoes, and monsoons and the landslides, mudslides, and floods that these heavy rainfall events cause (Fig. 3.8). The most important causes of the landslides, mudslides, and floods in the humid subtropics are the *monsoon* rains. The term monsoon is applied generally to tropical and subtropical seasonal reversals in both the atmospheric circulation and associated precipitation (Trenberth et al. 2000). During the summer monsoon, water-laden layers of air that form over the ocean move over large land masses and release the moisture as regular heavy rainfall over the land.

Flooding is common in the summer humid subtropics and is caused by heavy monsoon rains and other heavy rainfall events. About 80% of the annual precipitation falls during the summer monsoon. The large volumes of precipitation can quickly exceed the capacity of river channels, leading to flooding over large areas within short periods. These events result in flooding of rice fields and other crops, destruction or damage to houses and other infrastructure, and contamination of drinking water, leading to the spread of various diseases and diarrhoea (Rasmussen et al. 2015). On the other hand, the inundation of the floodplains is an important nutrient input for agriculture.

Mudslides and landslides can also be caused by monsoon rains. A landslide is the movement of a mass of rock, debris, or earth down a slope. Landslides can be initiated in slopes already on the verge of movement by rainfall, snowmelt, changes in water level, stream erosion, changes in groundwater, earthquakes, volcanic activity, disturbance by human activities, or any combination of these factors (USGS 2021). Mudslides are a type of landslide that tend to flow in channels. One important cause of landslides/mudslides is the monsoon rains. Because of the enormous amounts of rain, the soil quickly loses stability and is either washed away by the floods or detaches from the slope as a landslide or mudslide. The material transported in this way is usually deposited on valley bottoms, channels, or on the sides of the valley slopes. If the soil in these areas becomes water-saturated because of heavy rainfall, there is a risk that the deposited material will remobilize a mudslide with debris flow, thus posing a potential hazard in later monsoon periods (Collins and Jibson 2015).

Tropical cyclones (called 'hurricanes' in the Atlantic and northeast Pacific, 'typhoons' in the northwest Pacific, and just 'tropical cyclones' or 'cyclones' elsewhere) are tropical or sub-tropical large-scale low-pressure systems which generate winds reaching at least wind force 12 on the Beaufort scale ($\geq$117 km/h; 32.5 m/s). Hurricanes are tropical storms that occur in the seas west and east of the American continents. Hurricanes of the same strength in the southern Pacific Ocean and the Indian Ocean are called 'cyclones'. They form in the ITCZ where the trade winds converge and there is a water temperature of over 26.5 °C (Häckel 2012). The large area of warm water results in large areas of rising warm air laden with water vapour and the formation of huge cloud masses. This releases an enormous amount of latent energy, which heats the air inside the clouds. As a result, the air expands together with the remaining moisture that has not yet fallen as rain, leading to further rising of the air. Above the warm sea surface a negative pressure is created, to which moisture-saturated air from the surrounding area flows. The Coriolis force causes the air masses to rotate, creating a large-scale vortex. When a hurricane hits land, it is cut off from its supply of moist air and slowly loses strength. However, it carries an enormous mass of water and generates very strong winds, which cause devastating damage. The consequences of tropical cyclones are storm surges, windbreaks, coastal erosion, landslides, and floods (Kuttler 2013).

Tornadoes are smaller-scale low-pressure systems (often associated with thunderstorms) that result in rapidly rotating columns of air. Tornadoes occur all over the world (including the tropics and subtropics) but are most common in the central USA. They occur when large air masses of different temperatures and humidity collide over a large flat plain. The high dynamic wind pressure and wind speeds of ~430 km/h (~120 m/s) or more in the upper layers of the tornado are the main causes of the high destructive force. In spite of advances in technology, the occurrence of tornadoes is difficult to predict; early warning systems can provide advance warning of between 8 and 13 min before a tornado develops. This, combined with the very high wind speeds, makes tornadoes a particularly dangerous weather system.

In the subtropics, *climate change* is expected to be associated with changes in monsoon intensity and changes in the areas affected by monsoons. Fluctuations in sea surface temperature will have an impact on the amount of moisture the monsoon can absorb and thus influence the amount of precipitation. Monsoon rains will reach new areas, on the one hand, but, on the other hand will occur less frequently in previously typical monsoon regions, potentially with enormous effects on vegetation and the human population (Schiermeier 2006). Climate change is also expected to lead to more frequent and more intense tropical cyclones. Indeed, the global occurrence of category 4 and 5 tropical cyclones has already doubled since 1970 (Lin et al. 2012). It is also expected that there will be a higher frequency of tornadoes and a larger tornado catchment area in the future. As a consequence, so-called 'tornado outbreaks' will become more frequent (Tippett and Cohen 2016).

3.8   Summer Humid Tropics

3.8.1   *Characterization and Ecological Framework*

The zone of the summer humid tropics includes large areas south and north of the Equator with pronounced seasonal differences between rainy and dry seasons. The rain seasonality (850–2000 mm per year) is caused by the seasonal oscillation of the ITCZ between the Tropic of Capricorn (~23.4°N) and the Tropic of Cancer (~23.4°S). In this zone, there are – similar to the humid subtropics – evergreen tropical monsoon forests as well as tropical dry forests, xeric shrublands, and wet and dry savannahs. The zone covers about 16% of the land surface of the Earth, although currently only a fifth of this area can be considered near-natural (Schultz 2016). Many plant species have developed special strategies to survive the 2.5 and 7.5 months of the dry season. These plants include CAM plants[1], bottle trees, and poikilohydric plants. The further away the areas are from the Equator, the longer the dry season lasts. The area is generally frost free, with frosts only occurring in regions furthest away from the Equator and in mountain areas. The savannah areas consist of tropical grasslands, in which the tree growth is possible under year-round warm conditions, in contrast to the steppes of the middle-latitudes, where growth is paused during the cold season. Large areas of this zone have been shaped by overuse by humans for millennia.

---

[1] Crassulacean Acid Metabolism (CAM) plants photosynthesize during the day but only exchange gases at night when the stomata are closed, thereby enabling plants to increase water-use efficiency.

3.8.2   Typical Disturbance Regimes and Extreme Climatic Event

The typical disturbance regimes of the summer humid tropics include extreme rainfall events with soil erosion, droughts with fire, insect outbreaks, and mega-herbivory.

The *extreme rainfall events* (e.g. monsoon rains, tropical cyclones) are seasonal; the seasonality can be stronger, weaker, or occasionally completely absent (i.e. the rainfall is evenly distributed throughout the year). These events lead to severe soil erosion, especially in the savannahs and dry forests.

Fire is a natural disturbance factor in the tropics, especially in the dry seasons (even long before human influence). Many plant species are adapted to fire by sprouting more strongly directly after a fire (e.g. grasses) or seeds germinating after a fire (e.g. *Eucalyptus* spp.). Other species have developed thick bark (e.g. *Acacia* spp.) or survive fires by regrowing from vegetative organs that are present in the subsoil (e.g. some plants with tubers). Since the use of fire for hunting purposes, humans have significantly influenced the fire regime in these biomes (Walter and Breckle 1999). A ground fire usually only has a short-term effect on rapidly regenerating grasslands. In dry forests, on the other hand, ground fires can also pass through after several years of accumulation of dry litter (e.g. in eucalypt forests). If ground and bush fires have not occurred for some time and the amount of dry fuel has accumulated over the years, the danger of a crown fire increases. However, even this provides opportunities for the beginning of a new regeneration cycle and the establishment of new species.

There is an enormous diversity of *insect* species in the tropics. Many of them are leaf herbivores (locusts, butterflies, and moths) or suckers (cicadas, bugs). Flying insect species usually prefer dry weather conditions during their adult life. During the dry season, the reproduction and dispersion rates of herbivorous insects can also reach catastrophic proportions for humans (migratory locusts). In particular, many acacia species are well adapted to defend against herbivory, for example, through the presence of alkaloids in the leaves; or ant–plant mutualisms (Palmer et al. 2008).

The *El Niño–Southern Oscillation* (ENSO) causes disruption to the normal climate patterns, leading to irregular heavy rainfall events in some places (e.g. in the deserts and dry forests of the west coast of South America, Fig. 3.9) and droughts in other places. Because of the variable recurrence intervals of these events, some species have adapted to the irregular dry and wet periods (e.g. by evolving seeds that can remain dormant on the desert floor for long periods). In an El Niño year, oceanic and atmospheric currents change strongly in partly opposite directions for several months, thus creating intensively vegetated savannahs from deserts. Such a strong change from dry to humid is not found in any seasonal climate. Thus, the El Niño events act as time-limited windows of development and invasion in a landscape with usually only minor changes (Holmgren and Scheffer 2001).

The vegetation of the arid regions of the Indian subcontinent is adapted to the annual monsoon rains with, for example, deciduous species and species with specialized regeneration buds. In wide areas of the alternating humid tropics, there is a strong tendency towards succulent growth in order to survive the pronounced dry season – stem succulent cacti in South America, baobab (*Adansonia digitata* L.) in Africa, pawpaw (*Carica papaya* L.) in Australia, and leaf succulent *Agave* spp. and *Opuntia* spp. in Central America (Pfadenhauer and Klötzli 2015).

Mega-herbivory is an important factor in protecting savannahs from scrub encroachment and maintaining their ability to regenerate. The large savannah areas of the semi-humid tropics (eastern South America, East Africa, South Australia) are the migration areas of large herds of animals that follow the growth of grass at the beginning of the rainy season. Humans have influenced this regime by grazing farm animals.

Land-use change will manifest itself in the semi-humid tropics primarily through overgrazing by livestock, loss of savannah habitats owing to increasing large-scale cultivation of soybean, and loss of tree cover owing to excessive groundwater abstraction. Climate change is predicted to lead to a more pronounced rain–drought seasonality with increased fire frequency and an increase in monsoon and El Niño events (IPCC 2013; Cai et al. 2014).

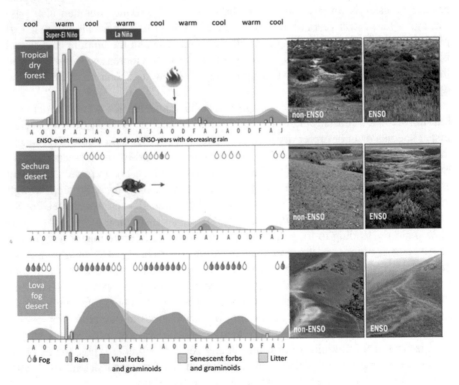

Fig. 3.9 ENSO (El Niño–Southern Oscillation)-induced heavy rainfall events on the west coast of Peru and Argentina have led to the greening of extensive dry forests and the Sechura Desert. (Photo: A. Jentsch)

3.9 Ever-Humid Tropics

3.9.1 Characterization and Ecological Framework

The ever-humid tropics occupy the land area around the ITCZ and account for about 8% of the total land area of the Earth (Schultz 2016). This zone experiences extremely high annual solar radiation (700–800 × 10^8 kJ/ha) and very high precipitation (up to 8000 mm/year) and a dry season that lasts less than 3 months. The only exceptions are areas in Southeast Asia, where there can be rare periods of up to 6 months of drought in El Niño years.

Because of the steep angle of the sun, the ever-humid tropics are characterized by a 'time of day climate'. The average daily temperature is 25–27 °C all year round and fluctuates only slightly during the year. The difference between day and night temperatures of 6–11 °C exceeds this annual amplitude (Kuttler 2013). There is no seasonal fluctuation of day length. The high availability of sunlight and water favours a high annual net primary production of 20–30 tons/ha.

The ever-humid tropics are covered almost everywhere by evergreen tropical rainforest; there are some tropical moorlands in upland areas of Africa, South America, Central America, North America, and New Guinea. The forests grow on relatively nutrient-poor, mostly strongly acidic topsoils with only a thin humus layer. Trees reach heights of up to 60 m and often a trunk diameter of 2 m. About 75–90% of the biomass grows above ground; the nutrient cycle is very short. An exception is the evergreen rainforest growing on peat, where the nutrient cycle is longer (Richter 2001). Because of the dense vegetation, only about 1–3% of the sunlight reaches the ground. Also, a large part of the precipitation is captured by the vegetation as interception. These two factors lead to large microclimatic differences in the various vertical layers within a tree stand and favour a large number of epiphytes. Almost 95% of all epiphyte species on Earth grow in the tropics.

3.9.2 Typical Disturbance Regimes and Extreme Climatic Event

Typical disturbance regimes in the humid tropics include insect herbivory, heavy rainfall, storm events with nutrient leaching, and occasionally volcanic eruptions.

Because of the enormous net primary production in the ever-humid tropics, there are almost unlimited resources for *herbivorous insects*. Insect herbivory can affect the leaf mass, the seeds, as well as the wood and the roots. A well-known example are leaf-cutting ants, which harvest large quantities of leaves from trees. By harvesting such quantities of leaves, the ants also play an important role in nutrient cycling and maintenance of soil fertility in tropical rainforests (e.g. Farji-Brener and Werenkraut 2017; Swanson et al. 2019). Most tropical plants react to leaf feeding by depositing secondary substances in the affected leaves, making them unattractive or

Fig. 3.10 Mudflows and flooding after heavy precipitation in Amazonia, Peru. (Photos: M. Richter)

toxic for further feeding. Other species perform a foliar flushing by quickly sprouting again with new leaves. Almost all herbivores are closely embedded within the stratigraphy of the tropical rainforest, many of them feeding exclusively in the crown of the trees, such as insects and herbivorous mammals (e.g. some monkeys).

Heavy rainfall and storm events lead to nutrient leaching of the already humus-poor soils (Fig. 3.10). The great rivers from the evergreen rainforests transport large amounts of fine sediments. This continuous leaching leads to open nutrient cycles of, for example, phosphorus, nitrogen, and carbon. To compensate for the losses, all unused dead biomass is quickly reintroduced into the cycle of living biomass (van Schaik and Mirmanto 1985).

Tropical cyclones, which lose much of their energy when they pass from the sea over land, can lead to severe mechanical disturbances such as crown fracture and uprooting. In semi-natural tropical forests, many tree species can react very flexibly to such disturbance and quickly close gaps that have arisen. Such storms offer the possibility for advanced regeneration, waiting on 'standby' in the undergrowth, to grow up into the canopy. Such rainforest dynamics and ecological instabilities are seen as important evolutionary opportunities in the otherwise species- and niche-saturated zone (Johns 1990).

Volcanic activity (Central Africa, Indonesia, Philippines) and strong tectonic activities (southern Central America to Colombia, Java, Sumatra) are common in some areas of the ever-humid tropics. From a global point of view, these disturbance regimes occur only rarely and selectively but are often enormously destructive for the affected region. Volcanic eruptions cause a variety of massive landscape-shaping disturbances, including lava flows and pyroclastic flows. Eruptions can also cause earthquakes (and vice versa), and the mass collapse of slopes caused by volcanic eruptions and earthquakes can cause tsunamis. Particularly in the tropical zone of Southeast Asia and Central/South America tectonic induction are most volcanically active regions ("Ring of Fire"). The largest eruptions in the last 200 years have occurred in Southeast Asia – Krakatoa, Indonesia (1883); Pinatubo, Philippines (1991); Tambora, Indonesia (1815) (Schmincke 2013). Depending on their chemical composition, volcanic lavas and ashes are of great importance for nutrient input.

Land-use change represents the greatest current and future threat to the ever-humid tropics. At the moment, only about 50% of the areas in this zone can be described as semi-natural. For example, the annual rate of deforestation in the Brazilian Amazon was calculated as 14,835 km^2 (±4706 km^2) in the period 2000–2015 (Silva Junior et al. 2020). Illegal logging and deforestation not only leads to a decrease in forest area, an increase in CO_2 emissions (van der Werf et al. 2009), and loss of biodiversity (Solar et al. 2015), but it also opens up the area for even further development and deforestation (Fearnside 2005). These developments, including the provision of transport infrastructure, mean that commercial production of commodities such as beef, soy, and palm oil become profitable on the cleared land (Pfaff et al. 2013; Nepstad et al. 2014). From the cleared forest areas, large companies financed by Western and Asian investors expand further into the untouched rainforest (Fearnside and Figueiredo 2015). The production of genetically modified soybeans is only possible with heavy herbicide use (especially glyphosate). This in turn poses a long-term threat to the surrounding forest areas and the local population (Cerdeira et al. 2007; Almeida et al. 2017).

References

Almeida VES, Friedrich K, Tygel AF, Melgarejo L, Carneiro FF (2017) Use of genetically modified crops and pesticides in Brazil: growing hazards. Ciencia Saude Coletiva 22:3333–3339

Bernes C, Bråthen KA, Forbes BC, Speed JDM, Moen J (2015) What are the impacts of reindeer/caribou (*Rangifer tarandus* L.) on arctic and alpine vegetation? A systematic review. Env Evid 4:4

Bliss LC (1971) Arctic and alpine plant life cycles. Annu Rev Ecol Syst 2:405–438

Buhk C, Meyn A, Jentsch A (2007) The challenge of plant regeneration after fire in the Mediterranean Basin: scientific gaps in our knowledge on plant strategies and evolution of traits. Plant Ecol 192:1–19

Burton PJ, Jentsch A, Walker LR (2020) The ecology of disturbance interactions: characterization, prediction and the potential of cascading effects. Bioscience 70:854–870

Cai W, Borlace S, Lengaigne M, van Rensch P, Collins M, Vecchi G, Timmermann A, Santoso A, McPhaden MJ, Wu L, England MH, Wang G, Guilyardy E, Fei-Fei J (2014) Increasing frequency of extreme El Niño events due to greenhouse warming. Nat Clim Chang 4:111–116

Cerdeira AL, Gazziero DLP, Duke SO, Matallo MB, Spadotto CA (2007) Review of potential environmental impacts of transgenic glyphosate-resistant soybean in Brazil. J Environ Sci Health B 42:539–549

Cerritos R (2011) Grasshoppers in agrosystems: pest or food. CAB Rev Perspect Agric Vet Sci Nutr Nat Resour 6(017). https://doi.org/10.1079/PAVSNNR20116017

Chmielewski F-M, Götz K-P, Weber KC, Moryson S (2018) Climate change and spring frost damages for sweet cherries in Germany. Int J Biometeorol 62:217–228

Collins BD, Jibson RW (2015) Assessment of existing and potential landslide hazards resulting from the April 25, 2015 Gorkha, Nepal earthquake sequence. U.S. Geological Survey: Open-File Report 2015–1142, 50 p

Deng L, Sweeney S, Shangguan ZP (2014) Grassland responses to grazing disturbance: plant diversity changes with grazing intensity in a desert steppe. Grass Forage Sci 69:524–533

Ernakovich JG, Hopping KA, Berdanier AB, Simpson RT, Kachergis EJ, Steltzer H, Wallenstein MD (2014) Predicted responses of arctic and alpine ecosystems to altered seasonality under climate change. Glob Chang Biol 20:3256–3269

Farji-Brener AG, Werenkraut V (2017) The effects of ant nests on soil fertility and plant performance: a meta-analysis. J Anim Ecol 86:866–877. https://doi.org/10.1111/1365-2656.12672

Fauria MM, Johnson EA (2008) Climate and wildfires in the North American boreal forest. Philos Trans R Soc B Biol Sci 363:2315–2327

Fearnside PM (2005) Deforestation in Brazilian Amazonia: history, rates, and consequences. Conserv Biol 19:680–688

Fearnside PM, Figueiredo AM (2015) China's influence on deforestation in Brazilian Amazonia: a growing force in the state of Mato Grosso. Global Economic Governance Initiative, University of Boston: Discussion Paper 2015-3, 51 p

Goulet F (1995) Frost heaving of forest tree seedlings: a review. New For 9:67–94

Häckel H (2012) Meteorologie, 7th edn. UTB, Ulmer, 447 p

Heavilin J, Powell J, Logan JA (2007) Dynamics of mountain pine beetle outbreaks. In: Johnson EA, Miyanishi K (eds) Plant disturbance ecology – the process and the response. Academic, London/New York, pp 527–553

Holdridge LR (1967) Life zone ecology. Tropical Science Centre, San Jose, 206 p

Holmgren M, Scheffer M (2001) El Niño as a window of opportunity for the restoration of degraded arid ecosystems. Ecosystems 4:151–159

Hugelius G, Strauss J, Zubrzycki S, Harden JW, Schuur EAG, Ping C-L, Schirrmeister L, Grosse G, Michaelson GJ, Koven CD (2014) Estimated stocks of circumpolar permafrost carbon with quantified uncertainty ranges and identified data gaps. Biogeosciences 11:6573–6593

IPCC (2012) Managing the risks of extreme events and disasters to advance climate change adaptation. A Special Report of Working Groups I and II of the Intergovernmental Panel on Climate Change [Field CB, Barros V, Stocker TF, Qin D, Dokken DJ, Ebi KL, Mastrandrea MD, Mach KJ, Plattner G-K, Allen SK, Tignor M, Midgley PM (eds)]. Cambridge University Press, Cambridge/New York, 582 p

IPCC (2013) Climate change 2013: the physical science basis. Contribution of Working Group I to the Fifth Assessment Report of the Intergovernmental Panel on Climate Change [Stocker TF, Qin D, Plattner G-K, Tignor M, Allen SK, Boschung J, Nauels A, Xia Y, Bex V, Midgley PM (eds)]. Cambridge University Press, Cambridge/New York, 1535 p. https://www.ipcc.ch/report/ar5/index.shtml

Jentsch A, Beierkuhnlein C (2011) Explaining biogeographical distributions and gradients: floral and faunal responses to natural disturbances. In: Millington A, Blumler M, Schickhoff U (eds) The SAGE handbook of biogeography. SAGE, London, pp 191–211

Jentsch A, White PS (2019) A theory of pulse dynamics and disturbance in ecology. Ecology 100:e02734

Johns R (1990) The illusionary concept of the climax. In: Baas P, Kalkman K, Geesink R (eds) The plant diversity of Malesia. Kluwer Academic Publishers, Dordrecht, pp 133–146

Johnson EA, Miyanishi K (2007) Plant disturbance ecology: the process and the response. Elsevier, Amsterdam, 698 p

Körner C (2003) Alpine plant life: functional plant ecology of high mountain ecosystems, 2nd edn. Springer, Berlin, 344 p

Kovats RS, Valentini R, Bouwer LM, Georgopoulou E, Jacob D, Martin E, Rounsevell M, Soussana J-F (2014) Europe. In: Climate change 2014: impacts, adaptation, and vulnerability. Part B: regional aspects. Contribution of Working Group II to the Fifth Assessment Report of the Intergovernmental Panel on Climate Change. Cambridge University Press, Cambridge/ New York, pp 1267–1326

Kroh El, Boonmofonb J, Dahov M, Dhuuu Whaat fl Thuuu f Vlhhuldhh i i H iijiiiiji i iiiiiiiiii K, Millbau A, Dorrepaal E (2018) Winter warming effects on tundra shrub performance are species-specific and dependent on spring conditions. J Ecol 106:599–612

Kreyling J, Stahlmann R, Beierkuhnlein C (2012) Spatial variation in leaf damage of forest trees and the regeneration after the extreme spring frost event in May 2011. Allg Forst Jagdztg 183:15–22

Kuttler W (2013) Klimatologie, 2nd edn. UTB, Ferdinand Schöningh, Paderborn, 306 p

Lees KJ, Quaife T, Artz R, Khomik M, Clark JM (2018) Potential for using remote sensing to estimate carbon fluxes across northern peatlands: a review. Sci Total Environ 615:857–874

Lin N, Emanuel K, Oppenheimer M, Vanmarcke E (2012) Physically based assessment of hurricane surge threat under climate change. Nat Clim Chang 2:462–467

Münchner Rück (1999) Naturkatastrophen in Deutschland: Schadenerfahrungen und Schadenpotentiale. Münchener Rückversicherungs-Gesellschaft, 98 p

Münchner Rück (2017) Zwischen Hoch und Tief – Wetterrisiken in Mitteleuropa. Münchener Rückversicherungs-Gesellschaft, 56 p

Nepstad D, McGrath D, Stickler C, Alencar A, Azevedo A, Swette B, Bezerra T, Digiano M, Shimada J, Da Motta RS (2014) Slowing Amazon deforestation through public policy and interventions in beef and soy supply chains. Science 344:1118–1123

Odenthal-Kahabka J (2004) Orkan "Lothar" – Bewältigung der Sturmschäden in den Wäldern Baden-Württembergs: Kap. 2: Ausmass und Ursache der Schäden. Schr.reihe Landesforstverwalt. Baden-Württ 83:37–90

Oteros-Rozas E, Ontillera-Sánchez R, Sanosa P, Gómez-Baggethun E, Reyes-García V, González JA (2013) Traditional ecological knowledge among transhuman pastoralists in Mediterranean Spain. Ecol Soc 18:33

Oxfam International (2013) Learning the lessons? Assessing the response to the 2012 food crisis in the Sahel to build resilience for the future. Oxfam Briefing Paper 168, 40 p

Palmer TM, Stanton ML, Young TP, Goheen JR, Pringle RM, Karban R (2008) Breakdown of an ant-plant mutualism follows the loss of large herbivores from an African savanna. Science 319:192–195

Pfadenhauer JS, Klötzli FA (2015) Vegetation der Erde: Grundlagen, Ökologie, Verbreitung. Springer Spektrum, Berlin, 643 p

Pfaff A, Amacher GS, Sills EO, Coren MJ, Streck C, Lawlor K (2013) Deforestation and forest degradation: concerns, causes, policies, and their impacts. In: Shogren JF (ed) Encyclopedia of energy, natural resource, and environmental economics. Elsevier, Amsterdam, pp 144–149

Ping CL, Jastrow JD, Jorgenson MT, Michaelson GJ, Shur YL (2015) Permafrost soils and carbon cycling. Soil 1:147–171

Price DT, Alfaro R, Brown K, Flannigan M, Fleming R, Hogg E, Girardin M, Lakusta T, Johnston M, McKenney D (2013) Anticipating the consequences of climate change for Canada's boreal forest ecosystems. Environ Rev 21:322–365

Rasmussen K, Hill A, Toma V, Zuluaga M, Webster P, Houze R (2015) Multiscale analysis of three consecutive years of anomalous flooding in Pakistan. Q J Roy Meteor Soc 141:1259–1276

Richter M (2001) Vegetationszonen der Erde. Klett-Perthes, Gotha, 441 p

Richter M, Ise M (2005) Monitoring plant development after El Niño 1997/98 in Northwestern Perú (Dauerbeobachtung der Pflanzenentwicklung nach El Niño 1997/98 in Nordwest-Peru). Erdkunde 59:136–155

Rundel PW (1999) Disturbance in Mediterranean-climate shrublands and woodlands. In: Walker LR (ed) Ecosystems of disturbed ground. Elsevier, Amsterdam, pp 271–285

Saccone P, Hoikka K, Virtanen R (2017) What if plant functional types conceal species-specific responses to environment? Study on arctic shrub communities. Ecology 98:1600–1612

Sala OE, Maestre FT (2014) Grass-woodland transitions: determinants and consequences for ecosystem functioning and provisioning of services. J Ecol 102:1357–1362

Scheffer M, Carpenter S, Foley JA, Folke C, Walker B (2001) Catastrophic shifts in ecosystems. Nature 413:591–596

Schiermeier Q (2006) Extreme monsoons on the rise in India. Nature News. https://doi.org/10.1038/news061127-12

Schmincke H-U (2013) Vulkanismus, 4th edn. WBG, Darmstadt, 264 p

Schultz J (2016) Die Ökozonen der Erde, 5th edn. UTB, Eugen Ulmer, Stuttgart, 332 p

Schuur EAG, Abbott B (2011) High risk of permafrost thaw. Nature 480:32–33

Schuur EAG, McGuire AD, Schädel C, Grosse G, Harden JW, Hayes DJ, Hugelius G, Koven CD, Kuhry P, Lawrence DM (2015) Climate change and the permafrost carbon feedback. Nature 520:171–179

Silva Junior CHL, Aragão, LEOC, Anderson LO, Fonseca MG, Shimabukuro YE, Vancutsem C, Achard F, Beuchle R, Numata I, Silva CA, Maeda EE, Longo M, Saatchi SS (2020) Persistent collapse of biomass in Amazonian forest edges following deforestation leads to unaccounted carbon losses. Sci Adv 6: eaaz8360. https://doi.org/10.1126/sciadv.aaz8360

Sklavou P, Karatassiou M, Parissi Z, Galidaki G, Ragkos A, Sidiropoulou A (2017) The role of transhumance on land use/cover changes in Mountain Vermio, northern Greece: a GIS based approach. Not Bot Hortic Agrobot Cluj-Napoca 45:589–596

Solar RRC, Barlow J, Ferreira J, Berenguer E, Lees AC, Thomson JR, Louzada J, Maués M, Moura NG, Oliveira VHF (2015) How pervasive is biotic homogenization in human-modified tropical forest landscapes? Ecol Lett 18:1108–1118

Swanson AC, Schwendenmann L, Allen MF, Aronson EL, Artavia-León A, Dierick D, Fernandez-Bou AS, Harmon TC, Murillo-Cruz C, Oberbauer SF, Pinto-Tomás AA, Rundel PW, Zelikova TJ (2019) Welcome to the *Atta* world: a framework for understanding the effects of leaf-cutter ants on ecosystem functions. Funct Ecol 33:1386–1399

Tarnocai C, Canadell JG, Schuur EAG, Kuhry P, Mazhitova G, Zimov S (2009) Soil organic carbon pools in the northern circumpolar permafrost region. Global Biogeochem Cycle 23:GB2023

Tippett MK, Cohen JE (2016) Tornado outbreak variability follows Taylor's power law of fluctuation scaling and increases dramatically with severity. Nat Comm 7:article 10668

Trenberth KE, Stepaniak DP, Caron JM (2000) The global monsoon as seen through the divergent atmospheric circulation. J Clim 13:3969–3993

Trenberth KE, Jones PD, Ambenje P, Bojariu R, Easterling D, Klein Tank A, Parker D, Rahimzadeh F, Renwick JA, Rusticucci M, Soden B, Zhai P (2007) Observations: surface and atmospheric climate change. In: The physical science basis. Contribution of working group I to the fourth assessment report of the Intergovernmental Panel on Climate Change. Cambridge University Press, Cambridge/New York, pp 236–336

Troll C (1944) Strukturböden, Solifluktion und Frostklimate der Erde. Geol Rundsch 34:545–694

Turner MG, Calder WJ, Cumming GS, Hughes TP, Jentsch A, LaDeau SL, Lenton TM, Shuman BN, Turetsky MR, Ratajczak Z, Williams JWA, Williams AP, Carpenter SR (2020) Climate change, ecosystems, and abrupt change: science priorities. Philos Trans B 375:20190105

USGS (2021) What is a landslide and what causes one? United States Geological Survey. https://www.usgs.gov/faqs/what-a-landslide-and-what-causes-one. Accessed 15 Sept 2021

van As D, Fausto RS, Steffen K, Ahlstrøm AP, Andersen SB, Andersen ML, Box JE, Charalampidis C, Citterio M, Colgan WT, Edelvang K, Larsen SH, Nielsen S, Veicherts M, Weidick A (2014) Katabatic winds and piteraq storms: observations from the Greenland ice sheet. Geol Surv Den Greenl:83–86

van der Werf GR, Randerson JT, Giglio L, Gobron N, Dolman A (2008) Climate controls on the variability of fires in the tropics and subtropics. Global Biogeochem Cycle 22:GB3028

van der Werf GR, Morton DC, DeFries RS, Olivier JGJ, Kasibhatla PS, Jackson RB, Collatz GJ, Randerson JT (2009) CO_2 emissions from forest loss. Nat Geosci 2:737–738

van Schaik CP, Mirmanto E (1985) Spatial variation in the structure and litterfall of a Sumatran rain forest. Biotropica:196–205

Walker LR (1999) Ecosystems of disturbed ground. Elsevier/Amsterdam, New York, 868 p

Walker DA, Epstein HE, Romanovsky VE, Ping CL, Michaelson GJ, Daanen RP, Shur Y, Peterson RA, Krantz WB, Raynolds MK, Gould WA, GG, Nicolsky DJ, Vonlanthen CM, Kade AN, Kuss P, Kelley AM, Munger CA, Tarnocai CT, Matveyeva NV, Daniëls FJA (2008) Arctic patterned-ground ecosystems: a synthesis of field studies and models along a North American Arctic Transect. J Geophys Res Biogeo 113:G03S01

Walter H, Breckle SW (1999) Vegetation und Klimazonen: Grundriss der globalen Ökozonen, 7th edn. UTB, Eugen Ulmer, Stuttgart

Werner-Gnann B, Minardi S, Ganninger-Hauck D (2017) Obstbauern in Schockstarre. BWagrar: published online 20 April 2017

White PS, Jentsch A (2001) The search for generality in studies of disturbance and ecosystem dynamics. Prog Bot 62:399–449

Yu Z (2012) Northern peatland carbon stocks and dynamics: a review. Biogeosciences 9:4071–4085

Yu C, Zhang J, Pang XP, Wang Q, Zhou YP, Guo ZG (2017) Soil disturbance and disturbance intensity: response of soil nutrient concentrations of alpine meadow to plateau pika bioturbation in the Qinghai-Tibetan Plateau, China. Geoderma 307:98–106

Zhang T, Barry R, Knowles K, Ling F, Armstrong R (2003) Distribution of seasonally and perennially frozen ground in the Northern Hemisphere. In: Permafrost: Proceedings of the 8th International Conference on Permafrost, 21–25 July, Zurich, Switzerland. AA Balkema Publishers, Lisse, pp 1289–1294

Part II
Concepts

Chapter 4
Disturbance and Biodiversity

Rupert Seidl ⑩, Jörg Müller ⑩, and Thomas Wohlgemuth ⑩

Abstract Biodiversity describes the variety of living creatures and habitats within ecosystems. It is the basis for the functioning of ecosystems but is currently strongly influenced by human activities. Natural disturbances promote biodiversity by creating landscape heterogeneity, releasing resources, reducing the dominance of competitive species, and increasing the diversity of niches. According to the intermediate disturbance hypothesis, the positive effect on biodiversity is at its peak at intermediate disturbance intensity, size, and frequency. In detail, however, the effect of disturbances on biodiversity depends on a multitude of factors that vary greatly with type of disturbance, ecosystem productivity, and spatial as well as trophic level of observation.

Keywords Biological diversity · Ecosystem dynamics · Intermediate disturbance hypothesis · Landscape heterogeneity · Spatial context

R. Seidl (✉)
Ecosystem Dynamics and Forest Management Group, School of Life Sciences, Technical University of Munich, Freising, Germany

Berchtesgaden National Park, Berchtesgaden, Germany
e-mail: rupert.seidl@tum.de

J. Müller
Chair of Animal Ecology and Tropical Biology (ZooIII), Biocenter, Julius-Maximilian University of Würzburg, Würzburg, Germany

T. Wohlgemuth
Forest Dynamics Research Unit, Swiss Federal Institute for Forest, Snow and Landscape Research WSL, Birmensdorf, Switzerland

T. Wohlgemuth et al. (eds.), *Disturbance Ecology*, Landscape Series 32,
https://doi.org/10.1007/978-3-030-98756-5_4

4.1 Introduction

The term biodiversity, which is used as an abbreviation for 'biological diversity', refers to the diversity of all living creatures and habitats. Biodiversity includes genetic diversity within species, interspecies diversity, and diversity of ecosystems (UN 1992). From an anthropogenic point of view, biodiversity has intrinsic value because it describes the totality of the biological heritage on our planet. However, it also has functional value for humans: a wide array of ecosystem functions and services, which contribute essentially to human well-being (see Chap. 18), are promoted by biodiversity (Hassan et al. 2005). Biodiversity is increasingly under pressure from land use, the exploitation of natural resources, and human-induced climate change (Butchart et al. 2010). The current rate of species loss exceeds the natural extinction rate by a factor of 100–1000, making the conservation of biodiversity a key challenge in the anthropogenic use of natural systems (Rockström et al. 2009). This chapter synthesizes the effects of disturbances on biodiversity. Not only do disturbances influence diversity, but diversity also affects disturbance (Hughes et al. 2007). The latter aspect will not be discussed here but will be addressed in Chaps. 7, 8, 9, 10, 11, 12, 13, 14, and 15 because of the strong dependence on the respective disturbance agent.

4.2 Disturbances Promote Biodiversity: But Not Always and Not Everywhere

Although disturbances cause the loss of live biomass and, in some extreme cases, even the local and temporary loss of species, they have a predominantly positive impact on biodiversity in general. This seemingly paradoxical relationship is inherent in the nature of disturbances: because of the abrupt loss of living biomass, the local dominance of some species, such as beech (*Fagus sylvatica* L.) in Central European lowland forests, gets disrupted (Wohlgemuth et al. 2002a). As a result, resources become temporarily available, which then can be utilized by other species. In addition, natural disturbances rarely lead to a total loss of live biomass in an ecosystem, which increases its structural heterogeneity and enhances the diversity of niches. In a global synthesis of publications on the effects of wind, fire, and bark beetle disturbances on diversity in temperate and boreal forest ecosystems, Thom and Seidl (2016) documented a predominantly positive effect of disturbances on biodiversity. Species richness responded positively to disturbances in 44.7% of the 262 studies examined. If studies that did not detect any disturbance-related change in biodiversity are also taken into account, more than two-thirds of the cases investigated show neutral or positive effects of disturbance on species richness (Fig. 4.1). These results are largely independent of the type of disturbance – that is, whether

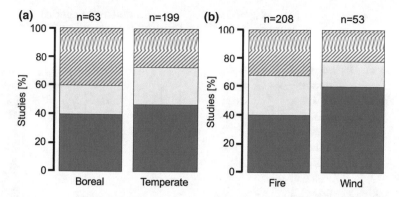

Fig. 4.1 The effect of disturbances on species richness in forest ecosystems (**a**) by biome and (**b**) disturbance agent based on a meta-analysis by Thom and Seidl (2016). Positive (dark grey), negative (hatched), and neutral effects (light grey) of disturbances on species richness are distinguished. The number of studies considered in each category is indicated by n

the disturbance was caused by wind, fire, or bark beetles – and of the biome. According to this meta-analysis, the number of species after a severe disturbance increases by +35.6% on average, relative to an undisturbed system (Thom and Seidl 2016).

However, there are also a considerable number of studies that document negative effects of disturbances on diversity. For a better understanding of the range of impacts that disturbances have on biodiversity, there is a need for a more in-depth look at the affected species, their habitats, and their habitat requirements: after the outbreak of the European spruce bark beetle (*Ips typographus* L.; see Chap. 12) in the Bavarian Forest National Park, Beudert et al. (2015) showed that especially those species that can take advantage of the resources freed up in the process (such as more light on the forest floor) benefitted from the disturbance. Lichens, for example, profit from the conditions after bark beetle infestation, while wood-decomposing fungi decreased as a result of the disturbance (Fig. 4.2). Although the amount of deadwood increased strongly as a consequence of the bark beetle outbreak, the deadwood dried out quickly after losing the bark in the absence of a closed canopy, which reduced the habitat quality for fungi living on wood. Furthermore, the dominance of one decomposer species (*Fomitopsis pinicola*) reduced diversity via competition (Hagge et al. 2019). However, the effects of disturbances must always be assessed in the context of the landscape as a whole: species temporarily affected negatively by disturbances can usually persist in undisturbed parts of a landscape. In the remainder of the chapter, the complex relationship between disturbances and biodiversity will be further examined by looking at the role of disturbance severity, temporal dynamics, and spatial scales.

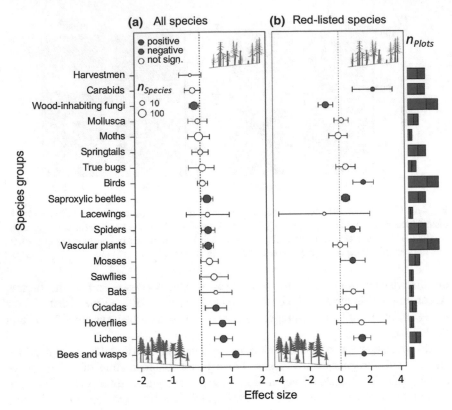

Fig. 4.2 Effect of a bark beetle outbreak on different species groups. (After Beudert et al. 2015)

4.3   How Disturbance Activity Affects Biodiversity

Connell (1978) formulated the Intermediate Disturbance Hypothesis (IDH) on the relationship between disturbance activity and its effects on biodiversity. It states that the diversity of species is highest under intermediate disturbance (size and frequency), while systems with frequent and/or severe disturbances as well as those with rare and/or low-severity disturbances show reduced diversity (Fig. 4.3). This hypothesis is based on the expectation that in systems with high disturbance activity primarily disturbance-adapted species ('pioneer species') prevail, whereas in less-disturbed systems, mainly climax species, that is, mostly long-lived and competitive species, dominate. Only intermediate levels of disturbance allow species from both groups to coexist.

Since its formulation four decades ago, the IDH has been intensively discussed and thoroughly tested in the ecological literature. Studies in different ecosystems have found results that support the IDH as well as results that contradict it (Mackey and Currie 2001; Molino and Sabatier 2001; Roxburgh et al. 2004; Shea et al. 2004; Zhang et al. 2014). This lack of generality of the IDH and the ambiguity in its

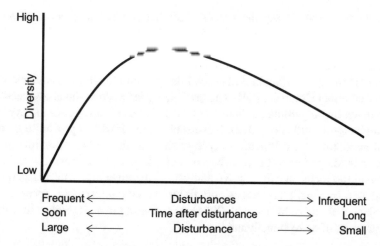

Fig. 4.3 The intermediate disturbance hypothesis states that the diversity of a system is highest under moderate levels of disturbance. (After Connell 1978)

evaluation could be attributed to a number of factors in further investigations. In experiments with microbes, Hall et al. (2012), for example, were able to show that the form of the relationship between disturbance and diversity depends not only on the general strength of the disturbance but also on the interaction between central parameters of the disturbance regime, for example, disturbance frequency and strength (see Chap. 2). Besides the characteristics of the disturbance regime, eco-system productivity is another factor that influences the relationship between distur-bance and diversity. Productivity and disturbance have opposing effects on the competition between species and thus also on species composition and biodiversity of an ecosystem: high productivity generally promotes highly competitive species, while severe disturbances and a high frequency of disturbances favour less competi-tive species that are, however, often responsive and fast growing. Consequently, low biodiversity is expected from combinations of either high productivity and low dis-turbance or low productivity and high disturbance (Huston 1994). The productivity of a system influences the position of the diversity optimum predicted by the IDH: at high productivity, the biodiversity optimum is reached under greater levels of disturbance (i.e. further to the right in Fig. 4.3) than in less productive systems (Kondoh 2001).

Besides these processes modifying the pattern predicted by the IDH, the influ-ence of the trophic level of investigation has been studied. Connell (1978) formu-lated the IDH for tropical rainforests and coral reefs, and the rationale for a unimodal relationship between disturbance and diversity was based on species interactions within one trophic level (i.e. within species that are on the same level in the food web, e.g. primary producers, consumers, or decomposers). Wootton (1998) demon-strated that part of the divergent observations regarding the IDH can be explained by the different trophic levels being considered. While the IDH is mostly accurate within a trophic level and especially holds for the basal levels of the trophic web (primary producers), it loses most of its predictive power when biodiversity across several trophic levels is considered. Another reason why observations of the relationship between disturbance and biodiversity are not consistent is the multitude of different biodiversity indicators used (Svensson et al. 2012). Recently, in addition to these methodological difficulties in its evaluation, more fundamental questions have been raised about Connell's IDH. The criticism goes beyond the above-mentioned lack of generality regarding predictions of the IDH and even suggests the hypothesis to be logically invalid (Fox 2012). Nevertheless, the IDH has proven to be a very fruitful hypothesis for disturbance ecology since in the course of its investigation the understanding of disturbance effects on ecosystems has progressed significantly. After an initial focus on a descriptive analysis of the diversity–disturbance relationship, more recent work has increasingly focused on the underlying processes of the IDH (Shea et al. 2004; Miller et al. 2011), contributing to an improved mechanistic understanding of ecosystem dynamics.

4.4 Disturbances, Biodiversity, and Changes in Ecosystems Over Time

When considering the effects of disturbances on biodiversity, the temporal dynamics of ecosystems play an important role. Ecosystems are subject to spatio-temporal fluctuations and can follow different development pathways. In forest ecosystems, for example, the natural sequence of species and communities over time is called succession, whereby each development stage is inherently diverse. Disturbances generally lead to a reset of vegetation development. Therefore, the effects of disturbance on diversity are affected by the system-dependent speed along such development trajectories (Aichinger 1951; Johst and Huth 2005). In the past decade, the biodiversity value of early-seral stages (i.e. communities emerging shortly after a disturbance) has been explicitly emphasized (Swanson et al. 2011). While earlier studies generally assumed a local homogenization through disturbance, more recent studies have shown that ecosystems are often comparable to late successional systems in terms of their complexity after disturbance (Donato et al. 2012). Usually, however, conservation efforts focus on late seral stages such as old-growth forests since these systems and their associated taxa are largely absent in landscapes with a long history of intensive human land use, as is the case in Central Europe. This absence of late successional stages is particularly problematic in the case of slow-growing and long-lived ecosystems such as forests, as the complex structures of old-growth forests take several centuries to develop. In addition to human activity, natural disturbances represent a risk for the few remaining late successional ecosystems and the species they contain. For example, a climate-induced increase in disturbances (Seidl et al. 2014b; see also Chap. 16) can lead to a further reduction of such habitats (Senf et al. 2021), which could negatively impact associated specialized taxa. To counteract this potentially negative effect of disturbances, Frelich

(2002) underlines the importance of large, contiguous protected areas: while smaller protected areas have a lower risk of being affected by disturbances, they often completely lose their late successional character when a disturbance occurs. Larger protected areas, on the other hand, are statistically more often affected by disturbances, but a total loss of late successional stages is unlikely (Frelich 2002). This facilitates the persistence of species dependent on late successional phases in the landscape. The discussion about the optimal size of protected areas is known as the SLOSS debate (Single Large or Several Small), whereby arguments for a few large as well as for many small protected areas have both been put forward (Diamond 1975; Honnay et al. 1999). A larger number of species is preserved in several small protected areas because of the species–area relationship. However, from the perspective of connectivity between habitats, and in the interest of conserving natural dynamics including disturbance (which can affect extensive areas), a certain minimum size of a protected area is warranted.

Disturbances are important catalysts of ecosystem dynamics and contribute to their continuous renewal (see Chap. 5). This role is of increasing importance since ecosystems are increasingly at disequilibrium with their environment because of global change. Disturbances can, for example, positively contribute to the adaptation of tree species to climate change (Thom et al. 2017; Scherrer et al. 2021; see Chap. 16). However, they also represent a challenge to local biodiversity since species from other ecosystems as well as introduced non-native species can often easily spread in disturbed stands (Zonneveld 1995). Burned areas in the insubric region of the south side of the Alps, for instance, are being invaded by various non-native plant species like the tree of heaven (*Ailanthus altissima* (Mill.) Swingle), black locust (*Robinia pseudoacacia* L.), and American pokeweed (*Phytolacca americana* L.) (Maringer et al. 2012). The introduction of alien species by humans can pose a threat to the naturally occurring biodiversity as non-native herbivores and diseases (see Chap. 10) often increase strongly because of missing antagonists, and thus can reduce or displace native species (Liebhold et al. 2017).

The temporal development of biodiversity after natural disturbances is often influenced by how people respond to these disturbances. A widespread example in Central Europe is the salvage logging of naturally disturbed forest areas (i.e. the removal of trees in areas affected by disturbance) (Lindenmayer et al. 2008). Commercial forests are mainly salvage logged to reduce the economic damage to landowners. However, even in protected areas of Central Europe, naturally disturbed areas are often cleared in order to prevent bark beetle outbreaks (see Chap. 12) in forest areas adjacent the original to bark beetle outbreaks. Such human modification of natural disturbances also has significant effects on biodiversity. For example, the removal of deadwood significantly reduces the number of saproxylic species (i.e. species living on and in deadwood, such as lichens and beetle species) (Thorn et al. 2018). Other species groups that do not depend on deadwood (such as several insect families and pioneer plants) can, however, benefit from human interventions in naturally disturbed areas (Wermelinger et al. 2002; Wohlgemuth et al. 2002b). As a consequence of widespread salvage logging, the biodiversity after disturbance in Central Europe is strongly influenced by humans.

4.5   Disturbances and Biodiversity in a Spatial Context

One aspect that allows many species to persist in the face of disturbance is the eco-logical memory or biological legacy persisting after a disturbance. These biological legacies include all organic remains from the ecosystem pre-disturbance that are carried over into the post-disturbance state, such as surviving plants, resprouting plant organs, storage organs surviving underground, or enduring seeds in the soil. Disturbances in natural systems are rarely so severe that all living organisms are destroyed by the disturbance. Even in severe forest fires, individual trees or groups of trees survive in topographically sheltered areas (Romme et al. 2011). Other forms of biological legacies in forests are regenerating trees not affected by windthrow and bark beetle infestation of the canopy trees, standing and downed deadwood (Fig. 4.4) which serves as a substrate for tree regeneration (Macek et al. 2017), and seed banks in the soil. An example of the latter is the leafy goosefoot (*Blitum virga-tum* L.), a pioneer species distributed in mountains around the Mediterranean and in Central Europe. Plants are found in fertile soils from where seeds are spread by, for example, goats or chamois. After a large stand-replacing fire in the Swiss Valais, individuals of this very rare species were observed much more commonly, indicat-ing that its seeds had probably survived in the soil for centuries (Moser et al. 2006). In general, the term biological legacy summarizes all organisms, structures, and patterns of an ecosystem that persist through a disturbance (Franklin et al. 2000). Biological legacies fulfil a lifeboat function for a variety of disturbance-sensitive species by providing resources, food, protection, and habitat in the wake of a

Fig. 4.4 Biological legacies after disturbance by windthrow and bark beetles in High Tatra National Park (Slovakia): surviving larches at the left margin of the picture, standing and downed deadwood, as well as tree regeneration established before the disturbance. (Photo: R. Seidl)

disturbance (Lindenmayer and Franklin 2002). Biological legacies also make an important contribution to the recovery of ecosystems after disturbances (Seidl et al. 2014). In the process of recovery, it is not only the legacies on the area directly affected by disturbance that are important but also the vegetation on adjacent undisturbed areas, that is, the spatial context of a disturbance, contributes significantly to post-disturbance recovery (Johnstone et al. 2016). An example is the dispersal of seeds of trees from surrounding areas into a disturbed area, a phenomenon that occurs after forest fires (Romme et al. 2011; Maringer et al. 2020), as well as after windthrows (Kramer et al. 2014).

The effect of the increased heterogeneity arising as a result of disturbances – which in turn has a positive effect on biodiversity – is not only a small-scale, local phenomenon in individual stands, but is also found at the landscape scale. The irregular temporal and spatial occurrence of disturbances as well as the combination of different types of disturbance on the landscape contribute significantly to habitat heterogeneity and therefore promote biodiversity (Warren et al. 2007; Fescenko and Wohlgemuth 2017). Thus, disturbances not only foster diversity within stands (alpha diversity), but also increase the diversity between stands in a landscape (beta diversity; Silva Pedro et al. 2016) and thus ultimately affect the overall diversity of a landscape (gamma diversity; Fig. 4.5). This effect exists not only for natural disturbances but is also observed for disturbances caused by human land use, such as

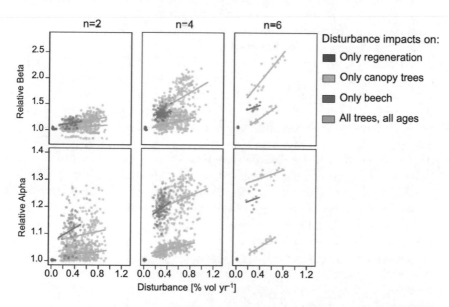

Fig. 4.5 Effect of disturbance type (colours) and disturbance rate (x-axis: average disturbance percentage per year, related to growing stock) on local tree species diversity (i.e. within 100 × 100 m stands, alpha diversity), as well as on tree species diversity between stands (beta diversity) of forest landscapes with a tree species pool of n = 2–6 species. The diversity response (y-axis) is given relative to undisturbed forest landscapes (value of 1). Values are based on 500-year simulations under typical environmental conditions for lowland broadleaved forest communities in Central Europe. (Data from Silva Pedro et al. 2016)

forest management (see Chaps. 14 and 15). Schall et al. (2018) showed, for example, that harvesting trees with varying intensity and size at different spatial scales has a stronger positive effect on biodiversity in beech forests of Central Europe than the emulation of late developmental stages across the landscape in continuous cover management. A combination of different disturbance sizes and severities is more similar to the natural disturbance regime in Central European forests, and therefore best suited to promote biodiversity also in managed forest areas (Nagel et al. 2014) (Boxes 4.1 and 4.2).

Box 4.1: Disturbance and Species Community Assembly

Anke Jentsch (iD)
Bayreuth Center of Ecology and Environmental Research (BayCEER),
University of Bayreuth, Bayreuth, Germany
e-mail: anke.jentsch@uni-bayreuth.de

Disturbance influences the development of plant species communities and the coexistence of different species in many ways. Assembly rules have been developed to understand how plant communities are formed, particularly with respect to the competition of species for abiotic resources. There is the concept of 'succession after disturbance' from vegetation ecology (Watt 1947; Pickett and Thompson 1978; Remmert 1991; van der Maarel 1993), as well as the concept of 'community assembly' from island biogeography and animal ecology (MacArthur and Wilson 1967; Diamond 1975). Succession processes in terrestrial ecosystems describe both the changes in species composition and the changes in the physical environment and resource availability – for example, with respect to inland dunes, the increase in organic matter and plant-available nitrogen as the dune succession progresses (Jentsch 2004). Succession is driven by the development of the biocoenosis, especially by the biomass growing after a disturbance, which then influences the temperature, humidity, light availability, production of organic detritus, and the allocation of soil resources between living organisms, organic material, and physical substrates. Consequently, the environment changes with the ongoing succession. Important aspects of species assembly and the formation of communities include the role of pioneer species for further successional processes. The type of disturbance determines how the initial resources are distributed after a disturbance event as well as which species can form a community and how species interact with each other.

Disturbance as a Filter

Many disturbance events act selectively on certain species, age groups, life stages, height growth, and other functional characteristics. To a large extent,

(continued)

Box 4.1 (continued)

this selectivity acts as a disturbance filter (Díaz et al. 1999) and determines the species composition and subsequent successional processes of the community. The prevailing environmental factors at a particular site are abiotic and disturbance filters. Similarly, the dispersal capabilities of species act as biotic filters. Biotic and abiotic filters determine which species reach the site, which species can tolerate the climate extremes, and which species can compete for resources. Disturbance filters in turn determine which species survive the dominant disturbance regimes, and therefore which species will be able to establish themselves and reproduce (Box Fig. 1).

Thus, disturbances take different roles (White and Jentsch 2004): (1) disturbance events may modify abiotic filters, for example, by changing the availability of resources such as nitrogen release after a fire or changing the availability of soil water during drought; (2) disturbance events may modify biotic filters, for example, by creating bare substrate free of competition through bioturbation of ground-dwelling rodents; and (3) disturbance events may work as filters themselves, influencing the functional characteristics of species within the community. Here, disturbance intensity acts selectively on the survival of species, disturbance frequency determines the window of

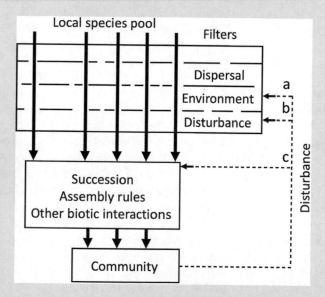

Box Fig. 1 The threefold role of disturbances in community assembly: (**a**) disturbances influence abiotic filters, e.g. nitrogen release after fire; (**b**) disturbances act directly as selective filters according to their properties, e.g. by mowing height, fire temperature, or wind energy; and (**c**) disturbances influence biotic interactions within a community, e.g. by selective feeding and shifting of competitive equilibria. (Modified after White and Jentsch 2004)

(continued)

Box 4.1 (continued)

opportunity for the reproduction of species, and timing of disturbance determines the success of recruitment and colonization of species. If disturbances occur suddenly and abruptly, they promote species that can quickly colonize areas where the vegetation is sparse because of the ability of these species to disperse seeds or propagate through vegetative means (e.g. rhizomes). When disturbance events suddenly release resources, they promote species with fast colonization and high growth rates. Seasonal patterns of disturbances promote species that spread, grow, and reproduce at a certain time. The higher the variability of disturbance regimes in a landscape, the greater the biodiversity. Direct, functional adaptations to disturbance are discussed in Chap. 6. Already several decades ago the functional traits of plant species had been classified into categories relevant to disturbances. The 'r–K continuum' distinguishes between two species characteristics, namely, 'r' for species with high reproduction rates and 'K' for species with high survival rates with only a few offspring (MacArthur and Wilson 1967). The 'CSR strategies' divide plant species into three types, with the C-type for competitive species, the S-type for stress-tolerant species, and the R-type for pioneer and ruderal species (Grime 1979). Noble and Slatyer (1980) distinguish three vital attributes of species: (1) the type of persistence during and after colonization after a disturbance; (2) the growth characteristics in a biocoenosis; and (3) the length of time until important life cycles are reached. With regard to human disturbances, the effects on species can be distinguished as concordant (i.e. the life cycle of the plants is not so disrupted by the disturbance) or discordant (i.e. the life cycle of the plants is disrupted by the disturbance) (Pavlovic 1994). Finally, the concept of plant functional traits can be used to group plants in terms of their growth forms, phenology, and other characteristics (Díaz et al. 1999, 2016). It is generally assumed that mutualism and facilitation are of great importance in a resource-limited or frequently disturbed environment, while inhibition and competition prevail in a favourable or rarely disturbed environment with abundant resources and favourable physical conditions. Because of the abiotic, biotic, and disturbance filters, the actual species numbers in plant communities are smaller than the species numbers of the local or regional species pool of a landscape. Thus, more stable communities can coexist in landscapes that are frequently disturbed than in landscapes in which there are only relatively small and infrequent disturbances (White and Jentsch 2004). Disturbances can accelerate successional dynamics, slow them down, return them to a previous state, or have no effect at all (Wilson and Agnew 1992).

(continued)

Box 4.1 (continued)

Acceleration of successional dynamics by disturbance usually happens through the elimination of inertia in species communities represented by long lifespans or imbalanced competitive interactions. Thus, disturbances act as catalysts of change.

References

Diamond JM (1975) The island dilemma: lessons of modern biogeographic studies for the design of nature reserves. Biol Conserv 7:129–146

Díaz S, Cabido M, Zak M, Martínez Carretero E, Araníbar J (1999) Plant functional traits, ecosystem structure and land-use history along a climatic gradient in Central-Western Argentina. J Veg Sci 10:651–660

Díaz S, Kattge J, Cornelissen JHC, Wright IJ, Lavorel S, Dray S, Reu B, Kleyer M, Wirth C, Prentice IC, Garnier E, Bönisch G, Westoby M, Poorter H, Reich PB, Moles AT, Dickie J, Gillison AN, Zanne AE, Chave J, Joseph Wright S, Sheremet ESN, Jactel H, Baraloto C, Cerabolini B, Pierce S, Shipley B, Kirkup D, Casanoves F, Joswig JS, Günther A, Falczuk V, Rüger N, Mahecha MD, Gorné LD (2016) The global spectrum of plant form and function. Nature 529:167–171

Grime JP (1979) Plant strategies and vegetation processes. Wiley, Chichester, 222 p

Jentsch A (2004) Disturbance driven vegetation dynamics: concepts from biogeography to community ecology and experimental evidence from dry acidic grasslands in Central Europe. Diss Bot 384:1–218

MacArthur RH, Wilson EO (1967) The theory of island biogeography. Princeton University Press, Princeton, 203 p

Noble IR, Slatyer R (1980) The use of vital attributes to predict successional changes in plant communities subject to recurrent disturbances. Vegetation 43:5–21

Pavlovic NB (1994) Disturbance-dependent persistence of rare plants: anthropogenic impacts and restoration implications. In: Boels ML, Whelan C (eds) Restoration of endangered species: conceptual issues, planning, and implementation. Cambridge University Press, Cambridge, pp 159–193

Pickett ST, Thompson JN (1978) Patch dynamics and the design of nature reserves. Biol Conserv 13:27–37

Remmert H (1991) The mosaic-cycle of ecosystems – an overview. In: Remmert H (ed) The mosaic-cycle concept of ecosystems, ecological studies, vol 85. Springer, Berlin/Heidelberg, pp 1–21

van der Maarel E (1993) Some remarks on disturbance and its relations to diversity and stability. J Veg Sci 4:733–736

Watt AS (1947) Pattern and process in the plant community. J Ecol 35:1–22

White PS, Jentsch A (2004) Disturbance, succession, and community assembly in terrestrial plant communities. In: Temperton VM, Hobbs RJ, Nuttle T, Halle S (eds) Assembly rules and restoration ecology: bridging the gap between theory and practice, pp 342–366

Wilson JB, Agnew ADQ (1992) Positive-feedback switches in plant communities. Adv Ecol Res 23:263–336

Box 4.2: A Unification of Many Diversity Hypotheses

Thomas Wohlgemuth (iD)
Forest Dynamics Research Unit, Swiss Federal Research Institute WSL,
Birmensdorf, Switzerland
e-mail: thomas.wohlgemuth@wsl.ch

Which factors determine the variation of biodiversity in ecosystems and communities, or more simply, which factors determine the coexistence of species? This topic has been dealt with in thousands of publications in the last decades, and more than a hundred hypotheses have been formulated. Michael W. Palmer listed 120 hypotheses in a review and discussed different approaches to categorize them (Palmer 1994). These categories concern the influence of genetics on the coexistence of species (Aarssen 1992), the aspect of scale (Auerbach and Shmida 1987), the focus on the difference between ecological balance and imbalance (Intermediate Disturbance Hypothesis; Connell 1978), or the relationship between resource availability and niche size (MacArthur 1965). This demonstrates the difficulty of classifying the significant hypotheses into a few categories to provide a clear overview of the topic. Therefore, in Palmer's search for a synthesis of the diversity hypotheses, he chose a fundamentally different approach: he linked one of the most central organizational concepts in ecology, the Competitive Exclusion Principle (CEP) (Grinnell 1904; Gause 1937; Begon et al. 1996), with the central principle of population ecology, the Hardy–Weinberg Equilibrium (HWE) (Hardy 1908), according to which geneticists explain why populations evolve, that is, why they genetically adapt to their environment. The HWE assumes that populations do not change if several conditions are simultaneously met, but as soon as one of the conditions does not apply, evolution takes place. This formulation was rephrased to meet the conditions for the CEP.

Amazingly, all conditions of the reformulated CEP are strongly bound to disturbances (Wohlgemuth et al. 2002):

1. Sooner or later, disturbance disrupts a species community in which few species have increased in abundance and structurally excluded many other species, for example, by shade or root competition. A 'reset' of the few species prevents the local disappearance of subordinate species.
2. This includes disturbance effects or increases in the temporal variation of environmental conditions.
3. It also means that disturbance changes and generally enhances the spatial environmental variation.
4. It follows that growth not only depends on one resource but rather on several resources. For example, in a forest, growth depends on the factors light, warmth, and nutrients, which change rapidly after disturbances.

(continued)

Box 4.2 (continued)

5. In turn, as availability of resources fluctuates (Davis et al. 2000) both abundant and rare species are affected disproportionally, in particular on a small scale to varying degrees, which promotes the coexistence of species.
6. As an effect of disturbances, spaces emerge across various scales that are temporarily free of competition, where many species and especially pioneer species can spread. In this way, the number of species around disturbed communities increases and will eventually decrease as a consequence of increasing competition.
7. Finally, open areas created by disturbances are not only colonized by species already in place, but also by new species, be it by wind-borne dispersal of plant seeds or by species attracted to the area as a result of a completely changing resource situation, for example, pyrophilous insects on fire-disturbed areas (Pradella et al. 2010) (Box Table 1).

Box Table 1 A reformulation of the Competitive Exclusion Principle

Principle
Given a suite of species, interspecific competition will result in the exclusion of all but one species.
Conditions of the principle
(1) There has been sufficient time to allow exclusion.
(2) Environment is temporally constant.
(3) Environment has no spatial variation.
(4) Growth is limited by a single resource.
(5) Rarer species are not disproportionately favoured in terms of survivorship, reproduction, or growth.
(6) Species have the opportunity to compete.
(7) There is no immigration.
Corollary
The greater the degree to which these conditions are broken, the greater the number of species that can coexist.

References

Aarssen LW (1992) Causes and consequences of variation in competitive ability in plant-communities. J Veg Sci 3:165–174
Auerbach M, Shmida A (1987) Spatial scale and the determinants of plant species richness. Trends Ecol Evol 2:238–242
Begon M, Harper GA, Townsend CR (1996) Ecology: individuals, populations and communities, 3rd edn. Blackwell Science, Oxford, 1068 p
Connell JH (1978) Diversity in tropical rain forests and coral reefs: high diversity of trees and corals is maintained only in a non-equilibrium state. Science 199:1302–1310

(continued)

Box 4.2 (continued)

Davis MA, Grime JP, Thompson K (2000) Fluctuating resources in plant communities: a general theory of invasibility. J Ecol 88:528–534
Gause GF (1937) Experimental populations of microscopic organisms. Ecology 18:173–179
Grinnell GF (1904) The origin and distribution of the chestnut-backed chickadee. Auk 21:364–382
Hardy GH (1908) Mendelian proportions in a mixed population. Science 28:49–50
MacArthur RH (1965) Patterns of species diversity. Biol Rev 40:510–533
Palmer MW (1994) Variation in species richness: towards a unification of hypotheses. Folia Geobot Phytotaxon 29:511–530
Pradella C, Wermelinger B, Obrist MK, Duelli P, Moretti, M (2010) On the occurrence of five pyrophilous beetle species in the Swiss Central Alps (Leuk, Canton Valais). Mitt Schweiz Entomol Ges 83:187–197
Wohlgemuth T, Bürgi M, Scheidegger C, Schütz M (2002) Dominance reduction of species through disturbance–a proposed management principle for central European forests. Forest Ecol Manag 166:1–15

References

Aichinger E (1951) Soziationen, Assoziationen und Waldentwicklungstypen. Angew Pflanezensoz 1:21–68
Beudert B, Bässler C, Thorn S, Noss R, Schröder B, Dieffenbach-Fries H, Foullois N, Müller J (2015) Bark beetles increase biodiversity while maintaining drinking water quality. Conserv Lett 8:272–281
Butchart SHM, Walpole M, Collen B, van Strien A, Scharlemann JPW, Almond REA, Baillie JEM, Bomhard B, Brown C, Bruno J, Carpenter KE, Carr GM, Chanson J, Chenery AM, Csirke J, Davidson NC, Dentener F, Foster M, Galli A, Galloway JN, Genovesi P, Gregory RD, Hockings M, Kapos V, Lamarque J-F, Leverington F, Loh J, McGeoch MA, McRae L, Minasyan A, Hernández Morcillo M, Oldfield TEE, Pauly D, Quader S, Revenga C, Sauer JR, Skolnik B, Spear D, Stanwell-Smith D, Stuart SN, Symes A, Tierney M, Tyrrell TD, Vié J-C, Watson R (2010) Global biodiversity: indicators of recent declines. Science 328:1164–1168
Connell JH (1978) Diversity in tropical rain forests and coral reefs: high diversity of trees and corals is maintained only in a non-equilibrium state. Science 199:1302–1310
Diamond JM (1975) The island dilemma: lessons of modern biogeographic studies for the design of nature reserves. Biol Cons:129–146
Donato DC, Campbell JL, Franklin JF (2012) Multiple successional pathways and precocity in forest development: can some forests be born complex? J Veg Sci 23:576–584
Fescenko A, Wohlgemuth T (2017) Spatio-temporal analyses of local biodiversity hotspots reveal the importance of historical land-use dynamics. Biodivers Conserv 26:2401–2419
Fox JW (2012) The intermediate disturbance hypothesis should be abandoned. Trends Ecol Evol:1–7
Franklin JF, Lindenmayer D, MacMahon JA, McKee A, Magnuson J, Perry DA, Waide R, Foster D (2000) Threads of continuity. Conserv Pract 1:8–17
Frelich LE (2002) Forest dynamics and disturbance regimes. Studies from temperate evergreen – deciduous forests. Cambridge University Press, Cambridge, 266 p
Hagge J, Bässler C, Gruppe A, Hoppe B, Kellner H, Krah F-S, Müller J, Seibold S, Stengel E, Thorn S (2019) Bark coverage shifts assembly processes of microbial decomposer communities in dead wood. Proc R Soc B 286:1912

Hall AR, Miller AD, Leggett HC, Roxburgh SH, Buckling A, Shea K (2012) Diversity-disturbance relationships: frequency and intensity interact. Biol Lett 8:768–771

Hassell H, Hahata H, Ash H (2005) Millennium Ecosystem Assessment and human well-being: current state and trends, vol 1. Island Press, Washington, 917 p

Honnay O, Hermy M, Coppin P (1999) Effects of area, age and diversity of forest patches in Belgium on plant species richness, and implications for conservation and reforestation. Biol Conserv 87:73–84

Hughes AR, Byrnes JE, Kimbro DL, Stachowicz JJ (2007) Reciprocal relationships and potential feedbacks between biodiversity and disturbance. Ecol Lett 10:849–864

Huston MA (1994) Biological diversity: the coexistence of species on changing landscapes. Cambridge University Press, Cambridge, 685 p

Johnstone JF, Allen CD, Franklin JF, Frelich LE, Harvey BJ, Higuera PE, Mack MC, Meentemeyer RK, Metz MR, Perry GLW, Schoennagel T, Turner MG (2016) Changing disturbance regimes, ecological memory, and forest resilience. Front Ecol Environ 14:369–378

Johst K, Huth A (2005) Testing the intermediate disturbance hypothesis: when will there be two peaks of diversity? Divers Distrib 11:111–120

Kondoh M (2001) Unifying the relationships of species richness to productivity and disturbance. Proc R Soc B-Biol Sci 268:269–271

Kramer K, Brang P, Bachofen H, Bugmann H, Wohlgemuth T (2014) Site factors are more important than salvage logging for tree regeneration after wind disturbance in central European forests. Forest Ecol Manag 331:116–128

Liebhold AM, Brockerhoff EG, Kalisz S, Nuñez MA, Wardle DA, Wingfield MJ (2017) Biological invasions in forest ecosystems. Biol Invasions 19:3437–3458

Lindenmayer DB, Franklin JF (2002) Conserving forest biodiversity: a comprehensive multiscaled approach. Island Press, Washington DC, 351 p

Lindenmayer DB, Burton PJ, Franklin JF (2008) Salvage logging and its ecological consequences. Island Press, Washington, 246 p

Macek M, Wild J, Kopecký M, Červenka J, Svoboda M, Zenáhlíková J, Brůna J, Mosandl R, Fischer A (2017) Life and death of Picea abies after bark-beetle outbreak: ecological processes driving seedling recruitment. Ecol Appl 27:156–167

Mackey RL, Currie DJ (2001) The diversity–disturbance relationship: is it generally strong and peaked? Ecology 82:3479

Maringer J, Wohlgemuth T, Neff C, Pezzatti GB, Conedera M (2012) Post-fire spread of alien plant species in a mixed broad-leaved forest of the Insubric region. Flora 207:19–29

Maringer J, Wohlgemuth T, Hacket-Pain A, Ascoli D, Conedera M (2020) Drivers of persistent post-fire recruitment in European beech forests. Sci Total Environ 699:134006

Miller AD, Roxburgh SH, Shea K (2011) How frequency and intensity shape diversity-disturbance relationships. Proc Natl Acad Sci USA 108:5643–5648

Molino J-F, Sabatier D (2001) Tree diversity in tropical rain forests: a validation of the intermediate disturbance hypothesis. Science 294:1702–1704

Moser B, Gimmi U, Wohlgemuth T (2006) Ausbreitung des Erdbeerspinats Blitum virgatum nach dem Waldbrand von Leuk. Wallis Bot Helv 116:179–183

Nagel TA, Svoboda M, Kobal M (2014) Disturbance, life history traits, and dynamics in an old-growth forest landscape of southeastern Europe. Ecol Appl 24:663–679

Rockström J, Steffen W, Noone K, Persson A, Chapin FS, Lambin EF, Lenton TM, Scheffer M, Folke C, Schellnhuber HJ, Nykvist B, de Wit CA, Hughes T, van der Leeuw S, Rodhe H, Sörlin S, Snyder PK, Costanza R, Svedin U, Falkenmark M, Karlberg L, Corell RW, Fabry VJ, Hansen J, Walker B, Liverman D, Richardson K, Crutzen P, Foley JA (2009) A safe operating space for humanity. Nature 461:472–475

Romme WH, Boyce MS, Gresswell R, Merrill EH, Minshall GW, Whitlock C, Turner MG (2011) Twenty years after the 1988 Yellowstone fires: lessons about disturbance and ecosystems. Ecosystems 14:1196–1215

Roxburgh SH, Shea K, Wilson JB (2004) The intermediate disturbance hypothesis: patch dynamics and mechanisms of species coexistence. Ecology 85:359–371

Schall P, Gossner MM, Heinrichs S, Fischer M, Boch S, Prati D, Jung K, Baumgartner V, Blaser S, Böhm S, Buscot F, Daniel R, Goldmann K, Kaiser K, Kahl T, Lange M, Müller J, Overmann J, Renner SC, Schulze E-D, Sikorski J, Tschapka M, Türke M, Weisser WW, Wemheuer B, Wubet T, Ammer C (2018) The impact of even-aged and uneven-aged forest management on regional biodiversity of multiple taxa in European beech forests. J Appl Ecol 55:267–278

Scherrer D, Ascoli D, Conedera M, Fischer C, Maringer J, Moser B, Nikolova PS, Rigling A, Wohlgemuth T (2021) Canopy disturbances catalyse tree species shifts in Swiss forests. Ecosystems 25:199–214

Seidl R, Rammer W, Spies TA (2014a) Disturbance legacies increase the resilience of forest ecosystem structure, composition, and functioning. Ecol Appl 24:2063–2077

Seidl R, Schelhaas MJ, Rammer W, Verkerk PJ (2014b) Increasing forest disturbances in Europe and their impact on carbon storage. Nat Clim Chang 4:806–810

Senf C, Sebald J, Seidl R (2021) Increasing canopy mortality affects the future demographic structure of Europe's forests. One Earth 4:1–7

Shea K, Roxburgh SH, Rauschert ESJ (2004) Moving from pattern to process: coexistence mechanisms under intermediate disturbance regimes. Ecol Lett 7:491–508

Silva Pedro M, Rammer W, Seidl R (2016) A disturbance-induced increase in tree species diversity facilitates forest productivity. Landsc Ecol 31:989–1004

Svensson JR, Lindegarth M, Jonsson PR, Pavia H (2012) Disturbance-diversity models: what do they really predict and how are they tested? Proc R Soc B Biol Sci 279:2163–2170

Swanson ME, Franklin JF, Beschta RL, Crisafulli CM, Dellasala DA, Hutto RL, Lindenmayer DB, Swanson FJ (2011) The forgotten stage of forest succession: early-successional ecosystems on forest sites. Front Ecol Environ 9:117–125

Thom D, Seidl R (2016) Natural disturbance impacts on ecosystem services and biodiversity in temperate and boreal forests. Biol Rev 91:760–781

Thom D, Rammer W, Seidl R (2017) Disturbances catalyze the adaptation of forest ecosystems to changing climate conditions. Glob Chang Biol 23:269–282

Thorn S, Bässler C, Brandl R, Burton PJ, Cahall R, Campbell JL, Castro J, Choi CY, Cobb T, Donato DC, Durska E, Fontaine JB, Gautier S, Hebert C, Hothorn T, Hutto RL, Lee EJ, Leverkus A, Lindenmayer D, Obrist MK, Rost J, Seibold S, Seidl R, Thom D, Waldron K, Wermelinger B, Winter B, Zmihorski M, Müller J (2018) Impacts of salvage logging on biodiversity: a meta-analysis. J Appl Ecol 55:279–289

UN (1992) Convention on biological diversity, Rio de Janeiro, 30 p

Warren SD, Holbrook SW, Dale DA, Whelan NL, Elyn M, Grimm W, Jentsch A (2007) Biodiversity and the heterogeneous disturbance regime on military training lands. Restor Ecol 15:606–612

Wermelinger B, Duelli P, Obrist MK (2002) Dynamics of saproxylic beetles (Coleoptera) in windthrow areas in alpine spruce forests. Forest Snow Landsc Res 77:133–148

Wohlgemuth T, Bürgi M, Scheidegger C, Schütz M (2002a) Dominance reduction of species through disturbance – a proposed management principle for central European forests. Forest Ecol Manag 166:1–15

Wohlgemuth T, Kull P, Wütrich H (2002b) Disturbance of microsites and early tree regeneration after windthrow in Swiss mountain forests due to the winter storm Vivian 1990. Forest Snow Landsc Res 77:17–47

Wootton JT (1998) Effects of disturbance on species diversity: a multitrophic perspective. Am Nat 152:803–825

Zhang Y, Chen HYH, Taylor A (2014) Multiple drivers of plant diversity in forest ecosystems. Glob Ecol Biogeogr 23:885–893

Zonneveld IS (1995) Vicinism and mass effect. J Veg Sci 6:441–444

Chapter 5
Disturbance Resilience

Rupert Seidl ⓘ, Anke Jentsch ⓘ, and Thomas Wohlgemuth ⓘ

Abstract Resilience is the capacity of ecosystems to recover from disturbance or to absorb disturbance without changing their structures and processes. While engineering resilience focuses solely on recovery from disturbance, ecological resilience also considers the possibility of a regime change after disturbance. A key element of resilience is the adaptive cycle in ecosystems, that is, the alternation of phases of growth, conservation, release, and renewal. Important mechanisms that make ecosystems resilient against disturbances are interactions over spatial and temporal scales, legacies of the pre-disturbance state, ecological stress memory, and the response diversity of plant communities.

Keywords Adaptive cycle · Cross-scale interactions · Disturbance legacies · Early warning indicators · Panarchy · Recovery · Regime shift · Response diversity · Stress memory

R. Seidl (✉)
Ecosystem Dynamics and Forest Management Group, School of Life Sciences, Technical University of Munich, Freising, Germany

Berchtesgaden National Park, Berchtesgaden, Germany
e-mail: rupert.seidl@tum.de

A. Jentsch
Bayreuth Center of Ecology and Environmental Research (BayCEER), University of Bayreuth, Bayreuth, Germany

T. Wohlgemuth
Forest Dynamics Research Unit, Swiss Federal Institute for Forest, Snow and Landscape Research WSL, Birmensdorf, Switzerland

© The Author(s), under exclusive license to Springer Nature Switzerland AG 2022
T. Wohlgemuth et al. (eds.), *Disturbance Ecology*, Landscape Series 32, https://doi.org/10.1007/978-3-030-98756-5_5

97

5.1 Introduction and Definitions

Disturbances are, by definition, discrete events in space and time which cause major changes in ecosystems, for example, in the form of plant mortality (see Chap. 2). Organisms, communities, and ecosystems are adapted to such changes and have a strong ability to recover from disturbances. This property is called resilience.

Resilience has received increasing attention in science in recent years, not least because of rapidly changing environmental conditions and new types of disturbances. The concept of resilience has become one of the most important research topics in the sustainability debate (Folke et al. 2004; Rockström et al. 2009). Fundamental research on mechanisms and limits of resilience is an active field that is developing rapidly. For example, achieving functional resilience is currently an important goal of risk research and experimental biodiversity research (Isbell et al. 2015; Kreyling et al. 2017). Furthermore, the concept of resilience is increasingly used as a guideline and target for the management of ecosystems (Biggs et al. 2012; Seidl 2014; see also Chap. 17). The diverse usages of the concept of resilience have led to a wide range of definitions (Brand and Jax 2007), which is why it is especially important to specify the meaning of resilience for the respective context or application (resilience of what? resilience to what? Carpenter et al. 2001). In general, the literature distinguishes three types of resilience: engineering resilience, ecological resilience, and social-ecological resilience (Nikinmaa et al. 2020).

5.1.1 Engineering Resilience

Engineering resilience focuses on the recovery after disturbance: the faster a system returns to its original state after a disturbance, the more resilient it is (Holling 1996). Engineering resilience assumes a predictable recovery path as well as constancy in the undisturbed state (ecological equilibrium; equilibrium assumption); systems, therefore, always recover along the same path and differ only in the speed of their recovery. As the name implies, this concept of resilience is often used in technical and engineering sciences, for example, to describe the development of material characteristics after stress. However, engineering resilience is also an important parameter in ecology: it can, for instance, be used to describe the recovery of tree growth after a drought period (Lloret et al. 2011; Zang et al. 2014). Tree growth is a narrowly defined indicator whose development has only one degree of freedom. The state of tree growth before a disturbance can be clearly defined, and therefore the requirements for the definition of engineering resilience are met in this example.

5.1.2 Ecological Resilience

The behaviour of communities and ecosystems is much more complex than materials, abiotic systems, or individual indicators such as tree rings. Many ecosystems are significantly affected by disturbance, but their characteristic functions remain intact despite disturbance or are restored relatively quickly after disturbance – thus, they are resilient. For example, after a fire, a forest stand will grow back into a forest stand, and after mowing a flowering mountain meadow will grow back into a flowering mountain meadow. Nevertheless, there are also disturbances which lead to regime shifts, especially if degradation has already occurred or if environmental conditions and resources change relatively quickly. If the ecological resilience of a system is exceeded, the system will change, for example, after a severe forest fire, trees may not regenerate sufficiently and there can be a transition towards open land. The concept of ecological resilience (Holling 1973, 1996; Gunderson 2000) considers this dynamic of alternative stable states, which are characterized by different structures and processes. Ecological resilience is thus the capacity of a system to absorb disturbances without changing the system's typical structures and processes. It is important to note that the preservation of structures and processes does not necessarily mean a deterministic return of the system to the state before a disturbance. In European primaeval forests, for example, there are a multitude of developmental paths after a disturbance (Meigs et al. 2017). However, characteristic structures (e.g. a complex canopy structure) and processes (e.g. carbon uptake) are recovered (to varying degrees) in all these paths of natural ecosystem development after disturbance – the system is thus ecologically resilient (Seidl et al. 2014). Ecological resilience can be seen as a mechanism of dynamic stability (Turner et al. 1993): disturbances and transient changes are an inherent part of many ecosystems, without fundamentally changing them (see Chap. 3).

If a disturbance exceeds the ecological resilience of a system, its regime and thus its structures and processes change fundamentally. In most cases, this regime change is not gradual but abrupt and occurs when a threshold is exceeded. For example, precipitation-induced boundaries between forest and savannah or temperature-induced boundaries between forest and tundra are not gradual but are rather expressed as relatively discrete tipping points of the biosphere (Hirota et al. 2011; Scheffer et al. 2012). An often-used metaphor to describe the concept of ecological resilience is the 'ball and cup' model, in which the ball describes the current system state and a landscape of valleys (i.e. stable system states, attractors of the system) and crests (i.e. unstable system states) describes the different possible states of the system (Fig. 5.1). A disturbance causes an impulse on the system and pushes it from its resting point in the current attractor. The deeper and narrower the valley, the faster the system returns to the centre of the attractor after the disturbance (and the greater its resilience). If the impulse from the disturbance is so strong that the ball is moved from one attractor to the next, the disturbance exceeds the resilience of the system and results in a regime shift. We note that the 'attractor landscape' of a system (i.e. its valleys and crests in the 'ball and cup' model) is in most cases not static

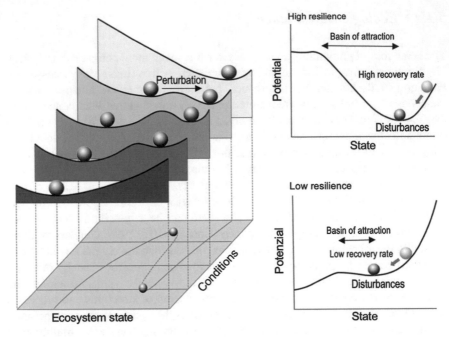

Fig. 5.1 Schematic representation of ecological resilience with different possible ecosystem states (valleys and crests) and the ball as an indicator of the current system state. (After Scheffer and Carpenter 2003; Scheffer et al. 2009, redrawn)

over time. Factors such as the extinction and immigration of species into a system or climatic changes may cause attractors (both in terms of strength and location) to change over time (Gunderson 2000; Seidl et al. 2016b). Different stable states of ecosystems are either alternating and thus reversible (e.g. the oligotrophic and eutrophic states of a lake) or they are largely irreversible (e.g. the nutrient enrichment from atmospheric deposition of ecosystems). Ecological resilience (i.e. the ability of a system to remain in a stable system state despite disturbance) is a neutral characteristic that is not per se good or bad. However, with regard to environmental changes such as climate change, the preservation of a certain system state is often the goal – here resilience is a desired property of the system, which can be further promoted by management measures. In contrast, restoration ecology often aims at restoring a system to its former state, possibly using disturbances to overcome the resilience of the current system. Examples are the liming of acidified lakes or the removal of topsoil from nutrient-polluted former nutrient-poor meadows (Fig. 5.2).

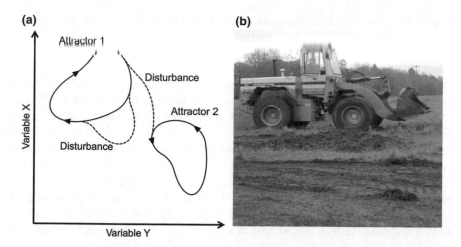

Fig. 5.2 (a) Schematic representation of different attractors for resilient ecosystem states. Disturbances act as triggers and catalysts for regime change. (After Scheffer and Carpenter 2003.) (b) An example of the use of disturbances in nature conservation is the removal of nutrient-rich topsoil to restore resource-limited sand ecosystems on inland dunes in southern Germany in order to promote rare and endangered pioneer species that are weak competitors

5.1.3   Social-Ecological Resilience

Social-ecological resilience takes up the concept of ecological resilience and extends it from ecological systems to social-ecological systems (Folke 2006). Resilience here means the ability of these systems to maintain their structures and processes in the face of disturbance and, for example, provide ecosystem services to society in a sustainable manner despite disturbance (Folke et al. 2002; Brand and Jax 2007; Biggs et al. 2012). An important aspect is social adaptive capacity, that is, the ability to respond to external stressors and disturbances with social or political change (Adger 2000). This type of resilience will not be discussed further in this chapter. However, it is of importance in the context of disturbance management (Seidl et al. 2016b) and is addressed in Chap. 17.

5.1.4   Panarchy

Closely related to the idea of ecological resilience is the concept of panarchy, which was introduced by Holling and Gunderson (2002). It is a model that describes the dynamic organization of complex systems in space and time, and can be used for the characterization and quantification of resilience in ecosystems. The panarchy model distinguishes between different levels of a system, for example, in the context of a forest ecosystem – leaf, single tree, stand, and landscape. For each of these levels, the dynamics of the system can be described by an adaptive cycle through the phases of growth, conservation, release, and reorganization (Fig. 5.3). From the growth to the conservation phase, the potential of the ecosystem increases; it accumulates biomass, energy, and other system components, such as species, growth forms, or functional groups. At the same time, the interconnectedness between the individual system components also increases, that is, system behaviour is increasingly determined by interactions such as competition or mutualism. In an old-growth forest, for example, the competition for light between trees determines the regeneration dynamics and species composition more than external factors. However, these strong system-internal interactions combined with a simultaneous increase in system potential (e.g. accumulated biomass) also lead to the system becoming increasingly inflexible and thus susceptible to disturbance. If a disturbance occurs (e.g. reduction of live biomass because of fire) the potential of the ecosystem is reduced in the release phase. In the subsequent phase of reorganization, the system components recombine. This can either lead to a growth phase along the previous system trajectory or to a regime change and the beginning of a qualitatively different adaptive cycle (see Fig. 5.3; Scheffer and Carpenter 2003; Allen et al. 2014). The duration of individual phases of the adaptive cycle can vary considerably. While in a forest ecosystem the growth phase typically lasts several decades, depending on the location, the conservation phase can span many centuries. The release phase, on the other hand, often lasts only a few hours (windthrow), days (forest fire), or years (bark beetle outbreak), and the reorganization of the system usually takes a few years to a few decades. Panarchy is a hierarchically nested arrangement of adaptive cycles (Fig. 5.4). Key components of the panarchy model are the cross-scale connections between the individual adaptive cycles. Thus, structures and processes on subordinate scales (e.g. individual trees on the landscape that survive a forest fire) contribute to the resilience of the system in the reorganization phase by acting as a systemic memory and promoting the reorganization towards the previous structures and processes (e.g. by seed dispersal). At the same time, if critical thresholds are

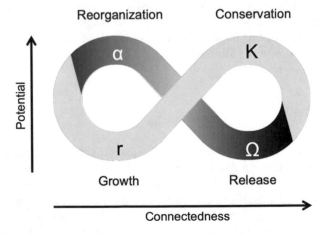

Fig. 5.3 The adaptive cycle forms the basis of panarchy in dynamic systems. (Holling and Gunderson 2002, redrawn)

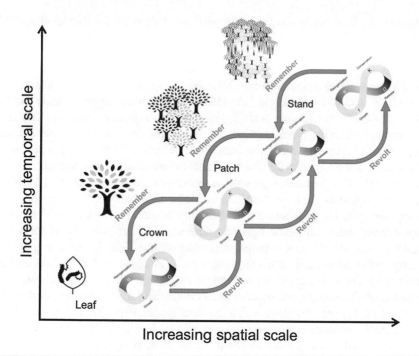

Fig. 5.4 Panarchy – A hierarchical arrangement of connected adaptive cycles at different spatial and temporal scales. (Modified from Allen et al. 2014)

exceeded, disturbance processes can spread to higher levels in the release phase and thus lead to a regime shift. For example, large-scale bark beetle disturbances in forests result from individual infestation spots, that is, from small groups of infested trees (Peters et al. 2004; see Chap. 12). These feedbacks and interactions across scales are an important mechanism of system resilience and will therefore be discussed in more detail in Sect. 5.2.

5.2 Mechanisms of Resilience

5.2.1 Feedbacks and Interactions Across Scales

Ecological resilience results from the interactions of different organizational levels. These include the direct stress response of individual organisms within short periods of time as well as acclimatization processes and adaptations at the population level throughout evolutionary periods. Feedbacks and interactions across temporal and spatial scales can contribute to resilience as well as lead to regime change (Jentsch et al. 2002). Because of the important role of different levels in the context of resilience, it is not sufficient to consider a system at a single scale to describe its resilience. At least one level above and one level below the focal level of analysis should be considered (Walker et al. 2004). Processes at higher levels often have a preserving effect on ecosystems (Meyn et al. 2007), that is, they make an important contribution to their resilience. In forests, intensive disturbances at the stand level (1–10 ha) can lead to a massive loss of live plant biomass. However, at the landscape level (1000–100,000 ha), single individuals or individual stands usually survive even extreme events (Romme et al. 2011). As a consequence, these individuals or stands contribute to the recolonization and revegetation of the system after a disturbance via seed dispersal. Since the immediate vicinity of a disturbed area is often similar in its composition to the disturbed area before the disturbance event (e.g. see Palmer 2005), this feedback represents a systemic memory (Franklin et al. 2000) and gives the system ecological resilience. Many disturbances first develop locally before they spread to populations and entire landscapes. Examples are fires (which often develop from individual lightning strikes) and plant diseases. If individual thresholds are exceeded, the propagation rate of the disturbance changes non-linearly and amplification occurs at higher levels (Peters et al. 2004). Fires above a certain size are self-reinforcing by influencing their surrounding climate (e.g. increased wind development, pyrocumulus clouds) and drying out the combustible material on the ground via the heat that precedes the fire front. Tree mortality caused by bark beetle infestation increases disproportionately after a local threshold population is reached as the defence mechanisms of trees, as well as the populations of antagonists, are overrun by the exponentially increasing beetle population (Raffa et al. 2008; see Chap. 12). Spatial connectivity in the landscape (e.g. between habitat for bark beetles or combustible material for fire) plays an important role in reaching critical thresholds (Meyn et al. 2007) and can contribute to a positive amplification across scales (Seidl et al. 2016a).

5.2.2   Legacies of the Pre-disturbance State

In most cases, disturbances do not result in the complete destruction of all organisms inhabiting an area, but organic remains (biological legacies) of the ecosystem before the disturbance and undisturbed islands of intact vegetation persist in a matrix of disturbed areas (Franklin et al. 2000; White and Jentsch 2001). Organic remains after a disturbance include, for example, surviving organisms, seeds surviving in the soil (seed banks), microorganisms, fungi and insects, organic material such as humus and deadwood, and structural remains such as tree stumps or freshly exposed mineral soil. The amount and distribution of biological legacies significantly influence the type and speed of recovery from a disturbance, and thus the resilience of the ecosystem. Examples of such legacies are surviving old trees after a forest fire and unmowed parts of a meadow. Even a small percentage of surviving individuals or stands can make a significant contribution to the resilience of the composition, structure, and functioning of a landscape. For example, in forests of the temperate zone, the survival of trees on only 12% of the landscape area increases the carbon storage by 33.8% in the first 100 years after a disturbance (compared to forests with total loss of live trees) At the same time, these legacies of the system before disturbance triple the recolonization with late successional species in the first century after disturbance (Seidl et al. 2014). Live tree legacies, therefore, contribute significantly to ecological resilience. Further forms of legacy are seed banks and the ability of certain species to resprout. Plants that can sprout from dormant buds after the death of aboveground plant parts often recover quickly after disturbance, as water and nutrients can be utilized by the established root network. Such plants oftentimes even benefit from disturbances as their competitors have been eliminated by the disturbance and more resources are available for their development (Buhk et al. 2007). In the Central Alps, aspen (*Populus tremula* L.) and downy oak (*Quercus pubescens* Willd.) are both able to resprout after fires, and thus benefit from fires in comparison with Scots pine (*Pinus sylvestris* L.), which is an obligate seeder (Wohlgemuth et al. 2018). Seeds can survive disturbances both in the soil (soil seed bank) and in tree crowns (canopy seed bank). The latter is called serotiny: seeds survive in closed, resinous cones in the canopy, which only open after the great heat produced by fire. This characteristic is an evolutionary adaptation to disturbances (see Chap. 6). For example, serotiny is shown by several pine species (*Pinus halepensis* Mill., *P. pinaster* Ait.) in the Mediterranean region and by black spruce (*Picea mariana* (Mill.) Britt.) in Alaska.

Frequently, not all developmental stages of plants are affected equally by disturbance, which means that less susceptible stages can serve as a system legacy and accelerate system recovery. Shade-tolerant tree species can, for example, regenerate under closed canopies even at low light levels and remain in a 'waiting position' for several decades until resources become available. This advanced regeneration is not affected by disturbances such as bark beetles or wind because the cambial layer of the young trees is still too thin to serve as breeding material for bark beetles and the susceptibility to wind is small owing to the high elasticity of young plants and low leverage because of low tree height. Consequently, advanced regeneration can play an important role in the recovery of forests following these disturbances. In the Bohemian Forest, for example, where mortality from bark beetle outbreaks in the last 25 years was up to 99% of canopy trees, this advanced regeneration led to a rapid recolonization (76% coverage 10 years after disturbance) by Norway spruce (*Picea abies* L.), which was already dominant before the disturbance (Zeppenfeld et al. 2015). The system thus proved to be very resilient to disturbance, which was largely due to the already existing regeneration under the canopy (Fig. 5.5).

5.2.3 Ecological Stress Memory and Acclimatization

Repeated disturbances can push ecosystems beyond a threshold so that a previous dynamic equilibrium can no longer be achieved. The consequences can be substantial changes in species composition and ecosystem services (see Chap. 18). However, some plant species have an individual 'ecological stress memory' and consequently

Fig. 5.5 Biological legacy after bark beetle infestation in the Bohemian Forest. Organic remains (here: standing and downed deadwood) as well as tree regeneration, some of which had already been established under the canopy before disturbance. (Photo: R. Seidl)

can react relatively quickly and efficiently to disturbances (Walter et al. 2011). Such a memory could be the mechanism by which ecological communities can remain stable even under extreme climatic conditions. Ecological stress memory is defined as a plant reaction which, after being exposed to stress, buffers the plant against the influence of future stress. This mechanism of individual resilience is observed, for example, as a consequence of drought, frost, or heat stress (see Chap. 6).

Ecological stress memory occurs either in the form of acclimatization processes or as damage as a result of stress. Acclimatization occurs, for example, when trees that are repeatedly exposed to dry periods invest a larger proportion of their assimilates into root growth and are therefore better adapted to water limitation (Kozlowski and Pallardy 2002). However, if the frequency of disturbances is increased, plants may not be able to recover sufficiently between disturbance events. Because of the reduced assimilation capacity in periods of drought, trees are often forced to reduce their leaf area, which leads to lower radiation absorption and thus reduced photosynthesis in the following years (e.g. Allen et al. 2015; Johnstone et al. 2016). Whether trees acclimatize to stress or not also affects their reaction to new disturbance events: either disturbances can reduce resilience through accumulation of stress, resulting in an increased sensitivity to subsequent disturbance events (Scheffer et al. 2001), or acclimatization may persist after stress, resulting in reduced sensitivity or faster recovery to new stress events (Walter et al. 2011). Delayed or belated stress effects, which only become visible after a considerable time, can have negative effects on the resilience of plants to further disturbances. A delayed reaction to past extreme events such as summer drought can be observed, for example, in trees, with increased mortality occuring several years after drought (Bigler et al.

2007). Late ground frosts can trigger increased mortality in dwarf shrubs in the vegetation periods following the frost (Kreyling et al. 2010). Such delayed responses to disturbances indicate significant carry-over effects in plant fitness; the responses may be barely noticeable immediately after a disturbance but explain part of the reduced resilience after repeated disturbances (Buma 2015).

5.2.4 Response Diversity and Trade-Offs Within Plant Communities

Many plant communities consist of species with complementary traits in terms of competitive strength and tolerance to disturbance (White and Jentsch 2001). The phenomenon of community resilience often arises from this diversity in traits. Often, species of a community complement each other with respect to their tolerances to disturbance (e.g. drought, plant diseases, and herbivory), resulting in systems that are resilient to different disturbances. In addition, a trade-off between disturbance tolerance and competitive capacity has often developed over the course of evolution: competitive species often show little resistance to disturbance, while less competitive species can benefit from disturbance and the resulting removal of superior competitors (dominance reduction; Wohlgemuth et al. 2002) and release of resources (Davis et al. 2000; White and Jentsch 2001). As discussed above, the resilience of forests is determined, for example, by characteristics such as seed dispersal, serotiny, the ability to resprout, or shade tolerance, which vary between species. In many cases, these properties are negatively correlated: seed dispersal over long distances is a typical property of early-successional species in forests, which are, however, very light-demanding species. Shade-tolerant climax species, on the other hand, often have heavy seeds and thus disperse only over short distances. Therefore, a key to the resilience of ecosystems is their diversity, that is, the coexistence of species with different characteristics, niches, and life history strategies. While disturbances influence diversity (see Chap. 4), the opposite is also true: diversity determines the resilience of ecosystems to disturbance.

A key element for resilience is response diversity, that is, the variability in the responses of species to fluctuations and disturbances (Mori et al. 2013). A landscape consisting of species with similar niches and characteristics has a lower response diversity than a landscape in which species with very different niches and strategies occur (e.g. a mixture of pioneer species and late-successional species). Silva Pedro et al. (2015), for example, showed that carbon uptake and storage in species-rich forest landscapes (European beech forests in Hainich National Park, Germany) are significantly more resilient to disturbances than species-poor systems. Further investigations of the same forest ecosystem showed that mixtures of species with different characteristics and strategies (i.e. early-successional, intermediate, and late-successional species) also show higher productivity and thus recover faster from disturbances (Silva Pedro et al. 2016). Beta diversity (i.e. mixtures of species between stands) is at least as effective as alpha diversity (i.e. mixtures within a stand) for buffering disturbance impacts (Sebald et al. 2021).

In grasslands, the positive impact of high biodiversity on productivity and stability is attributed to different mechanisms (see Chap. 15), including: (1) the degree of asynchronous behaviour of species in a plant community in the face of disturbances, (2) insurance effects through complementary plant strategies from fast-growing to stress-tolerant species, (3) overcompensation of individual species during disturbances by reducing competitive pressure, and (4) the likelihood that a particularly productive species will occur when a high number of species are present (Yachi and Loreau 1999; Lehman and Tilman 2000; Loreau and de Mazancourt 2008; Hautier et al. 2014).

We note that the mechanisms of resilience in ecosystems are still far from being fully understood and are a current topic of research. Therefore, the processes listed here should be seen as examples rather than as a comprehensive list.

5.3 Measuring and Describing Resilience

Despite the growing theoretical understanding of resilience and the steadily increasing scientific literature on resilience in different systems, practical applications of the concept (e.g. in ecosystem management) are still rare. One of the reasons for the slow transdisciplinary spread of the concept is the difficulty in measuring and describing resilience and the multi-scaled nature of the concept, as described above (Seidl 2014). Therefore, many studies are currently dealing with the quantification of resilience (Isbell et al. 2015; Kreyling et al. 2017; Ingrisch and Bahn 2018). In this section, we address some important aspects in this regard.

5.3.1 Recovery After Disturbances

The recovery after a disturbance is an important indicator of the resilience of a system. In the 'ball and cup' model, the faster the ball is at the bottom of the cup, the faster the system approaches the centre of the attractor, and the higher is its resilience (Fig. 5.1). This property is probably the most directly measurable indicator of resilience. Frequently performed productivity measurements can be used in this context to gain insights into the resilience of a system (Isbell et al. 2015). For example, the resilience of managed spruce forests (based on stand volume) decreases at the margins of the natural distribution of spruce in Central Europe, with stand age and stand structure also influencing resilience (Seidl et al. 2017). It should be noted, however, that the speed of recovery is in most cases an indicator of engineering resilience and therefore does not allow any inference about possible regime shifts.

5.3.2 *Position of the Attractor*

Another resilience indicator besides the speed of recovery is the recovery pathway of the system towards the dominant attractor (Seidl et al. 2016b). This requires a precise knowledge of the location of possible attractors. However, the estimation of attractor positions (Fig. 5.1) is complicated by the above-mentioned multitude of possible development paths of a system after disturbance (Fig. 5.3). How do we know whether an observed development pathway leads back to the former attractor of the system – indicating resilience – or leads to an alternative regime? In order to answer this question, it is necessary to quantitatively describe the position of the attractors of a system. The Historical Range of Variability (HRV) is suitable for the quantitative description of system attractors (White and Jentsch 2001). The HRV quantifies possible system states and specifies probabilities (over time) or frequencies (proportion of landscape) of system states for relevant variables (Keane et al. 2009). HRV thus quantitatively describes the past attractors of the system. A comparison of the current state or current recovery trajectory with the HRV provides information about the ecological resilience of a system. Human disturbances have, for example, led to the situation where Mediterranean ecosystems in Europe are currently characterized by maquis shrubland systems over large areas (Fig. 5.6; Tinner et al. 2009). Without human disturbances, these systems would approach their natural attractor again under the prevailing Mediterranean climate, that is, evergreen oak forests (*Quercus ilex* L.) would reappear (Henne et al. 2015). The HRV of a system can only be described for long periods of time because of the necessity to include the full variation inherent to the system. Since these periods of time are often not sufficiently covered by empirical data, simulation models are needed in many cases to describe the HRV quantitatively (Cyr et al. 2009; Seidl et al. 2013). However, the orientation on the HRV is increasingly criticized in the light of global change, as historical reference conditions can no longer be used as models for future conditions under changing environmental conditions (Seidl et al. 2016b; Jentsch and White 2019). It should, therefore, be noted that the attractor landscape is not static and can change over time. When estimating the resilience under future climatic conditions, it is necessary to consider not only the historical but also the future variability of the system (Duncan et al. 2010; Seidl et al. 2016b; Jentsch and White 2019). Nevertheless, analyses of the past provide an important basis for the contextualization of current disturbance events (Janda et al. 2017). Increasing availability of palaeo-data is providing better information about past climate-related changes in the attractor landscape (Tinner et al. 2009; Henne et al. 2015).

Fig. 5.6 Two alternative attractors of Mediterranean ecosystems in Southern Europe. (**a**) Open, shrubby maquis systems, which were created and are preserved by human disturbances. (Photo: M. Schweiss, Wikimedia Commons.) (**b**) Forests of Mediterranean oaks (e.g. *Quercus ilex* L.; Photo: F. Geller-Grimm, Wikimedia Commons) as they would appear without human disturbance (Henne et al. 2015)

5.3.3 Early-Warning Indicators

In applying the resilience concept, the question often arises of whether a regime shift is imminent or not. Many ecosystems recover relatively quickly after disturbance (high resilience), while others recover only over long periods of time (low resilience). In the case of ecosystems with low resilience, the question of a possible regime shift is increasingly being raised. If such a shift occurs, it can often have a large and long-lasting impact on ecosystems, which makes early-warning indicators of regime shifts particularly relevant.

The temporal autocorrelation of a system offers possible insights here. A system with high resilience (i.e. a deep valley in Fig. 5.1) quickly returns to the centre of the attractor after disturbance. The temporal autocorrelation of the system after disturbance (i.e. the temporal memory of a disturbance event) is low. At low resilience or near a tipping point, there are only weak gradients in the attractor landscape – the system returns only slowly to its starting point after disturbance. This slow return results in increasing temporal autocorrelation and is called 'critical slowing down'; it is a possible early-warning indicator of a regime shift (Scheffer et al. 2009).

The spatial structure of ecosystems can also provide information about their resilience. For example, the further a disturbance radiates into a system, or the greater its spatial effect on surrounding areas, the lower the resilience of the system (Dai et al. 2013). This phenomenon is also expressed in increasing spatial autocorrelation in systems near tipping points (Dakos et al. 2010). Furthermore, an ecosystem-related regime change can often be first announced by changes in limited areas of the system, which recover only slowly, or not at all, from disturbances. Increasing spatial variability may therefore also indicate a regime change in some systems (Kéfi et al. 2014).

References

Adger WN (2000) Social and ecological resilience: are they related? Prog Hum Geog 24:347–364

Allen CR, Angeler DG, Garmestani AS, Gunderson LH, Holling CS (2014) Panarchy: theory and application. Ecosystems 17:578–589

Allen CD, Breshears DD, McDowell NG (2015) On underestimation of global vulnerability to tree mortality and forest die-off from hotter drought in the Anthropocene. Ecosphere 6:article 129:1–55

Biggs R, Schlüter M, Biggs D, Bohensky EL, Burnsilver S, Cundill G, Dakos V, Daw TM, Evans LS, Kotschy K, Leitch AM, Meek C, Quinlan A, Raudsepp-Hearne C, Robards MD, Schoon ML, Schultz L, West PC (2012) Toward principles for enhancing the resilience of ecosystem services. Ann Rev Env Resour 37:421–448

Bigler C, Gavin DG, Gunning C, Veblen TT (2007) Drought induces lagged tree mortality in a subalpine forest in the Rocky Mountains. Oikos 116:1983–1994

Brand FS, Jax K (2007) Focusing the meaning(s) of resilience: resilience as a descriptive concept and a boundary object. Ecol Soc 12:23–23

Buhk C, Meyn A, Jentsch A (2007) The challenge of plant regeneration after fire in the Mediterranean Basin: scientific gaps in our knowledge on plant strategies and evolution of traits. Plant Ecol 192:1–19

Buma B (2015) Disturbance interactions: characterization, prediction, and the potential for cascading effects. Ecosphere 6:article 70

Carpenter S, Walker B, Anderies JM, Abel N (2001) From metaphor to measurement: resilience of what to what? Ecosystems 4:765–781

Cyr D, Gauthier S, Bergeron Y, Carcaillet C (2009) Forest management is driving the eastern north American boreal forest outside its natural range of variability. Front Ecol Environ 7:519–524

Dai L, Korolev KS, Gore J (2013) Slower recovery in space before collapse of connected populations. Nature 496:355–358

Dakos V, Nes EH, Donangelo R, Fort H, Scheffer M (2010) Spatial correlation as leading indicator of catastrophic shifts. Theor Ecol 3:163–174

Davis MA, Grime JP, Thompson K (2000) Fluctuating resources in plant communities: a general theory of invasibility. J Ecol 88:528–534

Duncan SL, McComb BC, Johnson KN (2010) Integrating ecological and social ranges of variability in conservation of biodiversity: past, present, and future. Ecol Soc 15:Article 5

Folke C (2006) Resilience: the emergence of a perspective for social-ecological systems analyses. Global Environ Chang 16:253–267

Folke C, Carpenter S, Elmqvist T, Gunderson L, Holling CS, Walker B (2002) Resilience and sustainable development: building adaptive capacity in a world of transformations. Ambio 31:437–440

Folke C, Carpenter S, Walker B, Scheffer M, Elmqvist T, Gunderson L, Holling CS (2004) Regime shifts, resilience, and biodiversity in ecosystem management. Annu Rev Ecol Evol 35:557–581

Franklin JF, Lindenmayer D, MacMahon JA, McKee A, Magnuson J, Perry DA, Waide R, Foster D (2000) Threads of continuity. Cons Pract 1:8–17

Gunderson LH (2000) Ecological resilience – in theory and application. Ann Rev Ecol Syst 69:473–439

Hautier Y, Seabloom EW, Borer ET, Adler PB, Harpole WS, Hillebrand H, Lind EM, MacDougall AS, Stevens CJ, Bakker JD, Buckley YM, Chu C-J, Collins SL, Daleo P, Damschen EI, Davies KF, Fay PA, Firn J, Gruner DS, Jin VL, Klein JA, Knops JMH, La Pierre KJ, Li W, McCulley RL, Melbourne BA, Moore JL, O'Halloran LR, Prober SM, Risch AC, Sankaran M, Schuetz M, Hector A (2014) Eutrophication weakens stabilizing effects of diversity in natural grasslands. Nature 508:521–525

Henne PD, Elkin C, Franke J, Colombaroli D, Calò C, La Mantia T, Pasta S, Conedera M, Dermody O, Tinner W (2015) Reviving extinct Mediterranean forests increases ecosystem potential in a warmer future. Front Ecol Environ 13:356–362

Hirota M, Holmgren M, Van Nes EH, Scheffer M (2011) Global resilience of tropical forest and savanna to critical transitions. Science 334:232–235

Holling CS (1973) Resilience and stability of ecological systems. Ann Rev Ecol Syst 4:1–23

Holling CS (1996) Engineering resilience versus ecological resilience. In: Schulze PC (ed) Engineering within ecological constraints. National Academy Press, Washington, DC, pp 31–44

Holling CS, Gunderson LH (2002) Resilience and adaptive cycles. Island Press, Washington, DC, pp 25–62

Ingrisch J, Bahn M (2018) Towards a comparable quantification of resilience. Trends Ecol Evol 33:251–259

Isbell F, Craven D, Connolly J, Loreau M, Schmid B, Beierkuhnlein C, Bezemer TM, Bonin C, Bruelheide H, De Luca E, Ebeling A, Griffin JN, Guo Q, Hautier Y, Hector A, Jentsch A, Kreyling J, Lanta V, Manning P, Meyer ST, Mori AS, Naeem S, Niklaus PA, Polley HW, Reich PB, Roscher C, Seabloom EW, Smith MD, Thakur MP, Tilman D, Tracy BF, van der Putten WH, van Ruijven J, Weigelt A, Weisser WW, Wilsey B, Eisenhauer N (2015) Biodiversity increases the resistance of ecosystem productivity to climate extremes. Nature 526:574–577

Janda P, Trotsiuk V, Mikoláš M, Bače R, Nagel TA, Seidl R, Seedre M, Morrissey RC, Kucbel S, Jaloviar P, Jasík M, Vysoký J, Šamonil P, Čada V, Mrhalová H, Lábusová J, Nováková MH, Ryavai M, Mateju L, Svoboda M (2017) The historical disturbance regime of mountain Norway spruce forests in the Western Carpathians and its influence on current forest structure and composition. For Ecol Manag 388:67–78

Jentsch A, White PS (2019) A theory of pulse dynamics and disturbance in ecology. Ecology 100:e02734

Jentsch A, Beierkuhnlein C, White PS (2002) Scale, the dynamic stability of forest eco-systems, and the persistence of biodiversity. Silva Fenn 36:393–400

Johnstone JF, Allen CD, Franklin JF, Frelich LE, Harvey BJ, Higuera PE, Mack MC, Meentemeyer RK, Metz MR, Perry GLW, Schoennagel T, Turner MG (2016) Changing disturbance regimes, ecological memory, and forest resilience. Front Ecol Environ 14:369–378

Keane RE, Hessburg PF, Landres PB, Swanson FJ (2009) The use of historical range and variability (HRV) in landscape management. For Ecol Manag 258:1025–1037

Kéfi S, Guttal V, Brock WA, Carpenter SR, Ellison AM, Livina VN, Seekell DA, Scheffer M, van Nes EH, Dakos V (2014) Early warning signals of ecological transitions: methods for spatial patterns. PLoS One 9:10–13

Kozlowski TT, Pallardy SG (2002) Acclimation and adaptive responses of woody plants to environmental stresses. Bot Rev 68:270–334

Kreyling J, Beierkuhnlein C, Jentsch A (2010) Effects of soil freeze-thaw cycles differ between experimental plant communities. Basic Appl Ecol 11:65–75

Kreyling J, Dengler J, Walter J, Velev N, Ugurlu E, Sopotlieva D, Ransijn J, Picon-Cochard C, Nijs I, Hernandez P, Güler B, von Gillhaussen P, De Boeck HJ, Bloor JMG, Berwaers S, Beierkuhnlein C, Arfin Khan MAS, Apostolova I, Altan Y, Zeiter M, Wellstein C, Sternberg M, Stampfli A, Campetella G, Bartha S, Bahn M, Jentsch A (2017) Species richness effects on grassland recovery from drought depend on community productivity in a multisite experiment. Ecol Lett 20:1405–1413

Lehman CL, Tilman D (2000) Biodiversity, stability, and productivity in competitive communities. Am Nat 156:534–552

Lloret F, Keeling EG, Sala A (2011) Components of tree resilience: effects of successive low-growth episodes in old ponderosa pine forests. Oikos 120:1909–1920

Loreau M, De Mazancourt C (2008) Species synchrony and its drivers: neutral and nonneutral community dynamics in fluctuating environments. Am Nat 172:E48–E66

Meigs GW, Morrissey RC, Bače R, Chaskovskyy O, Čada V, Després T, Donato DC, Janda P, Lábusová J, Seedre M, Mikoláš M, Nagel TA, Schurman JS, Synek M, Teodosiu M, Trotsiuk V, Vítková L, Svoboda M (2017) More ways than one: mixed-severity disturbance regimes foster structural complexity via multiple developmental pathways. For Ecol Manag 406:410–426

Meyn A, White PS, Buhk C, Jentsch A (2007) Environmental drivers of large, infrequent wildfires: the emerging conceptual model. Prog Phys Geogr 31:287–312

Mori AS, Furukawa T, Sasaki T (2013) Response diversity determines the resilience of ecosystems to environmental change. Biol Rev 88:349–364

Nikinmaa L, Lindner M, Cantarello E, Jump AS, Seidl R, Winkel G, Muys B (2020) Reviewing the use of resilience concepts in forest sciences. Curr For Rep 6:61–80

Palmer MW (2005) Distance decay in an old-growth neotropical forest. J Veg Sci 16:161–166

Peters DPC, Pielke RA, Bestelmeyer BT, Allen CD, Munson-McGee S, Havstad KM (2004) Cross-scale interactions, nonlinearities, and forecasting catastrophic events. Proc Natl Acad Sci USA 101:15120–15125

Raffa KF, Aukema BH, Bentz BJ, Carroll AL, Hicke JA, Turner MG, Romme WH (2008) Cross-scale drivers of natural disturbances prone to anthropogenic amplification: the dynamics of bark beetle eruptions. Bioscience 58:501–517

Rockström J, Steffen W, Noone K, Persson A, Chapin FS, Lambin EF, Lenton TM, Scheffer M, Folke C, Schellnhuber HJ, Nykvist B, de Wit CA, Hughes T, van der Leeuw S, Rodhe H, Sörlin S, Snyder PK, Costanza R, Svedin U, Falkenmark M, Karlberg L, Corell RW, Fabry VJ, Hansen J, Walker B, Liverman D, Richardson K, Crutzen P, Foley JA (2009) A safe operating space for humanity. Nature 461:472–475

Romme WH, Boyce MS, Gresswell R, Merrill EH, Minshall GW, Whitlock C, Turner MG (2011) Twenty years after the 1988 Yellowstone fires: lessons about disturbance and ecosystems. Ecosystems 14:1196–1215

Scheffer M, Carpenter SR (2003) Catastrophic regime shifts in ecosystems: linking theory to observation. Trends Ecol Evol 18:648–656

Scheffer M, Carpenter S, Foley JA, Folke C, Walker B (2001) Catastrophic shifts in ecosystems. Nature 413:591–596

Scheffer M, Bascompte J, Brock WA, Brovkin V, Carpenter SR, Dakos V, Held H, van Nes EH, Rietkerk M, Sugihara G (2009) Early-warning signals for critical transitions. Nature 461:53–59

Scheffer M, Hirota M, Holmgren M, Van Nes EH, Chapin FS (2012) Thresholds for boreal biome transitions. Proc Natl Acad Sci USA 109:21384–21389

Sebald J, Thrippleton T, Rammer W, Bugmann H, Seidl R (2021) Mixing tree species at different spatial scales: the effect of alpha, beta and gamma diversity on disturbance impacts under climate change. J Appl Ecol 8:1749–1763

Seidl R (2014) The shape of ecosystem management to come: anticipating risks and fostering resilience. Bioscience 64:1159–1169

Seidl R, Eastaugh CS, Kramer K, Maroschek M, Reyer C, Socha J, Vacchiano G, Zlatanov T, Hasenauer H (2013) Scaling issues in forest ecosystem management and how to address them with models. Eur J For Res 132:653–666

Seidl R, Rammer W, Spies TA (2014) Disturbance legacies increase the resilience of forest ecosystem structure, composition, and functioning. Ecol Appl 24:2063–2077

Seidl R, Müller J, Hothorn T, Bässler C, Heurich M, Kautz M (2016a) Small beetle, large-scale drivers: how regional and landscape factors affect outbreaks of the European spruce bark beetle. J Appl Ecol 53:530–540

Seidl R, Spies TA, Peterson DL, Stephens SL, Hicke JA (2016b) Searching for resilience: addressing the impacts of changing disturbance regimes on forest ecosystem services. J Appl Ecol 53:120–129

Seidl R, Vigl F, Rössler G, Neumann M, Rammer W (2017) Assessing the resilience of Norway spruce forests through a model-based reanalysis of thinning trials. For Ecol Manag 388:3–12

Silva Pedro M, Rammer W, Seidl R (2015) Tree species diversity mitigates disturbance impacts on the forest carbon cycle. Oecologia 177:619–630

Silva Pedro M, Rammer W, Seidl R (2016) A disturbance-induced increase in tree species diversity facilitates forest productivity. Landsc Ecol 31:989–1004

Tinner W, van Leeuwen JFN, Colombaroli D, Vescovi E, van der Knaap WO, Henne PD, Pasta S, D'Angelo S, La Mantia T (2009) Holocene environmental and climatic changes at Gorgo Basso, a coastal lake in southern Sicily. Italy Quaternary Sci Rev 28:1498–1510

Turner MG, Romme WH, Gardner RH, Neill RVO, Kratz TK, O'Neill RV (1993) A revised concept of landscape equilibrium: disturbance and stability on scaled landscapes. Landsc Ecol 8:213–227

Walker B, Holling CS, Carpenter SR, Kinzig A (2004) Resilience, adaptability and transformability in social-ecological systems. Ecol Soc 9:article 5

Walter J, Nagy L, Hein R, Rascher U, Beierkuhnlein C, Willner E, Jentsch A (2011) Do plants remember drought? Hints towards a drought-memory in grasses. Environ Exp Bot 71:34–40

White PS, Jentsch A (2001) The search for generality in studies of disturbance and ecosystem dynamics. Prog Bot 62:399–449

Wohlgemuth T, Bürgi M, Scheidegger C, Schütz M (2002) Dominance reduction of species through disturbance – a proposed management principle for central European forests. For Ecol Manag 166:1–15

Wohlgemuth T, Doublet V, Nussbaumer C, Feichtinger L, Rigling A (2018) Baumartenwechsel in den Walliser Waldföhrenwälder verstärkt nach grossen Störungen. Schweiz Z Forstwes 169.279–289

Yachi S, Loreau M (1999) Biodiversity and ecosystem productivity in a fluctuating environment: the insurance hypothesis. Proc Natl Acad Sci USA 96:1463–1468

Zang C, Hartl-Meier C, Dittmar C, Rothe A, Menzel A (2014) Patterns of drought tolerance in major European temperate forest trees: climatic drivers and levels of variability. Glob Change Biol 20:3767–3779

Zeppenfeld T, Svoboda M, Derose RJ, Heurich M, Müller J, Čížková P, Starý M, Bače R, Donato DC (2015) Response of mountain *Picea abies* forests to stand-replacing bark beetle outbreaks: neighbourhood effects lead to self-replacement. J Appl Ecol 52:1402–1411

Chapter 6
Adaptation to Disturbance

Georg Gratzer ⓘ and Anke Jentsch ⓘ

Abstract Plants began to colonize land more than 400 million years ago. Since then, they have been exposed to fire, wind, drought, frost, floods, herbivores, and other disturbances – sometimes simultaneously – and have developed a variety of strategies to reproduce despite their exposure to these disturbances. The length of the evolutionary period available for such adaptations has been long. Accordingly, the adaptations that have been developed are complex and diverse. On the other hand, recent studies have shown that plants can also adapt very quickly to new disturbance regimes. The adaptation potentials of organisms and biotic communities available in ecosystems represent an important basis for their future development in times of global change.

Keywords Plant adaptation to disturbance · Disturbances as selection factors · Drought · Fire · Frost · Herbivory · Secondary metabolites

6.1 Complex Interactions of Selection Factors Form Plant Communities

Like all other organisms, plants perform several tasks simultaneously. Growth and reproduction requirements result in combinations of plant traits that perform certain functions (photosynthesis, growth, reproduction) from the individual cell to the structure of the entire plant. Natural selection acts on the 'net result' – the performance of combinations of different plant characteristics rather than on individual

G. Gratzer (✉)
Institute of Forest Ecology, Department of Forest and Soil Sciences, University of Natural Resources and Life Sciences BOKU, Vienna, Austria
e-mail: georg.gratzer@boku.ac.at

A. Jentsch
Bayreuth Center of Ecology and Environmental Research (BayCEER), University of Bayreuth, Bayreuth, Germany

117
T. Wohlgemuth et al. (eds.), *Disturbance Ecology*, Landscape Series 32,
https://doi.org/10.1007/978-3-030-98756-5_6

plant traits (Lachenbruch and McCulloh 2014). Disturbances have the potential to cause organisms to die before they can reproduce, and thus have a major impact on plant fitness. This is particularly important in long-lived individuals such as trees: these organisms have an evolutionarily stable strategy (i.e. a strategy that cannot be displaced by another, better strategy) called the 'bang-bang strategy' (Falster and Westoby 2003). This term (borrowed from control engineering in which a controller switches abruptly between two states) describes a strategy in which the plants (trees) first invest their resources exclusively in height growth (and thus in competitive strength) and allocate resources to sexual reproduction only after this initial period of height growth. However, this results in high demand for survival up to the point where the switch is made, which often occurs only after decades, and thus a high pressure to adapt to disturbance.

The long evolutionary history of land plants and the immensely varying environmental conditions during this period have meant that the current shaping of plant traits in their respective environments is a combination of past selection pressures and also current adaptations (Reich et al. 2003).

6.1.1 Adaptation to Fire

'Serotiny is the prolonged storage of seeds in closed cones or fruits held within the crown of woody plants' (Lamont et al. 2020), and it is followed by the spreading of these seeds directly after a disturbance. This is a widespread strategy in fire-driven ecosystems. In Pinaceae, this strategy evolved about 90 million years ago (He et al. 2012). At this time (the older Upper Cretaceous), the oxygen content in the atmosphere and the temperature were high, and thus the fire frequency was also high; the Cretaceous is considered to be one of the periods with the highest fire frequency in the history of the Earth (He et al. 2012). Plants that display serotiny have a significantly higher seedling survival rate because the seeds fall to the ground directly after the fire, germinate, and then grow without competition because the competing species have been burned and are temporarily absent (Causley et al. 2016).

6.1.2 Adaptation to Drought

Plants can also adapt to repeated drought events and stress caused by dehydration by investing in their drought tolerance (Walter et al. 2013). For this purpose, they use mechanisms such as the accumulation of dehydrins (DHN), a protein group that protects against dehydration during osmotic processes (Lambers et al. 2008). Further adaptations are the enrichment of sugars (Walter et al. 2011) as well as a reduction of the photosynthetic apparatus or a specific incorporation of water-soluble organic compounds, so-called compatible solutes (small organic molecules that act as osmolytes and help organisms survive extreme osmotic stress) like proline and betaines. Drought adaptations are generally accompanied by changes in the gene expression of the plant hormone abscisic acid (ABA) (Lambers et al. 2008). In addition to physiological changes, anatomic and morphological adaptations to drought stress can also occur, for example, increased root to shoot ratio or increased rooting of deeper soil horizons.

6.1.3   Adaptation to Frost

In regions where temperatures below zero are reached, persistent plants show the potential to adapt to frost and cold spells. A typical example is oaks (*Quercus* spp.) that are able to sprout a second time after being damaged by late frosts (the so-called Lammas shoots). Frost generally damages cells by forming intracellular ice crystals and the resulting dehydration caused by mechanical damage to cell membranes. Such apoplastic ice formations lead (like drought stress) to a lack of water in the cells. Therefore, plants often use the same mechanisms for both frost and drought stress, such as the accumulation of soluble sugars or the transcription of DHN and LEA-genes [late embryogenesis abundant proteins (LEA); Lambers et al. 2008; Janská et al. 2010]. In order to maintain the fluidity of the cell membrane (membrane fluid) under increased dehydration pressure, plants increasingly store unsaturated fatty acids in their phospholipids as an adaptation to cold.

Plants can also synthesize so-called anti-frost proteins, especially as protection against dehydration as a result of frost. These proteins then prevent damage to the cell membrane. Frost adaptation is induced by low temperatures and a change in the photoperiod (Janská et al. 2010). While plants usually need several weeks to harden against cold spells, softening can take place within hours. For example, resistance to frost can disappear after a few hours if temperatures rise enough. The fact that adaptation to frost takes a long time and the loss of frost resistance happens relatively quickly presents many plants with a problem (Fig. 6.1). They are particularly susceptible to frost damage during short periods of frost during the growing season or after warm periods in winter.

6.1.4   Adaptation to Heatwaves

The reactions of plants to heatwaves are quite well understood at the cellular level: after plants have been exposed to extremely high temperatures, the expression of housekeeping genes (i.e. genes that are always expressed, except under extreme conditions, to maintain cellular functions) is stopped and increased heat shock proteins (HSPs; molecular chaperones in protein quality control) are synthesized. These protect other proteins from damage and repair already denatured proteins (Kotak et al. 2007). Furthermore, proteins can also be stabilized by compatible solutes such as proline or betaines (Schulze et al. 2002).

Fig. 6.1 Late frost damage to young beech (*Fagus sylvatica* L.) trees in May 2011, whose already sprouted leaves were no longer adapted to late cold spells. On a cloudless night, temperatures of around −11°C were reached in the early hours of the morning, which killed all spring shoots. (Photo: J. Kreyling)

6.1.5 Fitness Effectiveness of Disturbance Combinations

Usually, plants are exposed to different types of disturbance simultaneously, each of which exerts different selection pressures on plants. For example, predispersal seed predation and crown fire represent competing selection pressures for the lodgepole pine (*Pinus contorta* Douglas ex Loudon): seed predisposition by squirrels selects against perennial storage of seeds in a crown seed bank; therefore, only a small amount of serotiny is found in forests with high squirrel density (Talluto and Benkman 2013).

At the cellular level, multiple disturbances acting on a plant activate specific physiological responses that are different from responses to individual disturbance events. An increase in tolerance to one stressor can reduce susceptibility to another stressor or, in individual cases, increase it. Since heat, drought, and frost lead to dehydration of cells, the adaptation strategies are very similar. It is, therefore, possible that adaptation to a particular type of disturbance also protects against damage from other disturbances. This creates a multiple memory that leads to more diverse tolerance. For example, the frost tolerance of some populations is also associated with drought stress tolerance (Blodner et al. 2005). Specifically, the late frost tolerance and even the maximum frost resistivity of some plants increases if they have been exposed to extreme drought in the previous year (Kreyling et al. 2011, 2012a).

Another form of adaptation to combinations of disturbances is that after drought herbivore resistance increases. The reason for this is an increased production of carbon-based secondary metabolites and abiotic stress-induced growth inhibition. However, it is not entirely clear how long such a change in secondary compounds and metabolic processes will last. Agrawal (2002) showed that the offspring of plants attacked by herbivores also have a higher induced resistance to herbivores. This implies that stress memories with induced defence against herbivores are heritable. This has been corroborated by recent findings that show that herbivory induced changes in morphological and reproductive traits were retained in plants two generations after herbivory occurred (Kellenberger et al. 2018). However, it is still unknown whether this can also protect against damage caused by drought, heat, or frost.

Phenological and morphological adaptations of plant parts, which are preserved over a long period of time, are another possibility for resistance to disturbance. Changes in the ratio of root to shoot growth in response to drought or warming are obvious morphological reactions; they also make the plants more resistant to future disturbance events. Also, the specific leaf area (ratio of leaf area to leaf dry matter) can be reduced to minimize water losses in dry periods. In this way, the plant already creates a higher resistance to future drought events.

The reactions of plants to abiotic stress (drought, cold stress, salt stress) are largely controlled by ABA, which, for example, causes the stomata to close during dry periods. The defence against pathogens that need a living host, on the other hand, is usually triggered by salicylic acid (SA), while in the case of an attack by herbivores and pathogens that kill the host, jasmonic acid (JA) and ethylene (ET) dominate as primary signals for plant defence (Atkinson and Urwin 2012; Bruce 2014). ABA acts antagonistically, but also synergistically, on the signal pathways triggered by biotic stress. Interactions of different stressors show a very specific signature in the plant reaction patterns. The signalling pathways against biotic disturbance also interact, almost exclusively antagonistically. Plants that are infected by pathogens and trigger an SA signalling pathway suppress JA-dependent responses and vice versa (Ballaré 2011). However, this mutual suppression of the production and also the reception of plant hormones, each of which initiates the corresponding defence, can also be outsmarted: some generalist herbivores, such as *Spodoptera* spp., can trigger an SA signalling pathway in plants (*Arabidopsis*, *Nicotiana*) through an as yet unidentified mechanism during feeding, which reduces the herbivory defence responses of the host plant (Ballaré 2011). Similarly, substances in insect eggs of the cotton leafworm (*Spodoptera littoralis* (Boisduval)) trigger SA defence reactions and lead to better larval growth (Bruessow et al. 2010). Colorado beetles (*Leptinotarsa decemlineata* Say) 'infect' plants at their feeding sites with symbiotic bacteria that induce SA-induced defence, which in turn act antagonistically on the JA signalling pathways and reduce the defence against the Colorado beetle (Chung et al. 2013). Here, the insects have thus succeeded in outwitting the specific defence reactions of plants and inducing the 'false' defence. In any case, plants exposed to multiple types of disturbance show complex interaction patterns that differ from the reactions to an individual type of disturbance. This fact became obvious in agricultural breeding when, for example, manioc (*Manihot esculenta* Crantz) bred for resistance to the manioc mosaic virus became more susceptible to the silverleaf whitefly (*Bemisia tabaci* (Gennadius)) (Atkinson and Urwin 2012). In maize (*Zea mays* L.), on the other hand, cultivars with increased drought tolerance showed increased resistance to the hemiparasitic witchweed (*Striga hermonthica*

(Delile) Benth.) (Atkinson and Urwin 2012). Breeding for resistance to only one disturbance factor, as is commonly done in agriculture, can lead to undesirable results with respect to other disturbances.

Some types of disturbance only become a factor in determining fitness in combination with other types of disturbance. One example comes from the fir forests (*Abies densa* Griff.) in the Eastern Himalayas, which form closed, monospecific stands at the forest line (Fig. 6.2). This tree species has a broad-crowned tree architecture that is not adapted to heavy snowfall. As a result, in most winters strong crown breakage and branch breakage provide entry points for various wood-decomposing fungi, which eventually leads to rot in the majority of trees with trunks over 50 cm in diameter (Gratzer et al. 2002). However, the combination of disturbances caused by snow loading and pathogens does not have an effect on fitness because strong winds and storms are not frequent in these ecosystems. Trees affected by rot not only survive for centuries but also massively produce seeds, even at ages of more than 400 years. If there were a combination of snow breakage, pathogens, and storms in these regions, the reduced stability against stem breakage would represent a strong selection factor towards narrow crowning and thus reduced snow loading.

Plants have developed a number of different adaptations to the same disturbance, for example, thick bark and strong self-pruning (the shedding of dry and dead branches) and height against the influence of surface fires, but also the same adaptations for different disturbance agents. An example is the resprouting from roots

Fig. 6.2 The tree architecture of *Abies densa* in Bhutan is not adapted to snow loads because of its wide crowns with spreading branches. As a result, during snowfall, crowns and branches break off, allowing decay pathogens to enter the tree. However, because storms that would break such rotten trees are rare in these ecosystems, these trees can remain in the ecosystem for a long time and continue to produce seeds. This means that snow loads do not affect fitness and – if conditions remain the same – will not lead to adaptations like narrow tree crowns. (Photo: ©Mark Overgaard, Scots Valley, CA, USA)

Fig. 6.3 Epicormic resprouting in a Canary Island pine (*Pinus canariensis*) stand after a fire on La Palma, Canary Islands. (Photo: A. Jentsch)

(root sprout) or from adventitious buds after the aboveground plant parts have died as a result of wind, fire, drought, insect herbivores, or pathogens. Unique among the pine species is the Canary Island pine (*Pinus canariensis* C. Sm.), which sprouts new needles on both branches and trunks after a fire, insect infestation, and wood harvesting (Pausas and Keeley 2017; Fig. 6.3). Therefore, it is not always possible to ascribe adaptations in plants to a specific disturbance agent or to a disturbance agent that is currently affecting the plant.

6.1.6   Adaptations Are Grouped Geographically

Especially the mentioned disturbances by a fire show that there are many different adaptations, but surprisingly these are geographically coherently combined: plants in ecosystems with frequent fires of low intensity often show thick insulating bark (Schwilk and Ackerly 2001; Cavender-Bares et al. 2004), as well as high flammability of leaves (Engber and Varner III 2012) and self-pruning of the lower dead branches (Keeley et al. 2011). In ecosystems with destructive fires, however, plants often show thin bark, low investment in reserves for regrowth (Tng et al. 2012), and serotiny (Schwilk and Ackerly 2001). The disturbance regime (i.e. the sum of all disturbances acting on a site) is thus closely linked to certain plant traits, suggesting an evolutionary adaptation of plants to a disturbance regime (Pausas and Schwilk 2012).

However, evidence of fitness alone is not sufficient to show that a certain plant trait was caused by disturbance. The trait must also be heritable and thus have a genetic basis. In this context, the heritability of serotiny and thick bark in the Aleppo pine (*Pinus halepensis* Mill.) has recently been investigated. The heritability was 0.2 for serotiny and 0.15–0.24 for thick bark, that is, medium to relatively high heritability for these traits (Pausas 2015). This also explains the divergence of these plant traits for populations living in regions with different disturbance regimes: serotiny is much more common in populations with crown fires (Hernández-Serrano et al. 2013).

Biotic disturbance also produces spatially located and delimited genetic patterns in plant populations. With the 'favourite' plant of plant genetics, the thale cress (*Arabidopsis thaliana* (L.) Heynh.), which is widely distributed in Europe, Züst et al. (2012) demonstrated that the genetic expression of chemical plant defence (by glucosinolates) varies geographically and correlates with the respective distribution of two aphid species. The differential selection by the two aphid species was demonstrated in an experiment. An artificially mixed population of *Arabidopsis* from different regions of origin with the highest possible genetic variation in defence properties was exposed separately to the two aphid species. After only five generations, the same specific genetic differences in plant defence were found as in the different regions of origin of the plants. Plants that were not exposed to aphid infestation produced only small amounts of glucosinolate (Züst et al. 2012). Thus, it was demonstrated that the different causes of biotic disturbance, although functionally very similar, produce geographical–genetic variation in plant populations. Such geographical–genetic differences could potentially be used in agriculture and forestry, especially in the context of the predicted increasing frequency and severity of extreme weather events in the near future, by growing disturbance-resistant varieties of arable crops and provenances of forest trees (Fig. 6.4). With respect to the late frost tolerance of important forage grasses and tree species from different European countries, the tolerance is correlated with the May minimum temperatures of their regions of origin (Kreyling et al. 2012a, b, c).

6.1.7 Many Enemies, Much Resistance: Biotic Disturbance as a Driver of Adaptation and Diversity

The most diverse adaptations to disturbances in the plant kingdom are probably the defence strategies against herbivores. Herbivores have exerted a strong selection pressure on plants for over 400 million years. Of the total biomass produced by terrestrial plants, herbivores consume about 15%. Herbivores are one of the most important links between autotrophic plants and the rest of the food web. Plants have developed various strategies against this biotic disturbance: they increase their resistance by physical barriers (thorns, spines, trichomes, thickened cell walls, resin, latex) and chemical repellents and can compensate for lost biomass through

Fig. 6.4 European origins of the widespread common meadow-grass (*Poa pratensis* L.) from Germany, England, Croatia, Norway, Poland, Switzerland, and Hungary. The plants were cultivated from seeds in a north German research institution and were planted there. Depending on their origin, the grasses have different characteristics, which also lead to different sensitivity to late frost. (Photo: A. Jentsch)

compensatory growth and regrowth, that is, plants tolerate the effects of the biotic disturbance. Until recently, it was assumed that plants had either one or the other strategy – that is, resistance or tolerance – because these strategies would be functionally redundant, and both would ultimately lead to an increase in fitness towards herbivory. However, empirical studies have shown that plants very often invest resources in both strategies at the same time (Pilson 2000). Herbivory experiments with two different leaf beetle species, a generalist (*Epitrix parvula* (Fabricius)) and a specialist (*Lema daturaphila* Kogan & Goeden), on thornapple (*Datura stramonium* L.) showed that the two beetles each selected for different strategies, namely, resistance (the generalist) and tolerance (the specialist), and that a mixed strategy (both resistance and tolerance) resulted in the greatest gain in fitness for the plant (Carmona and Fornoni 2013). Given the large number of herbivores, it can be assumed that exclusivity in the interaction between plant and herbivore (i.e. only one herbivore species on one plant species at a time) occurs very rarely (if at all), and that mixed strategies of resistance and tolerance, therefore, represent a general adaptation pattern (Carmona and Fornoni 2013).

6.1.8   The Role of Secondary Metabolites in Herbivore Defence

The chemical arsenal of plants is extremely diverse with over 100,000 secondary metabolites already identified (Kessler 2015). These plant substances perform a number of tasks such as pigmentation, UV protection, and structure for the plant, but are in any case central to the plant's defence against herbivores. They have a direct toxic effect, such as protease inhibitors in pigeon pea (*Cajanus cajan* (L.) Huth) and its wild relatives that act against the cotton bollworm (*Helicoverpa armigera* (Hübner)) – a pest that causes global losses in legume yields of over $320 million per year (Parde et al. 2012), or indirectly by attracting natural enemies of herbivores (Bruce 2014). Some of these substances are constitutive, while others are induced by herbivory, that is, they are only formed when an attack by a herbivore occurs.

Which type of defence occurs in plants and which factors are important for the evolution of strategies is only slowly being better understood through the application of molecular methods. Kessler (2015) summarizes the functional hypotheses that explain the evolution of herbivory-induced secondary metabolites as follows: (1) the optimal defence hypothesis, (2) the moving target hypothesis, and (3) the information transfer hypothesis. The first hypothesis states that secondary metabolites are only formed when they are needed, that is, when the plant is eaten or eggs are laid. It is based on the assumption that constitutive defence results in higher costs for the plant than induced defence. In fact, the costs of constitutive defence are high and the plant can invest assimilates either in defence or in growth and propagation. Accordingly, in the absence of herbivores, the fitness of those plants that have a high level of constitutive defence in the form of secondary metabolites is reduced (Agrawal 2007; Ballaré 2011; Kessler 2015) – an allocation compromise called the 'dilemma of plants: to grow or defend' (Herms and Mattson 1992). However, is induced defence 'cheaper'? To answer this question, it is necessary to consider the costs of the signalling pathways that lead to the formation of plant substances. Results of recent molecular genetic studies suggest that the maintenance of endogenous signalling pathways results in costs to plants comparable to those of constitutive defence (Kessler 2015) – a basic assumption of the first hypothesis, therefore, does not seem to apply here. However, the assumption of equal costs of these two defence strategies applies to the hypothesis of the mobile target, the second hypothesis. This hypothesis states that the change in the phenotype of plants, which is associated with metabolic changes and resulting differences in food quality for herbivores, increases the fitness of plants because it makes adaptations of herbivores more difficult. This would be the case if the costs of defence were not too high, if herbivore rates varied, and if the success of plant phenotype defence varied between herbivores (Adler and Karban 1994; Kessler 2015). Finally, the information transfer hypothesis (the third hypothesis) states that plants exchange information with the environment through induced defence. The plants thus 'orchestrate' interactions with the recipients of this information. These can be neighbouring plants, other herbivores, and natural enemies of herbivores (such as parasitic wasps), which are attracted by this information and can greatly reduce the infestation of herbivores.

6.1.9 Herbivore-Induced Communication Between Plants

How do plants communicate with their environment? They release volatile organic compounds during attacks by herbivores or pathogens as well as during abiotic stress. These compounds initiate defence reactions in plant organs of the same plant, but also in neighbouring plants, which may not even have been in direct contact with the herbivores or pathogens. These volatile organic compounds, most of which are terpenes, are semiochemicals (i.e. messenger substances used by insects to find food or conspecifics). Another group of substances is plant fragrances, so-called green leaf volatiles (GLVs): C6 compounds, mainly aldehydes, alcohols, and esters (Fig. 6.5). They are probably among the first weapons available to plants after attacks and abiotic stress; only a few seconds are needed for their formation (Scala et al. 2013). They are antimicrobial, influence the production of plant hormones, and thus induce defence or preparation for defence within the plant and neighbouring plants. On the one hand, GLVs can be used by plants to directly fend off herbivores, and, on the other hand, they can also be used by herbivores as an attractant that leads them to the forage plant. Predators of insect herbivores also use them to find their prey (Scala et al. 2013). Such predators are other insects (e.g. ichneumon wasps).

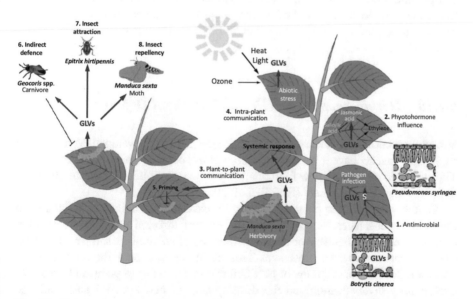

Fig. 6.5 Green leaf volatiles (GLVs) are emitted by plants during herbivory, pathogen infection, and abiotic stress. The GLVs have a variety of effects: (1) they are antimicrobial; (2) they influence phytohormonal networks; (3) they are involved in plant-to-plant communication; (4) they trigger defence reactions in unattacked leaves of the same plant; (5) they induce defence preparation in neighbouring plants; (6) some carnivores, e.g. *Geocoris* spp., show a preference for plants that produce GLVs – an example of indirect defence. GLVs can also (7) attract herbivorous insects, such as the tobacco flea beetle (*Epitrix hirtipennis* Melsheimer) or (8) repel herbivorous insects, e.g. the caterpillars of the tobacco hornworm *Manduca sexta* (L.) or the closely related tomato hornworm *Manduca quinquemaculata* (Haworth). (From Scala et al. 2013, redrawn)

Recent research has also shown that birds such as the great tit (*Parus major* L.) can also use purely olfactory stimuli to find insect caterpillars (Amo et al. 2013); previously it had been thought that these birds located their prey using visual stimuli only. It has also been proven recently that the indirect defence by the release of plant scents increases the fitness of plants. Wild tobacco (*Nicotiana attenuata* Torr. ex S.Watson) plants that released such volatile compounds had twice as many buds and flowers (as a measure of fitness) as genetically modified plants in which the ability to release these volatile compounds had been disabled; however, this effect only occurred when the tobacco hornworm (*Manduca sexta* L.), whose larvae feed on the leaves of wild tobacco, was reduced to half by predatory bugs of the genus *Geocoris* attracted by GLVs (Schuman et al. 2012). Thus, the fitness-enhancing effect is indeed caused by the indirect defence, that is, the attraction of the predatory bug.

Plants also use other communication pathways to obtain information about possible impending attacks. For example, plant hormones that prepare the plant defence are transferred from plant to plant by mycorrhizal networks (Babikova et al. 2013).

All these communication channels are used to exchange information across individual and species boundaries, which activates certain defence patterns in plants. The evolutionary significance of these sender–receiver relationships probably lies in the fact that the communication receivers 'eavesdrop' on their neighbouring plants and prepare their own defences when disturbance becomes likely. The benefit for the sender individuals lies in the fact that more distant plant parts can prepare the defence faster than would be possible through communication via the vessels (Heil 2014).

6.1.10   Herbivorous Insects Respond to Plant Defence

How do insects react to the arsenal of defence mechanisms developed evolutionarily by plants? Insects have a nervous system and the capacity to learn (Bruce 2014). They are able to adapt to new food plants by processing olfactory stimuli differentially and learning their scent (Riffell et al. 2013).

Another adaptation to plant defence is, for example, the ability of herbivores to digest toxic substances. Selection in plants is in turn directed towards the formation of new defence mechanisms or the intensification of existing defence mechanisms, that is, an increase in toxic substances. This is called escalation of the defence power (Agrawal 2007). The evolution of new defence mechanisms in plants in turn drives evolutionary diversification and the development of new species. This could be shown by the example of the defence mechanisms of herbivores on (tree) plants through latex, where latex-producing clades are more species-rich than sibling clades without latex (Agrawal 2007).

Evolution can be a very fast process: new studies have observed evolution in 'real time', for example, in plants of the common evening-primrose (*Oenothera biennis* L.), which were protected from herbivory. After several plant generations, this protection from herbivory led to a decrease in herbivore resistance and an increase in competitiveness. The genetic structure of the plant population had changed already after 5 years (Agrawal et al. 2012). In *Arabidopsis thaliana* studies that the immune response to pathogen infestation not only led to an increased resistance against the pathogens, but that this response was also genetically detectable in the next plant generation that had not been exposed to any pathogen (Luna et al. 2012). Thus, the plants 'remember' stress and biotic disorders in previous generations.

For such adaptations to occur, it is essential that sufficient intra-species genetic variation is present in plant populations. Here, a direct connection to the reproductive system of plants can be seen. In populations of self-pollinated species that are exposed to high herbivory pressure, the adaptive evolution of defence may therefore be limited as they often exhibit less genetic variation than outcrossing species (Campbell 2015). This example illustrates that reproduction and defence, which serves the maintenance of reproductive capacity, are coupled with each other and evolve in mutual dependence (Campbell 2015).

6.1.11 Adaptations at Community Level: Herbivores, Biomass Production, and Biodiversity

There are two main hypotheses that describe the interactions between herbivores and primary producers: the bottom-up hypothesis and the top-down hypothesis (Turkington 2009). The bottom-up hypothesis (Hunter and Price 1992) assumes that resources are made accessible by primary producers and gradually become available for higher trophic levels. Thus, plant biomass limits the herbivore biomass, which in turn limits the predator biomass. In contrast to the model of regulation by primary production, the top-down hypothesis postulates regulation by trophic interactions (Hairston et al. 1960; Menge and Sutherland 1976). This hypothesis assumes that consumers of higher trophic levels (e.g. herbivores) deterministically influence the organisms of lower trophic levels (e.g. plants) consumed by them. It is assumed that both bottom-up and top-down processes occur simultaneously, and also that there are interactions between the processes. By reducing herbivory, a short-term increase in stock biomass is achieved. However, the more decisive influence on the composition of species communities occurs through changes in the competitive relationships between the individuals and species concerned. When herbivores feed selectively, they impair their forage plants and thus reduce the competitive power of one plant species to the advantage of another; consequently, the relative abundance of the different plant species changes. In some cases, some plant species may disappear completely from a site as a result of selective herbivory (see Chap. 13).

In a study on Tenerife (Canary Islands), Cubas et al. (2017) reported on the impact of herbivores on two dominant endemic shrub species. While the legume *Spartocytisus supranubius* Christ ex G. Kunkel was heavily browsed by the non-native European rabbit (*Oryctolagus cuniculus* L.) and declined, the honeysuckle

Pterocephalus lasiospermus Link ex Buch was less affected and benefited from the dominance reduction of *S. supranubius*. A consequence of low herbivory, for example, on islands, is that plants grow larger because of the lack of feeding pressure and thus also demand more nutrients from the soil. An increasing number of herbivores, in contrast, cause an overall reduction of biomass. Among other things, this can result in a decrease in the number of herbaceous plants, which can affect endemic species on islands. Ultimately, it is a combination of herbivore pressure and soil resources that jointly control the vegetation. Herbivores with a broad range of food sources tend to influence the entire plant community, whereas specialized herbivores often only affect the abundance of singular species, impacting the overall diversity. Specialized herbivores and pathogens often regulate dominant plant species and thus in some cases lead to higher diversity. This phenomenon, first observed in tropical forests, is known as the Janzen–Connell effect (Connell 1971; Janzen 1970) and has been observed in other ecosystems, for example, temperate grasslands (Petermann et al. 2008).

6.1.12   Adaptation to Disturbance: Increasing Survival Probabilities of Affected Plants or Their Offspring

In terms of their function, the evolutionary adaptations of plants to disturbances can be divided into two major groups:

1. Probability of survival: Those adaptations that increase the survival of the plant under the influence of disturbances.
2. Propagation strategy: Those adaptations that promote the establishment of offspring of the plant under the influence of disturbances (Table 6.1)

An example of the first group of adaptations is the formation of mechanical and chemical defences against herbivores. The already described 'serotiny' (see Sect. 6.1.1) is an example of the second group of adaptations.

6.2   Disturbance as a Selection Factor for Plant Characteristics in Communities

6.2.1   Niche Partitioning in Habitats with High Levels of Disturbance

Functional approaches characterizing adaptations of different species to disturbances suggest that the functional properties and abundance of species are significantly adapted to the intensity, extent, duration, and frequency of disturbance (Grime 1979; Collins and Glenn 1988; Díaz et al. 2016). The resilience of plant

Table 6.1 Types of plant adaptation to abiotic disturbances

Increasing survival probability

Adaptation	Disturbance agent	Selected literature
Affected plant characteristic		
Bark thickness	Ground fire, rockfall, drought	Schwilk and Ackerly (2001), Cavender-Bares et al. (2004), and Lawes et al. (2011)
Sprouting ability	Ground and crown fires, wind, herbivory, avalanches	Bellingham and Sparrow (2000), Bond and Midgley (2001), Ellenberg and Leuschner (2010), and Vesk and Westoby (2004)
Large plant height	Ground fire	Cavender-Bares et al. (2004), He et al. (2012), Pausas and Schwilk (2012), and Dantas and Pausas (2013)
Self-pruning	Ground fire	Pausas et al. (2004) and Pausas (2015)
Plant architecture	Wind	Telewski (1995, 2012), Gardiner et al. (2013, 2016), and Hamant (2013)
Small plant height	Wind	Gardiner et al. (2016)
Reduction of exposed surface area through deflection of leaves, branches, and trunks (reconfiguration and streamlining)	Wind	Telewski (2012), Lopez et al. (2014), and Butler et al. (2011)
Reduction of exposed surface area through breakage of branches and leaves/needles	Wind	Telewski (2012)
Spiral grain in trees	Wind	Telewski (1995)
Small leaves with flexible petioles	Wind	Gardiner et al. (2016)
High flexibility of trunks and branches enables aerodynamic attenuation of oscillations	Wind	Hamant (2013), Lopez et al. (2014) and Gardiner et al. (2016)
Bending stiffness	Avalanches and landslides; wind	Gardiner et al. (2016)
Root adaptations (allocation pattern and root morphology, anchorage)	Wind	Nicoll and Ray (1996)
Allocation patterns for trunks and branches	Wind	Gardiner et al. (2016)
Reaction wood	Wind	Coutand et al. (2014)
Adaptations to flooding (lenticels, aerenchyma, adventitious roots, thin cuticle, chloroplasts in or close to epidermis)	Flooding	Glenz et al. (2006) and Catford and Jansson (2014)
Streamlined (elliptic) leaves	floods	Ellenberg (1996)
Elasticity of wood and predetermined breaking points	Flooding	Ellenberg and Leuschner (2010)

(continued)

Table 6.1 (continued)

Offspring		
Early reproduction, fast life cycle	Crown fires, avalanches, floods	Blom (1999), Ellenberg and Leuschner (2010), Keeley et al. (2011), Pausas and Schwilk (2012), and Reich (2014)
Large investment in seeds	Flooding	Blom (1999), Ellenberg and Leuschner (2010), and Catford and Jansson (2014)
Germination triggered by disturbances	Crown fires; generally seeds in soil diaspore banks	Thompson et al. (1997), Chiwocha et al. (2009), Nelson et al. (2012), and Long et al. (2015)
High flammability of plants and litter	Crown fires	Schwilk and Ackerly (2001), Schwilk (2003), Bond and Keeley (2005), and Crisp et al. (2011)
Serotiny	Crown fires	Schwilk and Ackerly (2001) and Causley et al. (2016)
Floating ability of seeds, high numbers of seeds	Flooding	Johansson et al. (1996), Gurnell et al. (2008), Ellenberg and Leuschner (2010), Nilsson et al. (2010), and Catford and Jansson (2014)

communities to disturbance is also characterized by the acquisition of ecological niches (see the resilience hypothesis, Walker et al. 1999; and the insurance hypothesis, Main 1982; Walker 1995; Naeem and Li 1997). In the case of disturbance, for example, less dominant species can contribute to the resilience of the system. These can replace the dominant species completely or partially to maintain ecosystem functions when the dominant species decline as a result of disturbance. Here, disturbances act as an evolutionary force to promote redundancy of certain ecosystem characteristics (such as the production of aboveground biomass) and at the same time to promote complementarity in other properties of species in the same community (such as tolerance to mowing and grazing).

Ecosystems respond differently to disturbances depending on whether the disturbances have been affecting the ecosystem for a long time or whether the disturbances are novel. If previous disturbances occur within an ecological timescale, species that have adapted to this disturbance may survive and thus participate in the response to a subsequent disturbance. Seen on an evolutionary timescale, the disturbance history of a habitat is decisive for the range of existing strategy types and the niche distribution between more or less adapted species (White and Jentsch 2001). Physiological tolerances, adaptation mechanisms, and genetic variations determine whether species and communities adapt to disturbances within an evolutionary timescale. Thus, it makes sense to measure the frequency, spatial extent, and intensity of disturbance events relative to the lifespan, regeneration time, and resource availability of the affected ecosystem. A gap in a tropical rainforest is closed much faster by regrowing individuals, because of higher growth rates, than a gap of the same size in a boreal forest. The number of species or functional groups available for regeneration also differs greatly between megadiverse tropical and relatively species-poor boreal forest ecosystems (Box 6.1).

Box 6.1: Disturbances as a Selection Factor: An Old Story

Georg Gratzer [iD]
Institute of Forest Ecology, Department of Forest and Soil Sciences,
University of Natural Resources and Life Sciences BOKU, Vienna, Austria
e-mail: georg.gratzer@boku.ac.at

Conditions in the Ordovician period (~485–444 million years ago) were extreme, and yet they contributed to one of the most significant evolutionary events on our planet: high temperatures caused by an atmospheric CO_2 content 18 times higher than today favoured the development of photosynthetically active microorganisms.

Before the Ordovician, tectonic events and sedimentation had created extensive shallow water areas, and on land during the Ordovician, acid rain accelerated rock weathering. This set the stage for plants to colonize the land. The photosynthetic activity of phytoplankton and microorganisms on land had finally reduced the strong greenhouse effect, and early soil formations provided sufficient nutrients for the great evolutionary step of plants leaving the water and colonizing land. Nevertheless, it took several evolutionary adaptations before plants could grow completely on land. A little more than 400 million years ago, plants began for the first time to spread quickly on land (Graham et al. 2000, Willis & McElwain 2014; Box Fig. 1). However, at that time, the land was already populated by microorganisms that had formed biofilms, exposing early land plants to biotic and abiotic disturbances from the beginning of their development: microbial attacks and the resulting mortality from such attacks may have been among the few biotic disturbances that affected the first land plants. Also, early terrestrial herbivores were present: small arthropods that sucked on the stems and consumed spores and sporangia more than 400 million years ago (Labandeira 2013a, Labandeira 2013b). It 'only' took a few million years for these early forms of herbivory to develop. Since then, herbivory has driven the co-evolution between plants and herbivores – plants have developed an arsenal of defence strategies and herbivores have developed strategies to overcome them.

When plants moved ashore, away from the permanently flooded shelf area, *drought* was a powerful driver of adaptation processes, affecting the development of drought-resistant spores and even the protection of sporophytes and gametophytes. In fossil records there are also traces of *fire* dating back over 400 million years (Glasspool et al. 2004, Bowman et al. 2009, He & Lamont 2017), illustrating the early exposure of plants to this disturbance agent 'soon' after the colonization of land. Hard coal from the Permian (299–252 million years ago) often shows high charcoal contents and indicates high fire frequency in an atmosphere with an oxygen content of over 30% (Berner et al. 2003).

(continued)

Box 6.1 (continued)

Box Fig. 1 In the middle Devonian (~380 million years ago) the first early forms of trees developed. Until the late Carboniferous (~300 million years ago) spore-producing trees, like the species of the extinct genus *Lepidodendron* (front right) shown here, coexisted with early seed plants, like the trees of the genus *Cordaites* shown on the left and in the centre of the picture. (© Walter B. Myers; Willis & McElwain 2014)

Wind-induced dynamic pressure on plant crowns increases exponentially with plant height. Therefore, land plants and especially trees, from the early hydrostatic cylinders with a height of some centimetres (Willis & McElwain 2014) to today's coastal redwoods with a height of more than 110 m, have been confronted with increasing stability requirements and wind as a disturbance agent in their evolutionary race for race for height and sunlight (sensu Falster & Westoby 2003).

In summary, disturbances have affected plants since their colonization of land and have probably also (partly) shaped their fitness (He & Lamont 2017). Protective mechanisms against herbivory, drought, fire, and windthrow are still part of the defence strategies of plants today.

References

Berner RA, Beerling DJ, Dudley R, Robinson JM, Wildman Jr. RA (2003) Phanerozoic atmospheric oxygen. Ann. Rev. Earth Pl. Sc. 31: 105–134

Bowman DMJS, Balch JK, Artaxo P, Bond WJ, Carlson JM, Cochrane MA, D'Antonio CM, Defries RS, Doyle JC, Harrison SP, Johnston FH, Keeley JE, Krawchuk MA, Kull CA, Marston JB, Moritz MA, Prentice IC, Roos CI, Scott AC, Swetnam TW, van der Werf GR, Pyne SJ (2009) Fire in the Earth system. Science 324: 481–484

Falster DS, Westoby M (2003) Plant height and evolutionary games. Trends Ecol. Evol. 18: 337–343

(continued)

> **Box 6.1** (continued)
>
> Glasspool IJ, Edwards D, Axe L (2004) Charcoal in the Silurian as evidence for the earliest wildfire. Geology 32: 381–383
> Graham LE, Cook ME, Busse JS (2000) The origin of plants: body plan changes contributing to a major evolutionary radiation. Proc. Nat. Acad. Sci. USA 97: 4535–4540
> He T, Lamont BB (2017) Baptism by fire: the pivotal role of ancient conflagrations in evolution of the Earth's flora. Natl. Sci. Rev. 5: 237–254
> Labandeira CC (2013a) Deep-time patterns of tissue consumption by terrestrial arthropod herbivores. Naturwissenschaften 100: 355–364
> Labandeira CC (2013b) A paleobiologic perspective on plant–insect interactions. Curr. Opin. Plant Biol. 16: 414–421
> Willis K, McElwain J (2014) The evolution of plants (2nd ed.). Oxford University Press, Oxford, 406 p

References

Adler FR, Karban R (1994) Defended fortresses or moving targets? Another model of inducible defenses inspired by military metaphors. Am Nat 144:813–832

Agrawal AA (2002) Herbivory and maternal effects: mechanisms and consequences of transgenerational induced plant resistance. Ecology 83:3408–3415

Agrawal AA (2007) Macroevolution of plant defense strategies. Trends Ecol Evol 22:103–109

Agrawal AA, Hastings AP, Johnson MTJ, Maron JL, Salminen J-P (2012) Insect herbivores drive real-time ecological and evolutionary change in plant populations. Science 338:113–116

Amo L, Jansen JJ, Dam NM, Dicke M, Visser ME (2013) Birds exploit herbivore-induced plant volatiles to locate herbivorous prey. Ecol Lett 16:1348–1355

Atkinson NJ, Urwin PE (2012) The interaction of plant biotic and abiotic stresses: from genes to the field. J Exp Bot 63:3523–3543

Babikova Z, Gilbert L, Bruce TJ, Birkett M, Caulfield JC, Woodcock C, Pickett JA, Johnson D (2013) Underground signals carried through common mycelial networks warn neighbouring plants of aphid attack. Ecol Lett 16:835–843

Ballaré CL (2011) Jasmonate-induced defenses: a tale of intelligence, collaborators and rascals. Trends Plant Sci 16:249–257

Bellingham PJ, Sparrow AD (2000) Resprouting as a life history strategy in woody plant communities. Oikos 89:409–416

Blodner C, Skroppa T, Johnsen O, Polle A (2005) Freezing tolerance in two Norway spruce (*Picea abies* L. karst.) progenies is physiologically correlated with drought tolerance. J Plant Physiol 162:549–558

Blom CWPM (1999) Adaptations to flooding stress: from plant community to molecule. Plant Biol 1:261–273

Bond WJ, Keeley JE (2005) Fire as a global 'herbivore': the ecology and evolution of flammable ecosystems. Trends Ecol Evol 20:387–394

Bond WJ, Midgley JJ (2001) Ecology of sprouting in woody plants: the persistence niche. Trends Ecol Evol 16:45–51

Bruce TJ (2014) Interplay between insects and plants: dynamic and complex interactions that have coevolved over millions of years but act in milliseconds. J Exp Bot 66:455–465

Bruessow F, Gouhier-Darimont C, Buchala A, Metraux JP, Reymond P (2010) Insect eggs suppress plant defence against chewing herbivores. Plant J 62:876–885

Butler DW, Gleason SM, Davidson I, Onoda Y, Westoby M (2011) Safety and streamlining of woody shoots in wind: an empirical study across 39 species in tropical Australia. New Phytol 193:137–149

Campbell SA (2015) Ecological mechanisms for the coevolution of mating systems and defence. New Phytol 205:1047–1053

Carmona D, Fornoni J (2013) Herbivores can select for mixed defensive strategies in plants. New Phytol 197:576–585

Catford JA, Jansson R (2014) Drowned, buried and carried away: effects of plant traits on the distribution of native and alien species in riparian ecosystems. New Phytol 204:19–36

Causley CL, Fowler WM, Lamont BB, He T (2016) Fitness benefits of serotiny in fire-and drought-prone environments. Plant Ecol 217:773–779

Cavender-Bares J, Kitajima K, Bazzaz F (2004) Multiple trait associations in relation to habitat differentiation among 17 Floridian oak species. Ecol Monogr 74:635–662

Chiwocha SD, Dixon KW, Flematti GR, Ghisalberti EL, Merritt DJ, Nelson DC, Riseborough J-AM, Smith SM, Stevens JC (2009) Karrikins: a new family of plant growth regulators in smoke. Plant Sci 177:252–256

Chung SH, Rosa C, Scully ED, Peiffer M, Tooker JF, Hoover K, Luthe DS, Felton GW (2013) Herbivore exploits orally secreted bacteria to suppress plant defenses. Proc Natl Acad Sci USA 110:15728–15733

Collins SL, Glenn SM (1988) Disturbance and community structure in North American prairies. In: During HJ, Werger MJA, Willems JH (eds) Diversity and pattern in plant communities. SPB Academic Publishing, The Hague, pp 131–143

Connell JH (1971) On the role of natural enemies in preventing competitive exclusion in some marine animals and in rain forest trees. In: den Boer PJ, Gradwell GR (eds) Proceedings of the advanced study institute on 'dynamics of numbers in populations', Oosterbeek, the Netherlands, 7–18 September 1970. Centre for Agricultural Publication and Documentation, Wageningen, pp 298–312

Coutand C, Pot G, Badel E (2014) Mechanosensing is involved in the regulation of autostress levels in tension wood. Trees 28:687–697

Crisp MD, Burrows GE, Cook LG, Thornhill AH, Bowman DM (2011) Flammable biomes dominated by eucalypts originated at the Cretaceous–Palaeogene boundary. Nat Commun 2:193

Cubas J, Martín-Esquivel JL, Nogales M, Irl SD, Hernández-Hernández R, López-Darias M, Marrero-Gómez M, del Arco MJ, González-Mancebo JM (2017) Contrasting effects of invasive rabbits on endemic plants driving vegetation change in a subtropical alpine insular environment. Biol Invas 20:793–807

Dantas VDL, Pausas JG (2013) The lanky and the corky: fire-escape strategies in savanna woody species. J Ecol 101:1265–1272

Díaz S, Kattge J, Cornelissen JHC, Wright IJ, Lavorel S, Dray S, Reu B, Kleyer M, Wirth C, Colin Prentice I, Garnier E, Bönisch G, Westoby M, Poorter H, Reich PB, Moles AT, Dickie J, Gillison AN, Zanne AE, Chave J, Joseph Wright S, Sheremet'ev SN, Jactel H, Baraloto C, Cerabolini B, Pierce S, Shipley B, Kirkup D, Casanoves F, Joswig JS, Günther A, Falczuk V, Rüger N, Mahecha MD, Gorné LD (2016) The global spectrum of plant form and function. Nature 529:167–171

Ellenberg H (1996) Vegetation Mitteleuropas mit den Alpen in ökologischer, dynamischer und historischer Sicht, 5th edn. Ulmer, Stuttgart, 1095 p

Ellenberg H, Leuschner C (2010) Vegetation Mitteleuropas mit den Alpen in ökologischer, dynamischer und historischer Sicht, 6th edn. Eugen Ulmer, Stuttgart, 1334 p

Engber EA, Varner JM III (2012) Patterns of flammability of the California oaks: the role of leaf traits. Can J For Res 42:1965–1975

Falster DS, Westoby M (2003) Plant height and evolutionary games. Trends Ecol Evol 18:337–343

Gardiner B, Schuck ART, Schelhaas M-J, Orazio C, Blennow K, Nicoll B (2013) Living with storm damage to forests. What science can tell us, 3. European Forest Institute Joensuu, 129 p

Gardiner B, Berry P, Moulia B (2016) Wind impacts on plant growth, mechanics and damage. Plant Sci 245:94–118

Glenz C, Schlaepfer R, Iorgulescu I, Kienast F (2006) Flooding tolerance of central European tree and shrub species. Forest Ecol Manag 235:1–13

Gratzer G, Rai P, Schieler K (2002) Structure and regeneration dynamics of *Abies densa* forests in Central Bhutan. Centralblatt für das gesamte Forstwesen 119:279–287

Grime JP (1979) Plant strategies and vegetation processes. Wiley, Chichester, 222 p

Gurnell A, Thompson K, Goodson J, Moggridge H (2008) Propagule deposition along river margins: linking hydrology and ecology. J Ecol 96:553–565

Hairston NG, Smith FE, Slobodkin LB (1960) Community structure, population control, and competition. Am Nat 94:421–425

Hamant O (2013) Widespread mechanosensing controls the structure behind the architecture in plants. Curr Opin Plant Biol 16:654–660

He T, Pausas JG, Belcher CM, Schwilk DW, Lamont BB (2012) Fire-adapted traits of Pinus arose in the fiery cretaceous. New Phytol 194:751–759

Heil M (2014) Herbivore-induced plant volatiles: targets, perception and unanswered questions. New Phytol 204:297–306

Herms DA, Mattson WJ (1992) The dilemma of plants: to grow or defend. Q Rev Biol 67:283–335

Hernández-Serrano A, Verdú M, González-Martínez SC, Pausas JG (2013) Fire structures pine serotiny at different scales. Am J Bot 100:2349–2356

Hunter MD, Price PW (1992) Playing chutes and ladders: heterogeneity and the relative roles aof bottom-up and top-down forces in natural communities. Ecology 73:724–732

Janská A, Maršík P, Zelenková S, Ovesná J (2010) Cold stress and acclimation – what is important for metabolic adjustment? Plant Biol 12:395–405

Janzen DH (1970) Herbivores and the number of tree species in tropical forests. Am Nat 104:501–528

Johansson ME, Nilsson C, Nilsson E (1996) Do rivers function as corridors for plant dispersal? J Veg Sci 7:593–598

Keeley JE, Pausas JG, Rundel PW, Bond WJ, Bradstock RA (2011) Fire as an evolutionary pressure shaping plant traits. Trends Plant Sci 16:406–411

Kellenberger RT, Desurmont GA, Schlüter PM, Schiestl FP (2018) Trans-generational inheritance of herbivory-induced phenotypic changes in *Brassica rapa*. Sci Rep 8:3536

Kessler A (2015) The information landscape of plant constitutive and induced secondary metabolite production. Curr Opin Insect Sci 8:47–53

Kotak S, Larkindale J, Lee U, von Koskull-Doring P, Vierling E, Scharf KD (2007) Complexity of the heat stress response in plants. Curr Opin Plant Biol 10:310–316

Kreyling J, Jentsch A, Beierkuhnlein C (2011) Stochastic trajectories of succession initiated by extreme climatic events. Ecol Lett 14:758–764

Kreyling J, Thiel D, Nagy L, Jentsch A, Huber G, Konnert M, Beierkuhnlein C (2012a) Late frost sensitivity of juvenile *Fagus sylvatica* L. differs between southern Germany and Bulgaria and depends on preceding air temperature. Eur J Forest Res 131:717–725

Kreyling J, Thiel D, Simmnacher K, Willner E, Jentsch A, Beierkuhnlein C (2012b) Geographic origin and past climatic experience influence the response to late spring frost in four common grass species in Central Europe. Ecography 35:268–275

Kreyling J, Wiesenberg GLB, Thiel D, Wohlfart C, Huber G, Walter J, Jentsch A, Konnert M, Beierkuhnlein C (2012c) Cold hardiness of *Pinus nigra* Arnold as influenced by geographic origin, warming, and extreme summer drought. Environ Exp Bot 78:99–108

Lachenbruch B, McCulloh KA (2014) Traits, properties, and performance: how woody plants combine hydraulic and mechanical functions in a cell, tissue, or whole plant. New Phytol 204:747–764

Lambers H, Chapin FS III, Pons TL (2008) Plant physiological ecology, 2nd edn. Springer, New York, 604 p

Lamont BB, Pausas JG, He T, Witkowski ETF, Hanley ME (2020) Fire as a selective agent for both serotiny and nonserotiny over space and time. Crit Rev Plant Sci 39:140–172

Lawes MJ, Adie H, Russell-Smith J, Murphy B, Midgley JJ (2011) How do small savanna trees avoid stem mortality by fire? The roles of stem diameter, height and bark thickness. Ecosphere 2:1–13

Long RL, Gorecki MJ, Renton M, Scott JK, Colville L, Goggin DE, Commander LE, Westcott DA, Cherry H, Finch-Savage WE (2015) The ecophysiology of seed persistence: a mechanistic view of the journey to germination or demise. Biol Rev 90:31–59

Lopez D, Eloy C, Michelin S, de Langre E (2014) Drag reduction, from bending to pruning. Eur J Entomol 108:48002

Luna E, Bruce TJ, Roberts MR, Flors V, Ton J (2012) Next-generation systemic acquired resistance. Plant Physiol 158:844–853

Main AR (1982) Rare species: precious or dross? In: Groves RH, Ride WDL (eds) Species at risk: research in Australia. Australian Academy of Science, Canberra, pp 163–174

Menge BA, Sutherland JP (1976) Species-diversity gradients – synthesis of roles of predation, competition, and temporal heterogeneity. Am Nat 110:351–369

Naeem S, Li SB (1997) Biodiversity enhances ecosystem reliability. Nature 390:507–509

Nelson DC, Flematti GR, Ghisalberti EL, Dixon KW, Smith SM (2012) Regulation of seed germination and seedling growth by chemical signals from burning vegetation. Annu Rev Plant Biol 63:107–130

Nicoll BC, Ray D (1996) Adaptive growth of tree root systems in response to wind action and site conditions. Tree Physiol 16:891–898

Nilsson C, Brown RL, Jansson R, Merritt DM (2010) The role of hydrochory in structuring riparian and wetland vegetation. Biol Rev 85:837–858

Parde VD, Sharma HC, Kachole MS (2012) Protease inhibitors in wild relatives of pigeonpea against the cotton bollworm/legume pod borer, *Helicoverpa armigera*. Am J Plant Sci 3:627–635

Pausas JG (2015) Evolutionary fire ecology: lessons learned from pines. Trends Plant Sci 20:318–324

Pausas JC, Keeley JE (2017) Epicormic resprouting in fire-prone ecosystems. Trends Plant Sci 22:1008–1015

Pausas JG, Schwilk D (2012) Fire and plant evolution. New Phytol 193:301–303

Pausas JG, Bradstock RA, Keith DA, Keeley JE, Network GF (2004) Plant functional traits in relation to fire in crown-fire ecosystems. Ecology 85:1085–1100

Petermann JS, Fergus AJF, Turnbull LA, Schmid B (2008) Janzen-Connell effects are widespread and strong enough to maintain diversity in grasslands. Ecology 89:2399–2406

Pilson D (2000) The evolution of plant response to herbivory: simultaneously considering resistance and tolerance in *Brassica rapa*. Evol Ecol 14:457

Reich PB (2014) The world-wide 'fast–slow' plant economics spectrum: a traits manifesto. J Ecol 102:275–301

Reich P, Wright I, Cavender-Bares J, Craine J, Oleksyn J, Westoby M, Walters M (2003) The evolution of plant functional variation: traits, spectra, and strategies. Int J Plant Sci 164:S143–S164

Riffell JA, Lei H, Abrell L, Hildebrand JG (2013) Neural basis of a pollinator's buffet: olfactory specialization and learning in *Manduca sexta*. Science 339:200–204

Scala A, Allmann S, Mirabella R, Haring MA, Schuurink RC (2013) Green leaf volatiles: a plant's multifunctional weapon against herbivores and pathogens. Int J Mol Sci 14:17781–17811

Schulze ED, Beck E, Müller-Hohenstein K (2002) Pflanzenökologie. Spektrum Akademischer Verlag, Heidelberg, 846 p

Schuman MC, Barthel K, Baldwin IT (2012) Herbivory-induced volatiles function as defenses increasing fitness of the native plant *Nicotiana attenuata* in nature. eLIFE 1:e00007

Schwilk DW (2003) Flammability is a niche construction trait: canopy architecture affects fire intensity. Am Nat 162:725–733

Schwilk DW, Ackerly DD (2001) Flammability and serotiny as strategies: correlated evolution in pines. Oikos 94:326–336

Talluto MV, Benkman CW (2013) Landscape-scale eco-evolutionary dynamics: selection by seed predators and fire determine a major reproductive strategy. Ecology 94:1307–1316

Telewski FW (1995) Wind-induced physiological and developmental responses in trees. In: Coutts MP, Grace J (eds) Wind and trees. Cambridge University Press, Cambridge, pp 237–263

Telewski FW (2012) Is windswept tree growth negative thigmotropism? Plant Sci 184:20–28

Thompson K, Bakker JP, Bekker RM (1997) The soil seed banks of North West Europe: methodology, density and longevity. Cambridge University Press, Cambridge, 276 p

Tng D, Williamson G, Jordan G, Bowman D (2012) Giant eucalypts – globally unique fire-adapted rain-forest trees? New Phytol 196:1001–1014

Turkington R (2009) Top-down and bottom-up forces in mammalian herbivore – vegetation systems: an essay review. Botany-Botanique 87:723–739

Vesk PA, Westoby M (2004) Sprouting ability across diverse disturbances and vegetation types worldwide. J Ecol 92:310–320

Walker B (1995) Conserving biological diversity through ecosystem resilience. Cons Biol 9:747–752

Walker B, Kinzig A, Langridge J (1999) Plant attribute diversity, resilience, and ecosystem function: the nature and significance of dominant and minor species. Ecosystems 2:95–113

Walter J, Nagy L, Hein R, Rascher U, Beierkuhnlein C, Willner E, Jentsch A (2011) Do plants remember drought? Hints towards a drought-memory in grasses. Environ Exp Bot 71:34–40

Walter J, Jentsch A, Beierkuhnlein C, Kreyling J (2013) Ecological stress memory and cross stress tolerance in plants in the face of climate extremes. Environ Exp Bot 94:3–8

White PS, Jentsch A (2001) The search for generality in studies of disturbance and ecosystem dynamics. Prog Bot 62:399–449

Züst T, Heichinger C, Grossniklaus U, Harrington R, Kliebenstein DJ, Turnbull LA (2012) Natural enemies drive geographic variation in plant defenses. Science 338:116–119

Part III
Abiotic Disturbances

Chapter 7
Fire in Forest Ecosystems: Processes and Management Strategies

Daniel Kraus ⓘ, Thomas Wohlgemuth ⓘ, Marc Castellnou ⓘ, and Marco Conedera ⓘ

Abstract There have been fires on this planet for 420 million years – especially since (a) the oxygen content of the atmosphere has exceeded 13% and (b) land plants have been available for combustion, with lightning strikes acting as the ignition source. Since then, vegetation fires have been the most frequent disturbance factor in terrestrial ecosystems worldwide. The long-term effect of fire thus became an evolutionary factor for many species. Where natural fires are frequent, species have developed characteristic traits of fire adaptation. Humans have greatly changed the geography of fire by burning vegetation to clear or fertilize the land; suppressing fires; and also land use and ecosystem fragmentation. A fire event depends not only on climatic and biological factors but also on cultural history. A thorough and systemic knowledge of fire and their medium- to long-term effects on the biosphere is a prerequisite for the appropriate management of vegetation fires.

Keywords Fire ecology · Fire regime · Fire history · Fire adaptation · Resilience · Pyrophilous · Fire-dependant · Integrated fire management

D. Kraus (✉)
University of Freiburg, Freiburg im Breisgau, Germany

Forstbetrieb Neureichenau, Bayerische Staatsforsten AöR, Neureichenau, Germany
e-mail: daniel.kraus@baysf.de

T. Wohlgemuth
Forest Dynamics Research Unit, Swiss Federal Institute for Forest, Snow and Landscape Research WSL, Birmensdorf, Switzerland

M. Castellnou
Direcció General de Prevenció i Extinció d'Incendis, Generalitat de Catalunya, Cerdanyola del Vallés, Spain

Departament de Ciència Forestal i Producció Vegetal, University of Lleida, Lleida, Spain

M. Conedera
Community Ecology Research Unit, Swiss Federal Institute for Forest, Snow and Landscape Research WSL, Cadenazzo, Switzerland

T. Wohlgemuth et al. (eds.), *Disturbance Ecology*, Landscape Series 32,
https://doi.org/10.1007/978-3-030-98756-5_7

7.1 History and Geography of Vegetation Fires

With the beginning of photosynthesis 2.3 billion years ago, oxygen entered the atmosphere for the first time. The oxygen content in the atmosphere rose sharply with the evolution of land plants, especially vascular plants, about 440 million years ago. Around 420 million years ago, the oxygen content in the Earth's atmosphere exceeded 13%, and lightning strikes could now cause the ignition of fuel in the form of standing or accumulated biomass. It is assumed that vegetation fires have occurred in most of the Earth's ecoregions since that time, although with large differences in frequency and fire characteristics (Krawchuk et al. 2009). From a physical point of view, fire activity basically depends on two mutually influencing factors: (i) the availability of fuel and (ii) the moisture content of the fuel. Both site productivity and desiccation of the accumulated biomass (i.e. the fuel) are influenced by climatic conditions (Krawchuk and Moritz 2011). Therefore, fire hardly ever occurs in arid and hot biomes (e.g. deserts), where dry weather conditions favourable for fire prevail but too little combustible biomass is present. On the other hand, fires hardly ever occur in tropical rainforests, because the abundant combustible material is almost constantly moist and difficult to ignite because of the constantly moist climatic conditions. Therefore, the highest degree of combustibility and corresponding fire susceptibility is found in biomes and ecosystems that are between the extremes with distinct wet and dry seasons (e.g. monsoon climate) and that produce sufficient biomass (Archibald et al. 2009; Pausas and Ribeiro 2017). The intermediate fire–productivity hypothesis thus serves as a model of global fire distribution (Kravchuk and Moritz 2011; Fig. 7.1). Worldwide, more than half of the land area covered by vegetation is potentially prone to burning (Fig. 7.2).

In prehistoric Europe, fire was a key factor in the dynamics of many ecosystems. In particular, the post-glacial climate warming played a key role in the global development of fire occurrences since the last ice age (Marlon et al. 2013). Corresponding to the climatic fluctuations during the Holocene, the conditions for vegetation fires varied. Paleoecological studies have shown that during dry-warm climate phases (e.g. the Allerød and Preboreal phases), lightning events and vegetation fires occurred more frequently (Granström 1993; Vázquez and Moreno 1998; Tinner et al. 1999). Studies from Mediterranean climates in other parts of the world suggest that fire has also strongly influenced the species composition of the vegetation in the Mediterranean Basin (Dodson 2001; Keeley and Fotheringham 2001; Atahan et al. 2004).

At least 500,000 years ago, humans began to control fire and use vegetation. Since then, the presence of humans has become an additional important factor affecting the occurrence of fire events in space and time. Depending on cultural and socio-economic development, human societies have used or suppressed fire. They

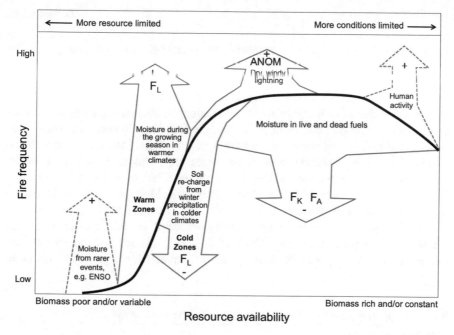

Fig. 7.1 Estimated long-term probability of vegetation fires or fire activity as a function of globally available resources, here net primary production (NPP, thick line). Fire is caused by the concurrence of fuel, drought, and ignition energy. The three components vary in space and time. Soil moisture (M) and weather anomaly (ANOM) were defined as indicators for the amount of fuel and its moisture content, with positive or negative relationships (arrows) to fire activity (y-axis) along a global gradient of available resources (x-axis). Soil moisture is differentiated between the current situation (Soil-M_C), the month just before (Soil-M_N, 1–2 months before; N = near-time conditions), or the month with the highest fire activity (Soil-M_A, 4–8 months before; A = antecedent conditions). The position of arrows is approximate; dotted arrows indicate additional drivers of fire activity. (Redrawn from Krawchuk and Moritz 2011)

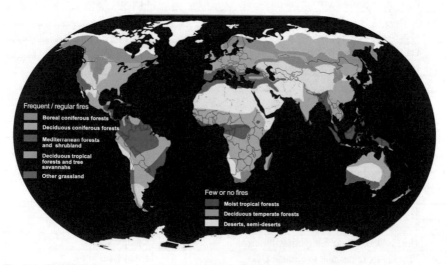

Fig. 7.2 Worldwide overview of ecosystems with different fire regimes

have changed the structure, composition, quantity, and continuity of the combustible vegetation through clearing, grazing, and other agricultural or industrial activities. Natural fire regimes have thus been prevented, weakened, or strengthened, and new regimes introduced (Table 7.1; Bowman et al. 2011).

The exact causes of the prehistoric fires cannot be determined accurately; however, charcoal particles and plant pollen in lake sediments indicate that during the post-glacial early warm period around 8000 BC (the Boreal stage), a moderate but regular charcoal accumulation took place, suggesting periodically recurring fires in both Central Europe and the Mediterranean region (Clark et al. 1989; Vázquez and Moreno 1998). After a brief decline, fire activity in the Mediterranean area increased again from about 4000 BC onwards (Carrión 2002; Franco-Múgica et al. 2005). In Central Europe, an increase in fire activity around 3700 BC is associated with Neolithic slash-and-burn activity (Clark et al. 1989), with another increase around 700 BC corresponding with the Iron Age settlement activity (Fig. 7.3; Tinner et al.

Table 7.1 Direct and indirect influence of human activity on fire activity

Fire regime attributes	
Parameters	
Impact of human activity	Time scale
Fire frequency	
Fire ignition	
Legislation	Short- / medium-term
Agricultural and silvicultural activity; socio-economic structure	Medium-term
Road networks, in particular agricultural and logging roads	Medium-term
Population density and education	Medium-term
Fire intensity	
Fuel amount and composition	
Preventing controlled burning	Short-term
Grazing	Medium-/long-term
Silviculture	Medium-/long-term
Land-use and landscape planning	Long-term
Fire spread	
Fuel continuity	
Firefighting by counterfires	Medium-term
Fire barriers, e.g. roads, fuel breaks	Medium-term
Landscape architecture/land use	Medium-term
Speed of fire front	
Organization of the fire service and operational readiness	Short- / medium-term
Fire suppression system, e.g. emergency water ponds	Medium-term
Quality of airborne firefighting systems	Medium-term
Weather extremes due to climate change	Long-term

Adapted from Bowman et al. (2011)

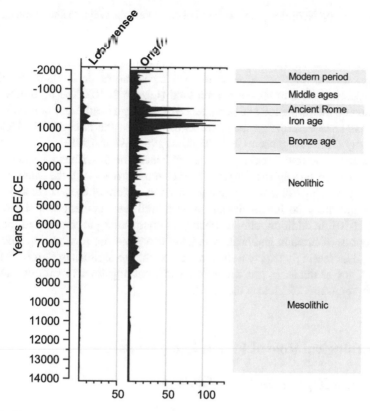

Fig. 7.3 Charcoal sedimentation rate in sediments at the Lobsigensee in the northern Alps and Lago d'Origlio in the southern Alps. (Adapted from Tinner et al. 2005)

2005; Colombaroli et al. 2007; Tinner et al. 2009; Conedera et al. 2017). While in the cooler and wetter climate of the Subatlantic (since 450 BC until today) natural fires were rather rare in Central Europe, they were clearly overlaid by anthropogenic fire activity in the Mediterranean area. Nevertheless, there are clear indications of natural fire regimes in large parts of Europe (Zackrisson 1977; Pausas and Fernándes-Muñoz 2012). In this context, it is interesting to note that early settlements on the Iberian Peninsula are very often located in areas where the wind-driven fire regimes are still associated with natural lightning fires. Actually, people preferred to settle precisely in areas where fire has created open forest landscapes (Castellnou 1996; Carrión 2002). Boreal forests in Scandinavia also developed under the influence of a regime of regularly recurring fires until the beginning of modern forestry (Zackrisson 1977; Niklasson and Granström 2000; Carcaillet et al. 2007). Tree ring analyses of conifers with multiple burn scars in the Białowieża forest in Poland indicate that fire events were probably more frequent even in some Central European forests, at least in the more continental regions, than would be expected from the written historical accounts of fires alone (Niklasson et al. 2010; Zin 2016).

Ever since the Neolithic, fire has been used in various ways to shape the environment and specifically for the purpose of clearing land (Clark et al. 1989). Where permanent farming was not possible for climatic and edaphic reasons, slash-and-burn/swidden agriculture was practised (Heikinheimo 1915; Kapp 1984). Fire has been used over a long period to remove plant waste or dead plant material, to improve pastures, and to preserve open land (Lázaro 2010). The influence of fire application on the formation and persistence of the large heathlands of northwestern Europe has been described, mainly for Great Britain, the Netherlands, Denmark, and Norway (Holst-Jørgensen 1993; Diemont 1996; Webb 1998).

The widespread traditional use of fire, which can be described as European fire culture, has been largely lost, at least in Western Europe, with the intensive rural exodus and growing urban development since the middle of the last century. Fire is therefore generally no longer perceived as a land management tool (Lázaro and Montiel 2010). In addition, climate change is contributing to longer and more frequent periods of drought and heat, which are changing fire regimes in some areas. This could increasingly lead to a shift in the natural fire regimes in Europe, which is indicated, for example, by the more frequent lightning fires in various European mountain regions (Conedera et al. 2006).

7.2 Ecological Role of Fire in Ecosystems

7.2.1 Fire Regime and Fire Ecology

Fire has an important regulatory and selective function in ecosystems. The selection pressure of fire on the survival of species strongly depends on the fire intensity, fire frequency, and time during the season when the fire occurs – in short the fire regime and its characteristics and processes (Table 7.2). The most important characteristics of fire regimes are fire intensity as a measurable variable of fire behaviour, fire frequency (Fig. 7.4), and fire type (Fig. 7.5).

Table 7.2 Characteristics of fire regimes and processes

Characteristics	Description
Regime	
Fire type	Ground (peat), surface, and crown fire
Frequency	Mean fire return interval for a given location
Fire behaviour	Fire intensity and rate of spread
Severity	Effect of fire on organisms and ecosystems
Seasonality	Differential susceptibility of organisms to fire damage due to time of year
Size	Area burned per time period or fire event
Process	
Combustion type	Complete/incomplete combustion, with or without visible flames
Heat transfer	Radiation, conduction, and convection

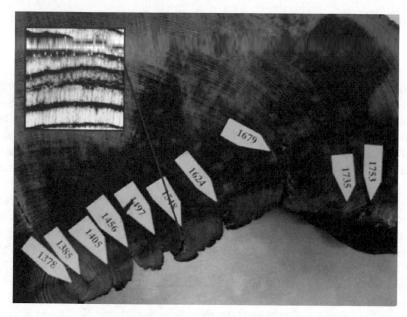

Fig. 7.4 Cross section through a 400-year-old pine trunk (near Haverö in Västernorrland, Sweden). The tree had survived eleven different fire events since 1378, which left clear marks in its burn scar as occluded charcoal inclusions. The mean return interval of the fire events was 47 years. (Dating: T. Rydkvist) (Photo: D. Kraus)

Fig. 7.5 Examples of different forms of fire type: (*i*) surface fire of low to medium intensity; (*ii*) passive crown fire, in which only single individuals completely burn up to the crowns (torching); (*iii*) active crown fire, in which the whole stand burns up to the crowns. (Adapted from Simard et al. 2011)

Organisms that are repeatedly exposed to fire have generally developed adaptations and survival strategies. In particular, plant species influence the development of fire regimes (Bond and van Wilgen 1996; Pausas and Paula 2012).

Research on fire-specific interactions and relationships is now a separate branch of ecology: fire ecology.

7.2.2 Fire Interactions with Other Disturbances

Although fire is often referred to as an abiotic disturbance, it clearly differs from other abiotic disturbances such as windthrows, avalanches, or late frosts. In fact, fire consumes biomass and is thus more comparable to herbivory, strongly resembling biotic disturbances in its effects (Bond and Keeley 2005). However, in contrast to herbivores, fire has a non-selective and broader 'food base' and can mineralize both dead and living plant parts and protein-rich as well as indigestible plant parts. Furthermore, fire metabolism has a neutralizing effect on various toxins accumulated in the soil, which can produce a negative feedback on species composition in ecosystems (Mazzoleni 1993).

Special interest has been paid – for a longer time – to the interactions between fire and other disturbances, as forest fires as well as windstorms and insect calamities are among the major drivers of landscape change. In particular, whether and how these disturbances influence each other in their effects is an important subject (see Chap. 17). It is often assumed that fire risk increases after large-scale bark beetle calamities because of the large amounts of deadwood produced. In the area of the devastating Yellowstone fires in 1988 (USA), the interplay between the infestation by the mountain pine beetle (*Dendroctonus ponderosae* Hopk.; see also Box 12.1 in Chap. 12 and forest fires has been investigated in detail. Bark beetle infestation initially occurs on a smaller scale than forest fires (Turner 2010; Simard et al. 2011). This results in a very heterogeneous infestation pattern (Fig. 7.6) with a mosaic of larger and smaller islands in different stages of decay: reddened crowns

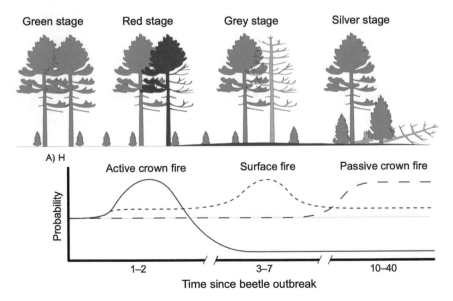

Fig. 7.6 Dependence of fire intensity/fire type on decay stages after bark beetle infestation in the Yellowstone National Park, USA. (After Simard et al. 2011)

after fresh infestation represent an increased fire hazard for 1–2 years owing to an increased supply of fine fuels for active crown fires ('red phase'), whereas after the needles fell off, there was only a small amount of fuel remaining in the crown ('grey phase') and thus only rather low-intensity surface fires can develop. It was only after herbs, shrubs, and trees have established ('silver phase') that more intensive fires, in the form of passive crown fires, could happen again. Bark beetle infestation can therefore affect the intensity of forest fires over a large area. Both fire weather and the consistency of available fuels, especially in the red phase, are the most important driving factors of natural fires (Harvey et al. 2013). Fire ecologists are aware that forest fires in the Yellowstone National Park did not represent an ecological disaster, because previous bark beetle outbreaks created such a heterogeneous mosaic of forest stands that a recolonization of the burned areas by the main tree species was possible in high densities and within a short time. For the European spruce bark beetle (*Ips typographus* L.), such interactions have not yet been sufficiently confirmed, but the processes appear to be transferable. With the pine processionary moth (*Thaumetopoea pityocampa* Denis and Schiffermüller) in northern Catalonia (Spain), on the other hand, similar effects to those observed in the Yellowstone National Park can be observed: after intensive infestation, an increased fire hazard can occur in the treetops at short notice, especially in dense stands of black pine (*Pinus nigra* J.F. Arnold) or Aleppo pine (*P. halepensis* Mill.). However, the defoliation after caterpillar feeding in spring results in a significantly reduced risk of forest fire in summer.

7.2.3   Mechanisms of Fire Adaptation

There is much evidence that plant species in fire-prone environments have developed adaptations to survive fire by selective pressure. Some of these are simply general disturbance adaptations, such as resprouting from stem and root, while others are so specific that species cannot survive or reproduce without the influence of fire. In general, fire adaptations may consist of *passive strategies* that affect the survival of individuals (resistance) and *active strategies* that are necessary for the emergence of a new generation (resilience). The passive resistance strategies of plant species consist of protecting living tissue, whereas the usually mobile animals tend to escape the fire front.

In woody species, a typical *passive* fire defence mechanism at the individual level is the formation of a thick bark to protect the cambium from high temperatures and flames, as is the case with larch (*Larix decidua* Mill.) and many oaks (*Quercus* spp.) and pines (*Pinus* spp.). Similarly, the rapid loss of dead branches in the lower crown can be interpreted as a passive avoidance strategy to prevent potential vertical fire bridges. However, when shaped by the fire regime itself, stand structure can also be considered as a survival strategy in this context. Regularly recurring fires of low intensity reduce the undergrowth, thus preventing the spatial continuity of the fuels up to the crown layer and preventing lethal heat transfer to the renewal organs and meristems. Under such circumstances, fire-resistant structures at the stand level can develop, which in the sense of Holling (Holling and Gunderson 2002) persist despite massive changes (disturbances) of the system (Fig. 7.7). Holling describes the ability of resilience in his adaptive cycle through phases of release, reorganization, growth/exploitation, and conservation. Intensive, stand-replacing fires in advanced succession are often referred to as representing resilience of forest stands to fire disturbances. A fire initiates the regeneration and reorganization of forests. This is followed by a long growth phase that can bind all available resources and nutrients of the ecosystem for a long time until the next fire event releases resources again. Examples are the Pacific coastal rainforests of North America (Agee 1993) or the *Araucaria–Nothofagus* forests of southern Chile (González et al. 2009). Single phases, however, often occur simultaneously in small patches, as clearly shown in the example of a black pine regeneration process dependent on low-intensity fire with intermediate return intervals (approximately 30–40 years) in Ports de Beseit, Catalonia (Fig. 7.8).

The regeneration processes after a fire event are called *active defence strategies*. On the one hand, this concerns the establishment of a new generation by vegetative propagation ('bud bank') and on the other hand reproduction via the formation of diaspores after the fire event or via germination from surviving seeds ('seed bank'; Table 7.3 and Fig. 7.9). Some fire adaptations of plant species are mistaken for 'exaptations' rather than true adaptive traits (Bradshaw et al. 2011). This is true, for example, for the vegetative propagation of plant species in non-fire-dominated habitats. In contrast, Keeley et al. (2011) describe adaptation mechanisms such as serotiny or fire-stimulated flowering and germination as highly fire-specific, since without these characteristics the regeneration process could not occur. An extreme fire adaptation in this sense is flammability. Some plants have developed characteristics that make them more susceptible to fire: they accumulate a lot of dead material – either as litter around the plant or as leaves that are kept on the branches – and store easily flammable oil

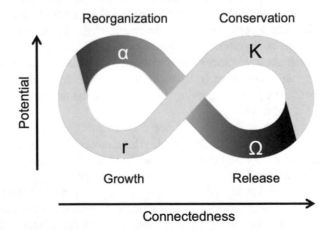

Fig. 7.7 Adaptation cycle and resilience in terms of the four ecosystem functions: growth (*r*), conservation (*K*), release (*Ω*), and renewal (*α*). (Redrawn after Holling and Gunderson 2002)

Fig. 7.8 Regeneration dynamics in black pine stands at Ports de Beseit, Catalonia (Spain), under the influence of fire: in a structurally rich, closed black pine (*Pinus nigra* J.F. Arnold) stand with sparse undergrowth of natural regeneration, holm oaks (*Quercus ilex* L.) and Mediterranean shrubs (*Cistus albidus* L., *Rosmarinus officinalis* L.), (*i*) a lightning strike leads to a low-intensity fire event (*ii*), in which individual trees of the lower and intermediate layer die and the entire undergrowth is removed. Under the new light conditions and owing to the released nutrients, black pine regenerates optimally, while other, more light-demanding pine species are shaded out. Holm oak regenerates by resprouting, which leads to a reorganization of the regeneration layer, followed by an undisturbed growth phase of the old stand. After a further fire event, the crown bases are pushed upwards (*iii*) by thermal (fire-induced) branching. By this process, most of the regeneration and undergrowth are reset, except for a few pre-growing black pines. Those pines that were able to escape the fire trap have a good chance to grow into the upper stand layer. (Photo: D. Kraus)

components in their leaves. During a fire event, these properties lead to increased mortality of their neighbours, creating space for seed germination and establishment, i.e. for their own offspring (the 'kill your neighbour' hypothesis).

Table 7.3 Strategies for active fire resistance

Strategy and process	Examples
Resprouting	
Root suckers	
Resprouting from belowground structures like lignotubers, bulbs, etc.	*Populus tremula* L.
	Cornus spp. L.
Epicormic Shoots	
Resprouting from dormant or adventitious vegetative buds protected by bark tissue from fire	*Castanea sativa* Mill.
	Quercus spp. L.
	Pinus canariensis C.Sm.
	Pinus banksiana Lamb.
	Sequoia sempervirens Endl.
	Eucalyptus spp. L'Her.
Flowering	
Fire-stimulated flowering	
Mast-flowering triggered by fire events creates ideal conditions for seed production and germination.	*Watsonia pyramidata* (Andrews) Stapf
	Cyrtanthus contractus N.E.Br.
Germination	
Smoke-induced germination	
Germination of seeds is triggered by combustion chemicals in smoke compounds	Proteaceae
Heat-induced germination	
Germination of seeds is triggered by heat-shock	*Calluna vulgaris* (L.) Hull
	Cistus spp. L.
	Acacia spp. Mill.
Serotiny	
Heat-induced seed release	
Seeds are stored in a canopy seed bank, which can be released by fire. Normally fruits or cone scales are sealed by a resin that melts when heated	*Pinus* spp. L.
	Sequoia sempervirens Endl.
	Banksia spp. L.F.
	Eucalyptus ssp. L'Her.
Flammability	
Improving conditions for recruitment	
Plant traits that increase flammability to free up space for recruitment by killing off less-competitive neighbours with lower resilience against fire events	*Pinus canariensis* C.Sm.
	Pinus contorta Dougl. ex Loud.
	Anthropogoneae Dumort.
	Eucalyptus spp. L'Her.

Fig. 7.9 In the Mediterranean area, serotinous cones of Aleppo pines (*Pinus halepensis* Mill.) remain closed until a fire event occurs. Stimulated by heat, they open, causing the seeds to fall onto the burned ground where they find optimal conditions for germination. (Photo: D. Kraus)

7.2.4 Pyrophilous Species, Fire-Dependent Species, and Fire Profiteers

Fires create new habitat conditions and substrates, which serve as ecological niches for specialized species. These include insects and fungi that prefer freshly burned areas, burned plant tissue, or charcoal. When flame temperatures exceed 300 °C, pyrolysis changes the chemical properties of wood (lignin content, evaporation of resins). This also changes decomposition processes, which give some species an advantage in colonizing deadwood. Fire also alters the competitive situation by destroying the pre-fire species communities, which enables new species to quickly establish. Other species benefit more from the changed environmental conditions after a fire, e.g. when deadwood is directly exposed to the sun and low humidity in contrast to pre-fire conditions (Wohlgemuth et al. 2010; Stokland and Siitonen 2012).

Fire-dependent and pyrophilous species are often found among the process-limited species (Jonsson and Siitonen 2013), i.e. species that occur predominantly or exclusively in certain stages of succession, which are created and maintained by disturbances such as fire. For example, some sac fungi (ascomycetes) colonize ash or charred wood residues immediately after fire (Fig. 7.10). In particular, species of fire-adapted coal fungi (*Daldinia* spp.) can remain invisible for decades but thrive abundantly after fire. *Daldinia loculata* (Lév.) Sacc. is a well-known specialist in this respect, which grows in Northern Europe and can also be found sporadically in Central Europe. It colonizes birches and beeches killed by fire (Conedera et al. 2007). In a study from central Sweden, a strong correlation was found between fire-dependent insects and birches (*Betula pendula* Roth) colonized by *D. loculata* (Wikars 2002). Details of this interaction are not yet sufficiently understood, but it has been found that many pyrophilous insects live in the fruiting bodies of the fungi,

Fig. 7.10 In the temperate zones of Europe and North America, the dwarf acorn cup (*Geopyxis carbonaria* (Alb. & Schwein.) Sacc.), a sac fungus, appears immediately after fires. Together with the dwarf acorn cup, also morels (*Morchella* spp.) can appear in great abundance in summer after a fire (Greene et al. 2010). (Photo: M. Moretti)

indicating a close connection. Other known fungal species that only colonize fire-created deadwood substrates are *Gloeophyllum carbonarium* (Berk. & M.A. Curtis) Ryvarden and *Phanerochaete raduloides* J. Erikss. & Ryvarden, as well as the lichen species *Hypocenomyce anthracophila* (Nyl.) Bendiksby & Timdal.

Another example of fire-adapted species is insects laying their eggs in trees killed by fire instantly after the flames have died down. Some of these insect species depend on burned areas to such an extent that they have developed infrared sensors to locate forest fires. A classic example is the black jewel beetle (*Melanophila acuminata* DeGeer) which has infrared receptors along each side of its body which allow the beetles to detect forest fires; the female beetles lay their eggs under the bark of freshly burned trees, and the larvae use the burned wood as food (Schmitz et al. 2009). Other species, such as flat bugs of the genus *Aradus*, colonize fire sites as soon as the fire is extinguished. These fire-dependent species often exhibit remarkable dispersal capabilities for colonizing forest fire sites. Thus, even tiny quantities of guaiacol (a phenolic compound formed when the lignin in wood is burned) and similar phenolic derivatives of smoke in the air are sufficient to attract many individuals of the rare pine jewel beetles over distances of several kilometres (Schütz et al. 1999). Resource-limited species can also benefit from fire events, as forest fires provide suitable substrates, e.g. in the form of deadwood. In forests with frequently recurring fires, Scots pine (*Pinus sylvestris* L.) trees may be subject to increased resin accumulation (impregnation) resulting from infection defence, which can lead to the death of woody tissue and ultimately the whole tree. The death of the trees can be slow and a lot of resin can accumulate in the wood. Consequently the decay of the wood is slow, and the dead trees can persist in forest stands for centuries. In Finland, these old dead trees are called 'kelo' trees; such trees were ᵁᵁᵁᵁᵁᵁᵁᵁᵁ ᵁᵁᵁᵁᵁᵁᵁᵁᵁ ᵁᵁ ᵁᵁ ᵁᵁᵁᵁᵁᵁ ᵁᵁᵁ ᵁᵁᵁᵁᵁ ᵁᵁ ᵁᵁᵁᵁᵁᵁ ᵁᵁᵁ ᵁᵁᵁᵁᵁᵁ ᵁᵁᵁᵁᵁᵁ but are now among the rarest types of deadwood (Fig. 7.11). Today, they are found

Fig. 7.11 So-called 'kelo' trees are dry and bark-less dead pine trees, which had been stimulated to produce resin by frequently recurring fires and have developed into particularly durable dead-wood structures in forests owing to this resin impregnation. As an outstanding example of a fire-specific substrate and as a representative of highly fire-affected pine forests, they are a special and now rare deadwood substrate for specialized species. (Photo: D. Kraus)

Fig. 7.12 Burn scar on an
old pine tree in Hede
Urskog, Härjedalen,
Sweden. (Photo: D. Kraus)

on a larger scale only in remote boreal areas (e.g. Karelia) and in a few lowland locations such as the Białowieża Forest (Niemelä et al. 2002).

Distinct burn scars on pine (*Pinus* spp.), larch (*Larix* spp.), and occasionally oak (*Quercus* spp.) species represent another important microhabitat for pyrophilous species (Kraus et al. 2016; Fig. 7.12). Only a few specialists like the beetles *Stephanopachys linearis* Kugelann or *Nothorhina punctata* Fabricius can tolerate the hostile conditions of a fresh burn scar, under which they are exposed to the defence mechanisms of the tree (Wikars 2002; Rydkvist and Kraus 2010).

Other species use the temporary vegetation-free conditions after fire events for predatory activities, e.g. the ground beetles *Sericoda quadripunctata* DeGeer (Moretti and Conedera 2003) and *Pterostichus quadrifoveolatus* Letzner (Pradella et al. 2010). A fire event temporarily creates competition-free habitats with a considerable nutrient supply. Arriving seeds or spores from adjacent undisturbed habitats germinate in large numbers if moisture conditions are suitable, which is why species diversity increases rapidly under such conditions (e.g. Wohlgemuth et al. 2002; Pausas and Ribeiro 2017). As a characteristic fire profiteer, we mention the common cord moss (*Funaria hygrometrica* Hedw.). After a fire in the Valais (Switzerland), this species was found everywhere 2 years after the event, and in one third of the 300 ha fire area, it even dominated (Wohlgemuth et al. 2010).

7.2.5 Fire Adaptations of Ecosystems

Depending on the frequency and intensity of the burns, fire-adapted species are part of the ecosystems (Moretti et al. 2010; Pausas and Paula 2012). These ecosystems can be divided into the following three categories (Hardesty et al. 2005; Kraus and Goldammer 2007):

(a) Fire-Independent Ecosystems

In fire-independent ecosystems, fire usually plays only a small role or is absent. Such ecosystems are either too cold, too humid, or too dry (fire is often rare in dry ecosystems because of a lack of vegetation) to burn. Examples are some high mountains, tundra, humid tropical forests, swamps and mires, as well as deserts and semi-deserts.

(b) Fire-Sensitive Ecosystems

Ecosystems that are sensitive to fire have not developed with fire as a significant and recurring process. Species in these ecosystems show few adaptations to the occurrence of fire. Even at low fire intensities, mortality of affected species can be high. Both the vegetation structure and species composition and traits tend to inhibit ignition and fire spread. In other words, such ecosystems are not very prone to fire. Under natural, unaffected conditions, fire can be such a rare event that these ecosystems can be considered to be almost like fire-independent ecosystems. Land use in these ecosystems leads to fragmentation, to changes in fuel load, or to increases in ignition sources. Together with the increase of climatic extremes, these features can result in fire becoming a problem. If fire events become more common, the ecosystem shifts towards a more fire-prone vegetation. Species composition often changes in tropical rain and cloud forests under these conditions, but it is still unknown how deciduous forests in the cool temperate zones will react to increasing fire frequency.

(c) Fire-Dependent Ecosystems

If fire returns at regular intervals in an ecosystem, species adapt in such a way that fire is a precondition for their existence and in particular may be necessary for reproduction and seed dispersal. Accordingly, ecosystems with such species are flammable and susceptible to fire. Therefore, they are often called fire-adapted or fire-maintained ecosystems. If fire is excluded or if the fire regime is changed from its natural variability, the ecosystem also changes, resulting in a loss of habitats and species. Examples of natural fire-dependent systems are the boreal coniferous forests of the western taiga and many Mediterranean pine forest communities. Atlantic heathlands in Europe and Mediterranean bush formations (e.g. maquis, garrigue) are also fire dependent, although they have developed under human influence (Mazzoleni 1993; Miles 1993).

7.3   Management of Vegetation Fires

7.3.1   Use of Fire in Silviculture and Ecosystem Restoration

Natural disturbance emulation (NDE) is often regarded as a basic principle for ecologically sustainable forest management (Bergeron et al. 2002; Kuuluvainen 2002;
Kuuluvainen 2012). The idea was developed in Europe already in the nineteenth
century to consider natural forest structures and development patterns as a model
for sustainable forest management. In his essay *Das Bodenfeuer als Freund des
Forstmannes* ("The surface fire as a friend of the forester"), Conrad (1925) described
experiences and hypotheses on the role of fire in inducing natural regeneration of
pine and spruce. On a trip to the primeval forests of the Rhodopes, Müller (1929)
coined the term 'fire thinning' (German, *Branddurchforstung*) after he had observed
the positive effect of low-intensity surface fires on the structure of uniform
pine stands.

Current approaches in forest management emphasize the preservation of ecosystem integrity, resilience, and biodiversity by adherence to ecological principles
(Drever et al. 2006). NDE strategically aims at implementing silvicultural practices
that reproduce and maintain stand structures and processes of forests with natural
disturbance regimes. Essentially, it develops and applies silvicultural practices that
are based on the fundamental ecological effects of disturbance. This assumes that in
the long run, the majority of forest processes become more stable by imitating
disturbance-induced forest structures than in forests that are conventionally managed. So far, only a few explicit guidelines have been developed to include disturbance dynamics in forest management. Examples for the consideration of fire
disturbances in silviculture are the ASIO model (Angelstam 1998) and the Multi-
cohort model (Bergeron et al. 2002; Fig. 7.13).

NDE models have often served as a basis for large-scale clear-cutting in boreal
forests, as the heterogeneity of boreal fire regimes was strongly underestimated.
However, recent studies have shown that even small to medium fire events are typical for boreal fire regimes, which is why forest models that imitate gap dynamics by
low-intensity fires are also considered (Kuuluvainen 2002).

Worldwide, the restoration of forest ecosystems is seen as an important instru-
ment to maintain ecosystem services and to reduce the loss of biodiversity. The
main focus is on restoring original habitats as far as possible, thereby allowing natural processes to take place and former ecosystem functioning to be restored. One
promising way of achieving such targets for the conversion of structurally poor,
uniform stands into diverse forest habitats is to allow or reintroduce disturbances
that naturally trigger the formation of diverse habitats (Arno and Fiedler 2005). Fire
is one of the most important structural factors in many forests, e.g. in the boreal
areas of Northern Europe. In historical times, forest utilization there led to an
increase in fire frequency. With the introduction of modern forestry, fire incidents
have greatly decreased. In particular, the naturally occurring fires are now suppressed, so that only a small proportion of the boreal forests in Finland and Sweden

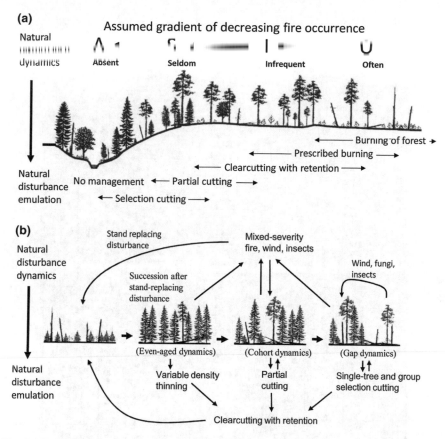

Fig. 7.13 (**a**) ASIO model after Angelstam (1998) and (**b**) Multi-cohort model after Bergeron et al. (1999)

are still affected by fire. The use of controlled fire is one of the most effective means of restoring and increasing the diversity of boreal forest communities (Rydkvist and Kraus 2010). Forest stands suitable for the application of controlled fire are burned between May and the end of August, with suitable burning conditions only on a few days a year (Similä and Junninen 2012). Natural forest fires are rarely so intense in these forest ecosystems that entire stands burn down by crown fires creating open habitats with large amounts of deadwood (Pitkänen and Huttunen 1999). Usually, in the frame of low-intensity surface fires, only a few trees are killed. Nevertheless, they change the stand structure in the long term, as tree growth after a fire event clearly changes. Controlled fires (Fig. 7.14) are used to imitate these properties and to exploit the positive effects of natural fires. Depending on the objective, 25–75% of the tree population should be kept alive. To create a high niche density, a mosaic of burned and unburned areas within the burn perimeter is achieved.

Fig. 7.14 Flow chart of a controlled fire in an open pine–spruce stand. (**a**) Along a natural barrier (course of a river, exposed rock, wet spots in the terrain), a backing fire is set, the spread of which is limited on the flanks by control lines (exposed mineral soil, forest roads). In the direction of propagation (against the wind), point fires are set to support and direct fire intensity; the number of point fires is adapted to the terrain and the rate of spread. (**b**) The fire intensity, which varies over a small area, leads to a varied fire mosaic in which some small trees under canopy die, though deadwood and undergrowth are preserved in other places. (Photo: D. Kraus)

7.3.2   Global Change and Forest Fires in Europe

In southern Europe in particular, the land use change in the second half of the twentieth century led to an increase in the frequency and intensity of forest fires (Valese et al. 2014). Today, forest fires in this region are predominantly classified as destructive and harmful, regardless of their actual size and impact. According to the statistics of the Joint Research Centre (JRC) of the European Union, the number of forest fires in the EU increased significantly between 1980 and 2006. Every year, an average area of about 600,000 ha of forest is affected by fire (San-Miguel-Ayanz and Camia 2009; Rego et al. 2010). On average, more than 50,000 forest fires are counted annually, with most fires occurring in Southern Europe, the eastern part of Central Europe, and Scandinavia (Fig. 7.15; San-Miguel-Ayanz and Camia 2009).

Most of these fires are caused by humans and they usually affect small areas. JRC (2007) reported that in the EU area in 2006, forest fires exceeding 50 ha were less frequent (only 2.6% of the total number of fires) but that these fires accounted for 75% of the total burned area.

The response to ever new generations of forest fires in Europe (see Table 7.4) has been for countries to increase their firefighting resources and fire-suppression efforts, with the result that small- and medium-sized fires have been suppressed with great success. However, despite the high investments in fire suppression measures, the size, rate of spread, and intensity of large forest fires are still increasing, with often uncontrollable consequences. This phenomenon is known as the 'fire

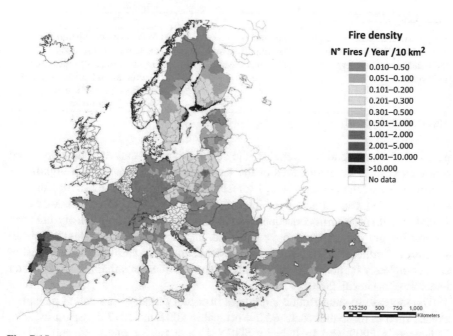

Fig. 7.15 Average annual number of fires in administrative units of European countries in the period 1998–2007. (Sources: San-Miguel-Ayanz and Camia 2009; Schuck et al. 2010)

Table 7.4 Generations of forest fires in Euro-Mediterranean landscapes since the 1950s

Description	Fire type	Response
1950s–1960s		
Continuity of vegetation cover leads to large fire perimeters, hardly interrupted by agricultural land	Between 1000 and 5000 ha, predominantly high-intensity surface fires, wind-driven	Local resources reinforced by seasonal firefighters, creation of linear infrastructure, water points and better access to remote areas
1970s–1980s		
Rate of spread. Fuel accumulation allows for fires that spread with high intensity and speed	Between 5000 and 10,000 ha, both wind and topography driven	Increase of water-based resources and aerial means for more forceful attack
1990s		
Fire intensity. Fuel accumulation reaches vertical fuel continuity. Crown fires and high convective potential are the consequence. Few opportunities for fire control	Crown fires with large distance spotting. Between 10,000 and 20,000 ha, extreme heat waves reinforce high-intensity fires	Risk modelling and fire analysis become more important, reintroduction of fire as a management tool, reinforcing aerial attack
Since 2000 (a)		
Wildland–urban interface fires with high intensities	Fires threatening residential areas larger than 1000 ha	New defensive situations need GIS and GPS applications to track resources in real time. Fire analysis becomes the most important tool
Since 2000 (b)		
Megafires. Simultaneous large fires in high-risk areas with extremely rapid, virulent fire behaviour	Simultaneous crown fires crossing urban and peri-urban areas	Necessity of cooperation and exchange of resources, information, and experiences. Coordination between regions. Continuous learning and exchange platforms

After Costa et al. (2010)

paradox'. According to this paradox, the suppression of small- and medium-sized forest fires led to the major fires of the last decades, which is why firefighting is equivalent to a negative selection of forest fire events (Costa et al. 2010; Curt and Frejaville 2018). Depending on the intensity of the forest fires, the effects vary considerably, with long-term consequences. While low- to medium-intensity fires can create a largely fire-resistant forest structure, high-intensity forest fires destroy all structures. Furthermore, the subsequent establishment of shrub vegetation makes such completely burned areas more vulnerable to further fires that return at shorter intervals (Costa et al. 2010).

The increase in catastrophic large fires in recent decades – so-called mega-fire events, such as those that occurred in Portugal in 2003 and 2017, in Spain in 2006, in Greece in 2009, and in Italy in 2017 – is a cause for major concern. These

devastating and almost uncontrollable fires represent a new generation of forest fires
unknown in European fire history until recently (Table 7.4).

Experiences from other parts of the world show that the origin of the mega fire
phenomenon in the Mediterranean region can only partly be explained by climate
change, in particular by altered precipitation regimes. The main reason for the
increase in the number of large fires has been a profound change in the landscape,
which has resulted in continuous and steadily growing fuel loads in the context of a
rural exodus and the related land abandonment. An increase in average temperatures
and dry periods only intensifies the symptoms of a phenomenon with a much longer
incubation period, which has its origin in extensive landscape changes that have
been occurring since the early twentieth century (Castellnou and Miralles 2009).

7.3.3 A Solution to the Problem: Integrated Fire Management

In the Mediterranean area, traditional fire use came to an end after many forms of
land use were abandoned and an exodus from the countryside to urban environments took place. An increasingly urban society was soon no longer able to recog-
nize the role of fire in the landscape. As a result, the rapid and effective suppression
of all fires became a high priority in the affected countries. The establishment of
powerful firefighting structures and airborne resources was also pushed forward. In
line with the 'fire paradox', this development increased the risk of fire. New and
innovative concepts were needed to counter this development. In order to reduce fire
risk, modern concepts must take into account landscape planning and targeted silvicultural measures. This preventive approach is also described by the term integrated
fire management (IFM) (Fig. 7.16). It mainly focusses on target-oriented landscape
planning and silvicultural measures with the aim of significantly reducing the fuel
load, i.e. the accumulated combustible biomass (Piñol et al. 2007; Costa et al. 2010).

The basic principles of IFM are briefly explained below (Silva et al. 2010):

- IFM: a concept for the planning and implementation of fire application with the
  aim of minimizing damage through the intensive use of fire, taking into account
  social, economic, cultural, and ecological aspects. IFM combines prevention and
  control strategies or techniques and regulates their application by qualified per-
  sonnel as well as by traditional fire use.
- Traditional fire use: fire use in rural areas for land and resource management,
  based on traditional knowledge. Appropriate applications follow legal regulations and are subject to good practice.
- Prescribed burning: application of fire under environmental conditions that allow
  the fire to be used within a limited, selected area to achieve planned management
  objectives. This reduces the fuel load outside the fire season or controls the differentiation of stand structures.
- Suppression fire (Fig. 7.16): the use of fire as a tool for fighting forest fires. In
  general, a distinction is made between the removal of fuel in front of the fire front

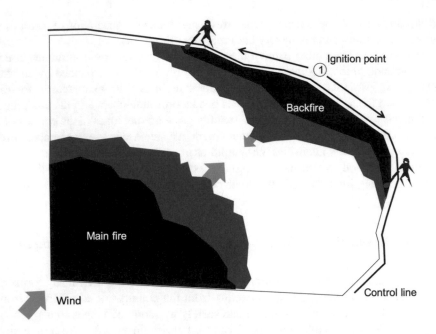

Fig. 7.16 Schematic illustration of backfiring: to prevent fire from further spreading, a counterfire is set from a safe line against the wind direction and the approaching fire front

by burning (burning out) and suppression fire, where the approaching fire front is opposed by a backfiring from a safe line. The time of ignition must be adjusted in such a way that the suppression fire is attracted by the developing convection (suction effect) at the fire front, thus making further fire spread impossible.

The goal of IFM is the combination of traditional and modern fire applications and the beneficial adoption of fire regimes of low- to medium-intensity disturbance to restore landscapes and make them largely resistant to intense fires (Fig. 7.17).

A comprehensive implementation with appropriate measures is difficult to realize, but decisive innovations in this direction are already being tested in various European countries. This includes the application of analytical methods to identify strategically important locations in the terrain (Castellnou and Miralles 2009; Costa et al. 2010; Molina et al. 2010), from where targeted measures such as controlled fire can start and where landscape planning measures may be taken to prevent the spread of fires. Additionally, methods and techniques of backfire application have been developed to decelerate and contain catastrophic fires with extreme fire behaviour (Castellnou et al. 2010; Miralles et al. 2010). In both cases, the focus is on the beneficial application of fire to solve the fire paradox.

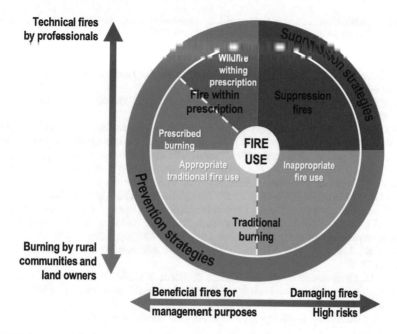

Fig. 7.17 Integrated fire management as a combination of prevention and control strategies with fire application as a central element. (Sources: Schuck et al. 2010; Silva et al. 2010)

References

Agee JK (1993) Fire ecology of Pacific Northwest forests. Island Press, Washington, DC. 493 p

Angelstam PK (1998) Maintaining and restoring biodiversity in European boreal forests by developing natural disturbance regimes. J Veg Sci 9:593–602

Archibald S, Roy DP, Van Wilgen BW, Scholes RJ (2009) What limits fire? An examination of drivers of burnt area in Southern Africa. Glob Change Biol 15:613–630

Arno S, Fiedler C (2005) Mimicking nature's fire: restoring fire-prone forests in the west. Island Press, Washington, DC, 247 p

Atahan P, Dodson JR, Itzstein-Davey F (2004) A fine-resolution Pliocene pollen and charcoal record from Yallalie, south-western Australia. J Biogeogr 31:199–205

Bergeron Y, Harvey B, Leduc A, Gauthier S (1999) Forest management guidelines based on natural disturbance dynamics: stand- and forest-level considerations. For Chron 75:49–54

Bergeron Y, Leduc A, Harvey BD, Gauthier S (2002) Natural fire regime: a guide for sustainable management of the Canadian boreal forest. Silva Fenn 36:81–95

Bond WJ, Keeley JE (2005) Fire as a global 'herbivore': the ecology and evolution of flammable ecosystems. Trends Ecol Evol 20:387–394

Bond WJ, van Wilgen BW (1996) Fire and plants. Chapman & Hall, London, p 263

Bowman D, Balch J, Artaxo P, Bond WJ, Cochrane MA, D'Antonio CM, Defries R, Johnston FH, Keeley JE, Krawchuk MA, Kull CA, Mack M, Moritz MA, Pyne S, Roos CI, Scott AC, Sodhi NS, Swetnam TW (2011) The human dimension of fire regimes on Earth. J Biogeogr 38:2223–2236

Bradshaw SD, Dixon KW, Hopper SD, Lambers H, Turner SR (2011) Little evidence for fire-adapted plant traits in Mediterranean climate regions. Trends Plant Sci 16:69–76

Carcaillet C, Bergman I, Delorme S, Hornberg G, Zackrisson O (2007) Long-term fire frequency not linked to prehistoric occupations in northern Swedish boreal forest. Ecology 88:465–477

Carrión JS (2002) Patterns and processes of Late Quaternary environmental change in a montane region of southwestern Europe. Quat Sci Rev 21:2047–2066

Castellnou M (1996) Reconstrucción de la progression de los incendios históricos y su efecto moldeador en la vegetación: Ribera d'Ebre. In: González JM (ed) Seminario sobre Incendios Forestales, 16–18 Decembre 1996. Centre Tecnològic Forestal de Catalunya, Solsona, pp 207–223

Castellnou M, Miralles M (2009) The great fire changes in the Mediterranean – the example of Catalonia. Spain. Crisis Response 5(4):56–57

Castellnou M, Kraus D, Miralles M, Delogu G (2010) Suppression fire use in learning organizations. In: Silva JS, Rego F, Fernandes P, Rigolot E (eds) Towards integrated fire management – Outcomes of the European project Fire Paradox, EFI Research Report, vol 23, pp 189–201

Clark JS, Merkt J, Müller H (1989) Post-glacial fire, vegetation, and human history on the northern Alpine forelands, south-western Germany. J Ecol 77:897–925

Colombaroli D, Marchetto A, Tinner W (2007) Long-term interactions between Mediterranean climate, vegetation and fire regime at Lago di Massaciuccoli (Tuscany, Italy). J Ecol 95:755–770

Conedera M, Cesti G, Pezzatti GB, Zumbrunnen T, Spinedi F (2006) Lightning-induced fires in the Alpine region: an increasing problem. For Ecol Manag 234(supplement):S68

Conedera M, Lucini L, Holdenrieder O (2007) Pilze als Pioniere nach Feuer. Wald Holz 11(07):45–48

Conedera M, Colombaroli D, Tinner W, Krebs P, Whitlock C (2017) Insights about past forest dynamics as a tool for present and future forest management in Switzerland. For Ecol Manag 388:100–112

Conrad A (1925) Das Bodenfeuer als Freund des Forstmannes. Forstl Wochenschr Silva 13:139–141

Costa P, Larrañaga A, Castellnou M, Miralles M, Kraus D (2010) Prevention of large wildfires using the Fire Types Concept. EU Fire Paradox Publication, Barcelona, 87 p

Curt T, Frejaville T (2018) Wildfire policy in Mediterranean France: how far is it efficient and sustainable? Risk Anal 38:472–488

Diemont WH (1996) Survival of Dutch heathlands. PhD thesis, Wageningen Agricultural University, Wageningen, The Netherlands, 80 p

Dodson JR (2001) Holocene vegetation change in the Mediterranean-type climate regions of Australia. Holocene 11:673–680

Drever CR, Peterson G, Messier C, Bergeron Y, Flannigan M (2006) Can forest management based on natural disturbances maintain ecological resilience? Can J For Res 36:2285–2299

Franco-Múgica N, García-Antón M, Maldonado-Ruiz J, Morla-Juaristi C, Sainz-Ollero H (2005) Ancient pine forest on inland dunes in the Spanish northern meseta. Quat Res 63:1–14

González JM, Veblen T, Sibold JS (2009) Influence of fire severity on stand development of *Araucaria araucana–Nothofagus pumilio* stands in the Andean cordillera of south-central Chile. Austral Ecol 35:597–615

Granström A (1993) Spatial and temporal variation in lightning ignitions in Sweden. J Veg Sci 4:737–744

Greene DF, Hesketh M, Pounden E (2010) Emergence of morel (*Morchella*) and pixie cup *(Geopyxis carbonaria)* ascocarps in response to the intensity of forest floor combustion during a wildfire. Mycologia 102:766–773

Hardesty J, Myers RL, Fulks W (2005) Fire, ecosystems, and people: a preliminary assessment of fire as a global conservation issue. George Wright Forum 22:78–87

Harvey BD, Donato DC, Romme WH, Turner MG (2013) Influence of recent bark beetle outbreak on fire severity and postfire tree regeneration in montane Douglas-fir forests. Ecology 94:2475–2486

Heikinheimo O (1915) Kaskiviljelyksen vaikutus Suomen metsiin. (Mit Deutschem Referat: Der Einfluss der Brandwirtschaft auf die Wälder Finnlands. Acta For Fenn 4/2, 263+59 p

Holling CS, Gunderson LH (2002) Resilience and adaptive cycles. In: Holling CS (ed) Gunderson LH. Island Press, Washington, DC, pp 25–62

Holst-Jørgensen B (1993) Erfahrungen beim Erhalt von Heideflächen im staatlichen Walddistrikt Ulfborg. Jütland. NNA-Berichte 6(3):67–79

Jonsson BG, Siitonen J (2010) Managing for biodiversity in boreal forest ... approaches as an opportunity for the conservation of forest biodiversity, European Forest Institute, pp 134–143

JRC (2007) Forest fires in Europe 2006. EC Joint Research Centre, Institute for Environment and Sustainability, Ispra, 77 p

Kapp G (1984) Agroforstwirtschaft in Deutschland. Der Waldfeldbau im 18. und 19. Jahrhundert. AFZ-Der Wald 155:266–270

Keeley JE, Fotheringham CJ (2001) Historic fire regime in Southern California shrublands. Cons Biol 15:1536–1548

Keeley JE, Pausas JG, Rundel PW, Bond WJ, Bradstock RA (2011) Fire as an evolutionary pressure shaping plant traits. Trends Plant Sci 16:406–411

Kraus D, Goldammer JG (2007) Fire regimes and ecosystems: an overview of fire ecology in tropical ecosystems. In: Schmerbeck J, Hiremath A, Ravichandran C (eds) Forest fires in India: workshop proceedings, Madurai, India, February 19–23, 2007. ATREE, Bangalore, India & Inst. Silviculture, Freiburg i.Br, pp 9–13

Kraus D, Bütler R, Krumm F, Lachat T, Larrieu L, Mergner U, Paillet Y, Rydkvist T, Schuck A, Winter S (2016) Catalogue of tree microhabitats – reference field list. Integrate+ Technical Paper. European Forest Institute, Freiburg i.Br, 16 p

Krawchuk MA, Moritz MA (2011) Constraints on global fire activity vary across a resource gradient. Ecology 92:121–132

Krawchuk MA, Moritz MA, Parisien MA, Van Dorn J, Hayhoe K (2009) Global pyrogeography: the current and future distribution of wildfire. PLoS One 4:e5102

Kuuluvainen T (2002) Disturbance dynamics in boreal forests: defining the ecological basis of restoration and management of biodiversity. Silva Fenn 36:5–11

Kuuluvainen T (2012) Natural disturbance emulation in boreal forest ecosystem management – theories, strategies, and a comparison with conventional even-aged management. Can J For Res 42:1185–1203

Lázaro A (2010) Development of prescribed burning and suppression fire in Europe. In: Montiel C, Kraus D (eds) Best practices of fire use – prescribed burning and suppression fire programmes in selected case-study regions in Europe. EFI Res Rep 24:17–31

Lázaro A, Montiel C (2010) Overview of prescribed burning policies and practices in Europe and other countries. In: Silva JS, Rego F, Fernandes P, Rigolot E (eds) Towards integrated fire management – Outcomes of the European project Fire Paradox. EFI Res Rep 23:138–150

Marlon JR, Bartlein PJ, Daniau AL, Harrison SP, Maezumi SY, Power MJ, Tinner W, Vanniere B (2013) Global biomass burning: a synthesis and review of Holocene paleofire records and their controls. Quat Sci Rev 65:5–25

Mazzoleni S (1993) Incendi e vegetazione mediterranea. In: Mazzoleni S, Aronne G (eds) Introduzione all'ecologia degli incendi. Liguori, Napoli, pp 43–72

Miles J (1993) Gli effetti degli incendi boschivi. In: Mazzoleni S, Aronne G (eds) Introduzione all'ecologia degli incendi. Liguori, Napoli, pp 13–27

Miralles M, Kraus D, Molina D, Loureiro C, Delogu G, Ribet N, Vilalta O (2010) Improving suppression fire capacity. In: Silva JS, Rego F, Fernandes P, Rigolot E (eds) Towards integrated fire management – outcomes of the European project Fire Paradox. EFI Res Rep 23:203–215

Molina D, Castellnou M, García-Marco D, Salgueiro A (2010) Improving fire management success through fire behaviour specialists. In: Silva JS, Rego F, Fernandes P, Rigolot E (eds) Towards integrated fire management – outcomes of the European project Fire Paradox, EFI Res Rep, vol 23, pp 105–119

Moretti M, Conedera M (2003) Waldbrände im Kreuzfeuer. Gaia 12:275–279

Moretti M, De Caceres M, Pradella C, Obrist MK, Wermelinger B, Legendre P, Duelli P (2010) Fire-induced taxonomic and functional changes in saproxylic beetle communities in fire sensitive regions. Ecography 33:760–771

Müller KM (1929) Aufbau, Wuchs und Verjüngung der südosteuropäischen Urwälder: eine wald-bauliche Studie über den Urwald unserer Zone überhaupt. Schaper, Hannover, 322 p

Niemelä T, Wallenius T, Kotiranta H (2002) The Kelo tree, a vanishing substrate of specified wood-inhabiting fungi. Polish Bot J 47:91–101

Niklasson M, Granström A (2000) Numbers and sizes of fires: long-term spatially explicit fire history in a Swedish boreal landscape. Ecology 81:1484–1499

Niklasson M, Zin E, Zielonka T, Feijen M, Korczyk AF, Churski M, Samojlik T, Jędrzejewska B, Gutowski JM, Brzeziecki B (2010) A 350-year tree-ring fire record from Białowieża Primeval Forest, Poland: implications for Central European lowland fire history. J Ecol 98:1319–1329

Pausas JC, Fernándes-Muñoz S (2012) Fire regime changes in the Western Mediterranean Basin: from fuel-limited to drought-driven fire regime. Clim Change 110:215–226

Pausas JG, Paula S (2012) Fuel shapes the fire-climate relationship: evidence from Mediterranean ecosystems. Glob Ecol Biogeogr 21:1074–1082

Pausas JG, Ribeiro E (2017) The global fire–productivity relationship. Glob Ecol Biogeogr 22:728–736

Piñol J, Castellnou M, Beven KJ (2007) Conditioning uncertainty in ecological models: assessing the impact of fire management strategies. Ecol Model 207:34–44

Pitkänen A, Huttunen P (1999) A 1300-year forest-fire history at a site in eastern Finland based on charcoal and pollen records in laminated lake sediment. Holocene 9:311–320

Pradella C, Wermelinger B, Obrist MK, Duelli P, Moretti M (2010) On the occurrence of five pyrophilous beetle species in the Swiss Central Alps (Leuk, Canton Valais). Mitt Schweiz Entomol Ges 83:187–197

Rego F, Rigolot E, Fernandes P, Montiel C, Silva JS (2010) Towards integrated fire management. EFI Policy Brief 4:16

Rydkvist T, Kraus D (2010) Prescribed burning for nature conservation in Västernorrland, Sweden. In: Montiel C, Kraus D (eds) Best practices of fire use – prescribed burning and suppression fire programmes in selected case-study regions in Europe, EFI Research Report, vol 24, pp 47–60

San-Miguel-Ayanz J, Camia A (2009) Forest fires at a glance: facts, figures and trends in the EU. In: Birot Y (ed) Living with Wildfires: What Science Can Tell Us. EFI Discussion Paper 15. European Forest Institute, pp 11–18

Schmitz H, Norkus V, Hess N, Bousack H (2009) The infrared sensilla in the beetle *Melanophila acuminata* as model for new infrared sensors. In: Bioengineered and Bioinspired Systems IV. Presented at the Bioengineered and Bioinspired Systems IV, 73650A. International Society for Optics and Photonics. SPIE Europe Microtechnologies for the New Millennium, 2009, Dresden, Germany. https://doi.org/10.1117/12.821434

Schuck A, Kraus D, Rego F, Montiel C, Castellnou M (2010) Reduzierung von Waldbrandgefahren durch Integratives Feuermanagement. AFZ/DerWald 9(2010):36–37

Schütz S, Weisbecker B, Hummel HE, Apel KH, Schmitz H, Bleckmann H (1999) Insect antenna as a smoke detector. Nature 398:298–299

Silva JS, Rego F, Fernandes P, Rigolot E (2010) Towards integrated fire management: outcomes of the European project Fire Paradox. European Forest Institute Research Report 23, Joensuu, 228 p

Simard M, Romme WH, Griffin JM, Turner MG (2011) Do mountain pine beetle outbreaks change the probability of active crown fire in lodgepole pine forests? Ecol Monogr 81:3–24

Similä M, Junninen K (2012) Ecological restoration and management in boreal forests – best practices from Finland. Metsähallitus, Natural Heritage Services, Vantaa, 50 p

Stokland J, Siitonen J (2012) Mortality factors and decay succession. In: Stokland J, Siitonen J, Jonsson BG (eds) Biodiversity in dead wood. Cambridge University Press, Cambridge. 521 p

Tinner W, Hubschmid P, Wehrli M, Ammann B, Conedera M (1999) Long-term forest fire ecology and dynamics in southern Switzerland. J Ecol 87:273–289

Tinner W, Conedera M, Ammann B, Lotter AF (2005) Fire ecology north and south of the Alps since the last ice age. Holocene 15:1214–1226

Tinner W, Van Leeuwen JFN, Colombaroli D, Vescovi E, Van Der Knaap WO, Henne PD, Pasta S, D'Angelo S, La Mantia T (2009) Holocene environmental and climatic changes at Gorgo Basso, a coastal lake in southern Sicily, Italy. Quat Sci Rev 28:1498–1510

Turner MG (2010) Disturbance and landscape dynamics in a changing world. Ecology 91:2833–2849

Valese E, Conedera M, Held AC, Ascoli D (2014) Fire, humans and landscape in the European Alpine region during the Holocene. Anthropocene 6:63–74

Vázquez A, Moreno JM (1998) Patterns of lightning-, and people-caused fires in peninsular Spain. Int J Wildland Fire 8:103–115

Webb NR (1998) The traditional management of European heathlands. J Appl Ecol 35:987–990

Wikars LO (2002) Dependence on fire in wood-living insects: an experiment with burned and unburned spruce and birch logs. J Insect Cons 6:1–12

Wohlgemuth T, Bürgi M, Scheidegger C, Schütz M (2002) Dominance reduction of species through disturbance – a proposed management principle for central European forests. For Ecol Manag 166:1–15

Wohlgemuth T, Brigger A, Gerold P, Laranjeiro L, Moretti M, Moser B, Rebetez M, Schmatz D, Schneiter G, Sciacca S, Sierro A, Weibel P, Zumbrunnen T, Conedera M (2010) Leben mit Waldbrand. Merkbl Prax 46:1–16

Zackrisson O (1977) Influence of forest fires on the north Swedish boreal forest. Oikos 29:22–32

Zin E (2016) Fire history and tree population dynamics in Białowieża forest, Poland and Belarus. Act Univ Agri Sueciae 2016(105):1–57

Chapter 8
Wind Disturbances

Thomas Wohlgemuth (iD), Marc Hanewinkel (iD), and Rupert Seidl (iD)

Abstract Depending on wind speed and wind field, storms generate small- to large-scale disturbances, which affect forests more than other ecosystems because of the height and extent of forests. Wind is the most significant disturbance agent for forests. In Europe, western, central, and northern areas have the greatest storm damage in terms of growing stock affected. In Central Europe, coniferous trees – in particular Norway spruce – are more vulnerable to winter storms than broadleaved deciduous trees, which have shed their leaves. Stem breakage and uprooting also depend on factors regarding species identity, site conditions, and stand structure. Another important effect of recurring wind loads – e.g. at forest edges – is that trees adapt by forming compression wood, making them more robust. With respect to regeneration and regrowth after windthrow, trees regrow faster at low than at high elevations. Also, studies have shown that forest regeneration occurs at a similar rate in no-intervention and cleared areas. In the context of climate change, forest disturbance from winter storms is likely to increase.

Keywords Climate change · Deadwood · Economic damage · European winter storms · Management strategies · Natural regeneration · Stem breakage · Tree stability · Wind adaptation · Windthrow

T. Wohlgemuth (✉)
Forest Dynamics Research Unit, Swiss Federal Institute for Forest, Snow and Landscape Research WSL, Birmensdorf, Switzerland
e-mail: thomas.wohlgemuth@wsl.ch

M. Hanewinkel
Albert Ludwig University of University of Freiburg, Freiburg im Breisgau, Germany

R. Seidl
Ecosystem Dynamics and Forest Management Group, School of Life Sciences, Technical University of Munich, Freising, Germany

Berchtesgaden National Park, Berchtesgaden, Germany

© The Author(s), under exclusive license to Springer Nature Switzerland AG 2022
T. Wohlgemuth et al. (eds.), *Disturbance Ecology*, Landscape Series 32,
https://doi.org/10.1007/978-3-030-98756-5_8

8.1   Cause and Development of Winds

Worldwide, winds cause both the largest number of damaging events and the greatest financial loss compared with other types of natural disturbances. Focusing on forests, wind ranks third at the global scale regarding financial loss, behind forest fires and pests. Most damage is caused by storms, i.e. winds with a Beaufort number of 12 or more (corresponding to a wind speed of $\geq$117 km/h; 32.5 m/s). Large areas of wind disturbance occur when peak wind speeds exceed 35 m/s (126 km/h). Such strong winds can have different causes. Quine and Gardiner (2007) list five types of strong winds: tropical cyclones (hurricanes), extratropical cyclones (e.g. European winter storms), thunderstorms, tornadoes, and orographically generated winds (e.g. foehn winds). In Europe, the following wind event types are the most important causes of disturbance:

(a) Cyclonic Storms

Cyclonic storms occur as a result of large-scale weather conditions and within the framework of low-pressure systems that move over hundreds or thousands of kilometres in the form of hurricanes or cyclones. In the tropics and subtropics, these storms are often several hundred kilometres wide (Fig. 8.1) and are called hurricanes, typhoons, cyclones, or tropical storms when they occur in summer. In the mid-latitudes, the extratropical cyclones are called blizzards or 'Nor'easters' in North America and winter storms in Europe as their main occurrence is in the winter season. The maximum wind speeds vary between 40 and 71 m/s (150 and 250 km/h) and decrease towards the edge of the storm system. Areas along the coasts are most severely affected by these storms.

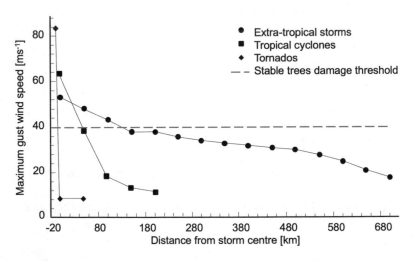

Fig. 8.1 Profile of storm winds and critical wind speed (From Quine and Gardiner 2007, redrawn)

(b) (Summer) Thunderstorms

Thunderstorms develop within the framework of large vertical temperature ⟨⟨⟨⟨⟨⟨⟨⟨⟨⟨ in the atmosphere and instable stratification. For Central Europe, this includes the local thunderstorms that almost always occur during the vegetation period (summer) and are caused by strong solar radiation and considerable evapotranspiration (evaporation and transpiration, i.e. release of water from plant stomata) or by large temperature differences in the atmosphere as a weather front passes over an area. They can produce large local wind peaks (gusts) and lead to small-scale damage to forests and settlements. Tornadoes are another type of strong wind that can form in association with summer thunderstorms. Tornadoes are destructive vortices of violently rotating winds having the appearance of a funnel-shaped cloud and advancing beneath large storm systems. Their radius of action is even smaller than that of thunderstorms, but their destruction power is greater because wind peaks of up to 140 m/s (500 km/h) can occur. Although most common in the Midwest of the USA, tornadoes also occur in other regions of the world including Europe.

(c) Warm Downslope Winds (Foehn Storms)

A foehn results from the ascent of moist air up the windward slopes; as this air climbs, it expands and cools until it becomes saturated with water vapour, after which it cools more slowly because the moisture in the air condenses as rain or snow, releasing latent heat. By the time it reaches the peaks and stops climbing, the air is quite dry. The ridges of the mountains are usually obscured by a bank of clouds known as a foehn wall, which marks the upper limit of precipitation on the windward slopes. As the air makes its leeward descent, it is compressed and warms rapidly all the way downslope because there is little water left to evaporate and absorb heat; thus, the air is warmer and drier when it reaches the foot of the leeward slope than when it began its windward ascent (www.britannica.com). The resulting winds, known as foehn winds, can reach high wind speeds and cause considerable damage. For instance, the damage caused by the foehn storm Uschi in the province of Salzburg, Austria, on 14–17 November 2002 totalled at least €100 million (Salzburger Nachrichten [Salzburg News], 4 October 2012). Corresponding to the different wind conditions, the south foehn occurs on the northern side of the Alps and the north foehn on the southern side. Foehn phenomena are also found in the Spanish Pyrenees ('Fogony'), in the Scandes in Norway ('fønvind'), on the east side of the Rocky Mountains ('Chinook'), as well as along the Andes ('Puelche' in Chile, 'Zonda' in Argentina).

(d) Cold Downslope Winds (Bora, Mistral)

Cold air masses form over large cold areas such as glaciers, mountains, or high plateaus. The denser cold air sweeps downhill as a cold wind (driven by the air pressure gradient) towards the low-pressure system located over the sea (katabatic, flowing down). The strongest katabatic winds are in Antarctica and Greenland ('Piteraq') where they can reach speeds of over 83 m/s (300 km/h). In Europe they are called 'bora' on the Adriatic coast and 'mistral' in the French Mediterranean region, and they can cause significant damage to vegetation. For example, the bora wind on 19

November 2004 destroyed 12,600 ha (2.6 million m³) of coniferous forest on the southern slopes of the High Tatra Mountains, Slovakia (Jonášová et al. 2010).

8.2 Direct Effects on Vegetation

In comparison to other vegetation types, forests are usually affected more by strong winds and hurricanes because they are taller and have a larger crown surface. Individual trees, entire tree populations, or even large forest areas can be disturbed over very short time periods. Tornadoes can also impact meadows and pastures, while gusts, heavy rain, or hail during summer thunderstorms is able to flatten grain fields. Such effects, however, mainly impact annual crops, many of which recover quickly after exposure to wind. In contrast, hurricane winds leave distinct patterns in forest stands in the form of flattened patches or corridors, which modify structure and species composition for several decades or even centuries (e.g. Čada et al. 2016; Kulakowski et al. 2017). Which are the factors that lead to even healthy trees or entire forest areas breaking or being uprooted in strong winds? Various factors are described and briefly discussed below and depicted in Fig. 8.2. There is a wide range of scientific papers which can serve as further literature on the topic of wind influence on ecosystems (e.g. Everham and Brokaw 1996; Johnson and Miyanishi 2007; Gardiner et al. 2010; Fischer et al. 2013).

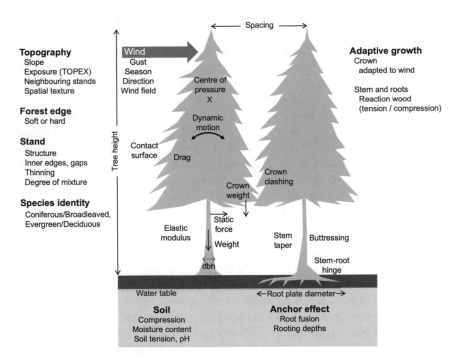

Fig. 8.2 Components of tree stability. (From Quine and Gardiner 2007, modified)

8.2.1 Stem Breakage or Uprooting

Whether a tree breaks along the stem when the wind load is too high (stem breakage) or is torn from its anchorage with the entire root plate (uprooting) depends on various factors. The most important factor here is the anchoring of a tree in the ground. Species-specific differences affect root growth under given soil conditions. In particular, soil moisture, groundwater levels, and nutrient conditions limit root and stem growth. Shallow-rooted tree species, such as many spruce species (*Picea* spp.) and Douglas fir (*Pseudotsuga menziesii* (Mirbel) Franco), are less well anchored in the soil and tend to uproot more easily; important factors include the ratio of the size of the root plate to that of the stem including crown (leverage effect) and the weight of the tree (counterforce). Tree pulling experiments have shown that shallow roots are less resistant to tension than deep roots (Peltola and Kellomäki 1993; Peltola et al. 2000). When soils are frozen, shallow-rooted tree species are better anchored than in unfrozen soil, but the trees are then more prone to stem breakage (Peltola et al. 2000). European winter storms are large-scale weather situations in which temperatures rise over several days before the storm event. Frozen soils may thus thaw, making uprooting of trees during storms more likely (Usbeck et al. 2010a). In winter, broadleaved deciduous trees are generally less susceptible to damage than Norway spruce (*Picea abies* (L.) H. Karst.) because of the lack of foliage and the correspondingly smaller area of exposure (Dobbertin et al. 2002). During summer thunderstorms with strong gusts, deep-rooted broadleaved deciduous trees are broken more often than they are uprooted.

Good nutrient and light conditions promote rapid growth, which result in wider tree rings and tend to make stems more susceptible to breakage (Meyer et al. 2008). Other studies point to an inverse relationship between stem uprooting and base saturation in soil (Braun et al. 2003; Mayer et al. 2005). High nitrogen inputs not only promote acidification but also lead to (i) a reduction of fine root biomass; (ii) the promotion of *Phytophthora* root disease and root rot (*Heterobasidion annosum*); and (iii) reduced storage of carbohydrates in roots. All these factors lead to a weakening of the roots on eutrophic and acidified soils. Braun et al. (2003) found large proportions of uprooted Norway spruce and European beech (*Fagus sylvatica* L.) on acidic soils, a result confirmed by an extensive sample of windthrow damage in Germany, France, and Switzerland (Mayer et al. 2005).

8.2.2 Non-lethal Damage and Adaptation to Wind

As long as wind forces do not lead to stem breakage or uprooting, tree individuals can adapt through growth, a process that has been addressed generally as thigmomorphogenesis, i.e. the effect of motion on growth (Jaffe 1973; Telewski 2006). Plants tune their height and slenderness biologically as a response to the mechanosensing of wind (de Langre 2008). A known characteristic of wind-exposed trees is that they are shorter than individuals sheltered from wind. In cases of recurring wind loads, exposed tree crowns align themselves in such a way that a smaller crown surface area is created (Vollsinger et al. 2005). In response to frequent pressure from snow or wind, conifers form so-called compression wood with a higher lignin content and stronger cell walls. Broadleaved deciduous trees, on the other hand, form tensile wood as a reaction to tensile stress. Compression and tension wood stabilizes tree stems against external stresses (Fournier et al. 1994). Roots experience wind pressures as compression and tension movements. Roots also react to these forces by forming compression wood, which leads to the formation of buttress roots with I- or 8-shaped cross sections (with profiles comparable to those of I-beams) (Matthek and Breloer 1995).

During storms, a tree first loses its weakest parts, i.e. needles/leaves and branches, which reduces the surface area exposed to wind and the risk of stem breakage (Spatz and Bruechert 2000). Likewise, roots are damaged by storm-induced vibrations of the tree. Therefore, trees that survive a storm experience a loss of growth in the first years after the event and use part of the assimilated carbon to restore the damaged plant parts. These reductions in growth are well documented for the first 3 to 5 years following a storm event and can be detected at both the individual tree and the landscape levels (Busby et al. 2008; Seidl and Blennow 2012).

8.3   Economic Damage to Forests

Over 50% of all forest damage in Europe is attributed to winds, occurring mostly in the form of winter storms (Schelhaas et al. 2003). In Europe, these storms first hit the countries facing the Atlantic, and they then travel several hundred to thousand kilometres inland (Fig. 8.3). Depending on the size of the winter storm, the disturbed area can be substantial (Fig. 8.3 and Table 8.1). The damage to forests, and also to infrastructure and buildings, causes widespread concern among the general public, which often leads to state support for particularly badly affected areas (Baur et al. 2003).

The simultaneous clearing of many wind-damaged areas leads to a temporary drop in timber prices (Fig. 8.4 for price dynamics in Germany). In Switzerland after storm Lothar (26 December 1999), there was around 15 million m³ of damaged timber. By 2001, prices for spruce and fir logs had fallen to around 70% of the price before the storm. Prices then slowly recovered but declined again with the arrival of more wood salvaged from the bark beetle infestations during the drought of 2003. It was not until 2007 that timber prices returned to the pre-storm level of 1999.

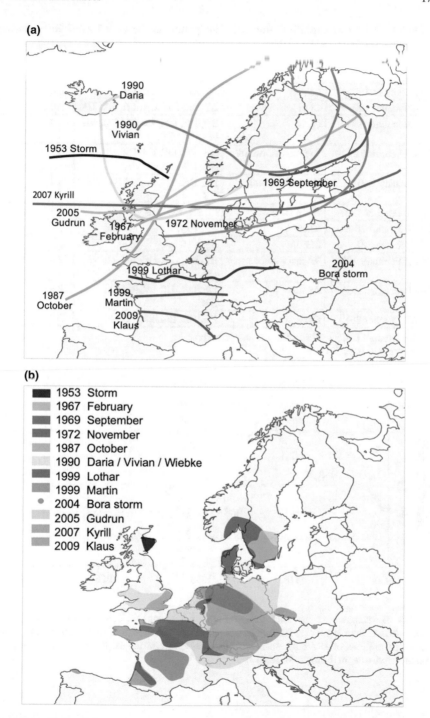

Fig. 8.3 (**a**) Paths of low-pressure systems (storms) in the winter months from 1961 to 2010. (**b**) Estimated areas affected by storms in the winter months from 1961 to 2010. (From Gardiner et al. 2010, modified)

Table 8.1 The 'most expensive' storms in Europe during the period 1980–2016. Country abbreviations according to ISO 3166-1

Date	Storm	Affected countries	Total losses in billion €[a]
26 December 1999	Lothar	AT, BE, CH, DE, FR	10.4
18–20 January 2007	Cyril	AT, BE, DE, DK, FR, NL, UK	9.0
25–26 January 1990	Daria	BE, DE, DK, FR, IE, LU, NL, PL, UK	6.2
26–28 February 2010	Xynthia (storm tide)	BE, CH, DE, ES, FR, LU, NL, PT	5.5
7–9 January 2005	Erwin (Gudrun)	DE, DK, EE, FI, LV, NO, SE, UK	5.4
15–16 October 1987	Not named	IT, FR, NO, UK	4.8
24–27 January2009	Klaus	ES, FR, IT	4.6
27 December 99	Martin	ES, CH, FR	3.7
25–27 February 1990	Vivian/Wiebke	AT, BE, CH, DE, FR, IT, NL, UK	2.8
3–4 December 1999	Anatol	DE, DK, LT, LV, RU, PL, SE, UK	2.8
27–30 October 2013	Christian	BE, DE, DK, FR, NL, SE	2.2

Source: Munich RE, http://natcatservice.munichre.com
[a]Exchange rate June 2014: $1 = €0.9

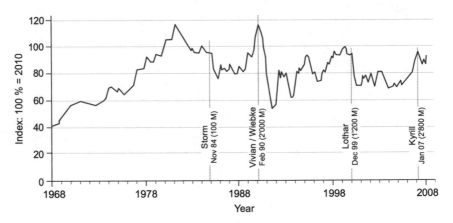

Fig. 8.4 Development of the wood price for spruce long logs (B quality) in Germany (index: 100% = 2010). Sources: Wood: [German] Federal Statistical Office: total damage caused by winter storms, in millions of DM/€: (Münchener Rück 1999; https://de.wikipedia.org/wiki/Liste_von_Wetterereignissen_in_Europa; 31.1.2017)

8.4 Influence on Ecosystem Dynamics

8.4.1 Root Plates and Root Pits

Frequently described and investigated phenomena in windthrown forest ecosystems are uprooted trees, with root plates often reaching several metres into the air, and the corresponding pits, which can vary in moisture depending on soil condition and slope. Shortly after a windthrow event, this pit-and-mound topography (Ulanova 2000) with exposed mineral soil offers favourable germination conditions for a number of species, and in particular European larch (*Larix decidua* Mill.) but also silver fir (*Abies alba* Mill.) and Norway spruce. In comparison, germination for these species in undisturbed stands with dense vegetation cover is often more difficult (Wohlgemuth et al. 2002). Root plates and mounds weather over time, with exposed roots drying out quickly and losing their function and the exposed soil being colonized by plants (herbs, tree seedlings). Regeneration of trees in these microsites is quite variable (Ulanova 2000). For mosses, the newly created microtopology offers many new locations, which has a positive effect on species diversity (Jonsson and Esseen 1990).

8.4.2 Succession After Windthrow

Since windthrows do not occur regularly but episodically, studies of the succession process are also often one-time efforts that are limited with respect to space or sample size. Another complicating factor is the long observation period necessary to make inferences on succession. In Central Europe, the high frequency of wind disturbances towards the end of the last century led to broadly discussed questions about the treatment of wind-disturbed areas, with three main topics: (i) the effects on bark beetle development (Bouget and Duelli 2004; Müller et al. 2008); (ii) the quantity and quality of natural reforestation; and (iii) the effect on biodiversity (Chap. 4, Biodiversity; see reviews by Thom and Seidl 2016; Thorn et al. 2018). In the Alps, the storm Vivian (1990) started a debate on the maintenance of avalanche and rockfall protection in windthrown mountain forests (Schönenberger et al. 2002), investigating effects of clearing vs. no intervention. While in North America the effects of windthrow and forest fires on natural regeneration had already been investigated in the 1980s and discussed in various textbooks and review articles (Chap. 2, Definitions; White 1979; Pickett and White 1985; Attiwill 1994; White and Jentsch 2001), such reviews followed much later for the boreal forests of Northern Europe (Kuuluvainen 1994; Kuuluvainen and Aakala 2011) and for the temperate forests of Central and Eastern Europe (Ulanova 2000; Fischer et al. 2013).

8.4.3   Natural Regeneration in Central Europe

In a study of 90 windthrow areas in Switzerland, Kramer et al. (2014) determined the regeneration densities and tree heights 10 years after the storm Lothar (1999) and 20 years after the storm Vivian (1990). The sample sites ranged from the lowlands (370 m a.s.l.) north of the Alps to higher-elevation sites in the Northern and Central Alps, where windthrow areas up to 1800 m a.s.l. were considered. Overall, the patterns of recovery were very heterogeneous; this is shown by the large differences in densities of beech regeneration with minimum densities of 1000 stems/ha (stems >20 cm height) and maximum densities of almost 80,000 stems/ha) (Fig. 8.5). In the higher-elevation Vivian windthrow areas, average densities of 2600 and 4600 stems/ha were found in no-intervention and cleared areas, respectively. Higher densities were found in the areas disturbed by storm Lothar, with averages of 7600 and 10,800 stems/ha in no-intervention and cleared 10-year-old areas, respectively. As the monitoring of Vivian and Lothar areas showed, higher stem density can result in slower colonization dynamics at high-elevation sites as a result of light limitation and shorter vegetation periods (Brang et al. 2015). At lower elevations, the larger trees in an area recovering from windthrow reach a height of around 6 m after 10 years, whereas at higher elevations a period twice as long is required for trees to grow to this size. The proportion of regeneration that was already present before the windthrow event was smaller at higher elevations (approximately 10%), as fewer shade-tolerant tree species are able to survive in closed stands and as competing tall herbs in more open subalpine forests hamper the establishment of trees. In contrast, at lower elevations an average of 30% of the regeneration present

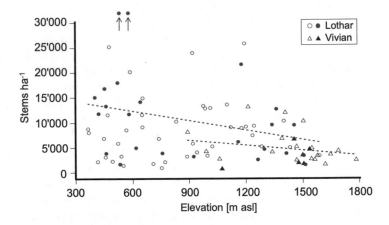

Fig. 8.5 Natural regeneration in windthrow areas in Switzerland. (Data from Kramer et al. 2014), 10 years after storm Lothar (circles) and 20 years after storm Vivian (triangles). Filled symbols represent no-intervention areas and open circles represent cleared areas. Regression lines for Vivian and Lothar areas show the differences between lower- and higher-elevation windthrow areas. However, for both groups, the relationship between elevation and regeneration density is not significant

10 years after a storm event germinated before the windthrow. This is especially true for shade-tolerant beech and sycamore (*Acer pseudoplatanus* L.; Wohlgemuth and Wohlgemuth 2002).

8.4.4 Limiting Factors of Regeneration

Kramer et al. (2014) cite elevation, competing vegetation, and acidity of the substrate as the most important factors limiting regeneration in windthrow areas. At higher elevations, not only is there less time for growth because of the short vegetation periods, but also fewer tree species are present. Tall herbs, reed grasses (*Calamagrostis* spp.), and raspberries (*Rubus idaeus* L.) can prevent tree regeneration over a long period of time. Conifers that grow slowly in the first 10 years after establishment (e.g. fir and spruce) can, if germinated after the storm, initially grow in the shade of pioneers or avoid tall herbs by establishing on decaying deadwood ('nurse logs'). Especially at higher elevations, an already existing regeneration shortens recovery times. Also at low elevations, rapidly emerging herbs, shrubs, or grasses can prevent tree regeneration. Thus, on about one-third of the investigated 10-year-old Lothar areas below 800 m a.s.l., regeneration did not reach a density of 4000 stems/ha. On such sites blackberry (*Rubus fruticosus* aggr. L.) is the most important competitor of young trees. On acidic soils, bracken (*Pteridium aquilinum* (L.) Kuhn) can also prevent tree regeneration (Brang et al. 2015). The higher the pH of the mineral soil, the greater the regeneration density in windthrow areas. Further, the number of tree species on carbonate-rich substrate is greater than on carbonate-poor or acidic substrates (Kramer et al. 2014). Van Couwenberghe et al. (2010) found a similar correlation in windthrow areas in Lorraine, France. The resource situation improves temporarily with the sudden opening of the forest. On barren sites in the limestone Alps, for example, accelerated decomposition dynamics of organic matter in the soil were observed after disturbance by wind. This effect was caused by higher soil temperatures and lower litter input (Mayer et al. 2014); these factors can have a long-term influence on forest development after windthrow. A recent study in France found evidence in windthrow gaps for forest adaptation to global warming due to a loss of cold-adapted species and a gain of warm-adapted species (Dietz et al. 2020).

The regeneration densities in Vivian and Lothar windthrow areas in Switzerland did not differ significantly between cleared and no-intervention areas. Although existing regeneration of small- and medium-sized trees may be destroyed during clearing, the disturbed soils of cleared areas provide favourable conditions for the germination and growth of many tree species, which is why stem densities can be high a few years after the intervention. In contrast, in no-intervention areas, optimal places for seed germination are restricted to root plates and pits, where vegetation is temporarily absent. These results from storms Vivian and Lothar contrast with various case studies that highlight the importance of microhabitats for the development of natural structures and forest biodiversity in uncleared windthrow areas (Fischer and Fischer 2012; Michalová et al. 2017).

Various studies have investigated regeneration as a function of distance to the intact forest and thus to potential seed sources. In a large sample in Lorraine, France, the frequency of small beech, hornbeam (*Carpinus betulus* L.), and oak (*Quercus robur* L., *Q. petraea* (Matt.) Liebl.) decreased in the first 20 m from the edge of the forest 5 years after the storm, whereas larger specimens of the same species increased in density with increasing distance from the edge (Van Couwenberghe et al. 2010). Although there are clear limits to the dispersal of heavier seeds, young trees that germinated before the disturbance generally grow better in open areas in comparison to close to the edge of an intact forest canopy (White and Jentsch 2001). In lowland forests in Switzerland, Kramer et al. (2014) found a negative correlation between regeneration density and distance to the forest edge for the first 80 m 10 years after windthrow events. This was mainly explained by the predominance of beech on these stands and the limited dispersal distance of beech seeds (heavy weight). In forests at higher elevations, distances to the forest edge of 20–80 m were not found to influence regeneration density. This may be related to the much higher seed dispersal distance of Norway spruce, the most common tree species in European mountain forests. On the other hand, in a windthrow area at 1200 m a.s.l. in Schwanden, Canton Glarus (Switzerland), 30% of the seed rain occurred within a distance of 60 m from the edge and only 10% at distances ≥80 m from the forest edge (Lässig et al. 1995).

8.4.5   Regeneration on Downed Deadwood

Competitive exclusion by herbaceous vegetation is a major reason why forest regeneration in disturbed areas is often slow, especially in high-elevation forests where the vegetation period is short (Brang et al. 2015; Wohlgemuth and Kramer 2015). Consequently, downed deadwood is an important substrate for tree regeneration in the mountain forests of Central Europe (Imbeck and Ott 1987; Svoboda and Pouska 2008) and Scandinavia (Hofgaard 1993). The importance of slowly decaying deadwood for regeneration (i.e. so-called nurse logs) was demonstrated, for instance, in a study of windthrow areas in Switzerland (Wohlgemuth and Kramer 2015): 20 years after the windthrow, the regeneration density on deadwood was equal to that on the forest floor. In areas where the regeneration of conifers on bare soil is impeded by the rapid emergence and vigorous growth of tall herbs and grasses (often reed grass, i.e. species of the genus *Calamagrostis*), slowly decaying wood remains an alternative substrate where conifers, in particular Norway spruce, can establish without serious competition (Kramer et al. 2014). In these high-elevation forests, where regeneration is generally less dense than in the lowlands (Wohlgemuth et al. 2017) and competition from forest floor vegetation prevents or delays tree regeneration, establishment on nurse logs is an important plant strategy for disturbance recovery (Imbeck and Ott 1987; Priewasser et al. 2013). In addition, the growing season on the deadwood substrate is extended as deadwood is free of snow earlier in spring than the adjacent forest floor. In contrast to the prominent role of nurse wood at elevations with shorter vegetation season and lower temperatures, nurse logs are relatively less important in lowland forests because of the rapid decomposition of deadwood, in particular of broadleaved tree species such as European beech (Kahl et al. 2017).

8.5 Forest Management Strategies to Minimize Storm Damage

How can the damage caused by strong winds be reduced? This question has concerned forestry for more than a century (e.g. Wagner 1923). Annual data on the loss of timber by windthrow are available for the canton of Zurich, Switzerland, for instance, extending over an area of 1729 km² and spanning a period of almost 120 years. The variation in timber loss can be explained surprisingly well with only two variables (Usbeck et al. 2010b): (i) the wind gust peaks measured in Zurich and (ii) the growing stock. Both variables are both strongly positively correlated with the timber damaged in winter storms (Fig. 8.6).

A storm is a physical factor that affects both individual trees and forest stands. From a process-oriented perspective, the physical variables of relevance for wind breakage and uprooting are the intensity of the storm and parameters of tree or stand

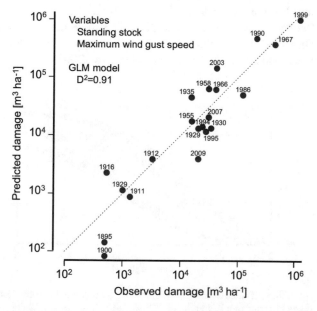

Fig. 8.6 Observed and predicted timber damage of 22 windthrow years from 1891 to 2014 in the canton of Zurich, based on various sources, with growing stock and gust wind peaks measured in Zurich as explanatory variables. (Data source: Usbeck et al. 2010b)

stability, which are discussed in more detail below. Wind gusts, which act over a longer period of time (wind force) with values of 35 m/s (126 km/h) or more, can be destructive for stands regardless of their susceptibility and prior silvicultural treatment. However, as the wind field within a storm is heterogeneous, winds do not affect all locations within a forest landscape with the same force. Consequently, there are a number of other factors that are important for determination of the level of wind damage at the stand level. Many of these drivers were analysed regarding their comparative effect on windthrow damage (Hanewinkel et al. 2015). The short summary that follows is largely based on this work.

8.5.1 Species Identity

The different susceptibilities of species to windthrow are, among other things, related to the greater air resistance of the crowns of evergreen conifers to winter storms as compared to the crowns of broadleaved deciduous species (see Fig. 8.7). Broadleaved deciduous trees are much less at risk than, for example, Norway spruce, as shown by several studies in Switzerland (Dobbertin et al. 2002), Germany (Hanewinkel et al. 2008), Austria (Thom et al. 2013), and Central Europe (Mayer et al. 2005). In Baden-Württemberg (Germany), Norway spruce was the species most severely affected by storm Lothar, followed by Douglas fir, Scots pine (*Pinus sylvestris* L.), and European beech (Fig. 8.7). Based on new analyses, Douglas fir, together with Norway spruce, is assigned as being at the greatest risk of stormthrow in Central Europe (Albrecht et al. 2013, 2015). The smaller windthrow susceptibility of silver fir in comparison to Norway spruce relates to different root structures, with silver fir developing a deep rooting system in contrast to Norway spruce which has a shallower root system (Stokes et al. 2007).

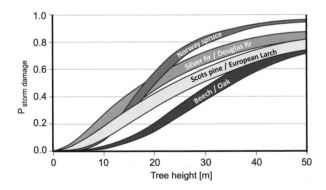

Fig. 8.7 The probability of stem breakage or uprooting as a result of storm Lothar (1999) in Baden-Württemberg as a function of tree height, separated by tree species (according to Schmidt et al. 2010). The shaded areas indicate the (95%) confidence intervals. Because of a lack of data, individual tree species (e.g. Douglas fir and fir) were grouped together. (From Schmidt et al. 2010, redrawn)

8.5.2   Tree Height and Stem Taper

In general, the storm susceptibility of trees in Europe increases with tree height (Dobbertin et al. 2002), with the increase varying from sigmoidal (Norway spruce) to near linear (European beech) depending on species (Fig. 8.7). Parameters such as stem diameter, average stand height, and stand age are closely correlated to tree height and are also often used as alternative explanatory variables of windthrow risk.

According to many studies, stem taper (i.e. h/d ratio; the ratio of tree height to tree diameter at breast height) is positively correlated with susceptibility to storm damage, because of the lower mechanical stability of thin and tall stems. Yet, trees with a lower h/d ratio often have larger crowns, which in turn results in a larger surface area exposed to wind. At the same time, however, such trees also have a larger root volume, which in turn suggests better anchoring. Stem taper should therefore be used with caution in the context of wind susceptibility. As a general rule, a meaningful interpretation of stem taper requires a joint assessment with stand or tree height (Albrecht et al. 2012).

8.5.3   Silvicultural Interventions and Management History

Shortly after a thinning intervention, older stands in particular have a temporarily elevated risk of windthrow because the forest canopy has been disrupted, thereby exposing the trees to increased wind. Silvicultural interventions were the third most important variable (after tree species identity and tree/stand height) explaining damage from storm Lothar (1999) in Baden-Württemberg (Albrecht et al. 2012). In particular, damage was greater where more dominant trees had been removed from the stands (thinning from above).

In areas with clear-cutting as silvicultural strategy, the newly exposed stand edges are particularly vulnerable to windthrows, because the trees that were previously sheltered within the stand are suddenly exposed to wind and are not adapted to this exposure. Such effects can be reduced if the newly created stand edges are not oriented facing the main wind direction (Blennow and Olofsson 2008). The risk of storm damage at these stand edges can also be reduced by retaining individual trees or groups of trees on the clear-cut areas. However, the latter trees are particularly vulnerable to storm events.

In Germany, large-scale Norway spruce plantations following the intensive timber extraction during the first half of the twentieth century resulted in uniform, single-layer pure stands often dominated by Norway spruce. Towards the end of the twentieth century, these stands became increasingly vulnerable to storms. Part of the large-scale storm damage in Germany since the 1980s (Table 8.1) can be attributed to this particular management history. This example shows that changes in forest structure and composition can have a long-term effect on the susceptibility to wind disturbance.

8.5.4 Soil Conditions

Various studies have shown that forest stands on wet soils are particularly susceptible to windthrow (Hanewinkel et al. 2008; Schmidt et al. 2010; Albrecht et al.
2012). The Europe-wide increase in storm damage in recent decades has also been
associated with the increased cultivation of Norway spruce on such soils in the first
half of the twentieth century (Schelhaas et al. 2003). In the Swiss Central Plateau,
major windthrow damage has occurred in winter when the soil was not frozen but
wet (Usbeck et al. 2010a), leading to a higher probability of uprooting. This may be
linked to the fact that winter storms in Europe are usually weather events lasting
several days, during which temperature first increases considerably, followed by the
periods with high wind speeds. Higher temperatures under climate change are likely
to exacerbate this situation. For the Lothar (1999) storm damage in Switzerland,
France, and Germany, also a statistically significant negative relationship with soil
acidity was found, i.e. greater damage occurred on acidic soils (Mayer et al. 2005).
Likewise, in the Black Forest (Germany), major Lothar damage was found on
waterlogged/wet soils (Hanewinkel et al. 2008).

8.5.5 Topographic Exposure

Storm Lothar, which had a dominant wind direction of west or northwest and was
thus typical for cyclonic winter storms in Europe, hit forests most severely in
exposed west-facing areas as well as in west–east oriented valleys (Mayer et al.
2005; Schmoeckel and Kottmeier 2008; Stadelmann et al. 2014). This occurred
despite the fact that trees develop wind-adapted growth forms in locations with
frequent exposure to strong winds and can thus withstand stronger storms than trees
on less exposed sites (Hanewinkel et al. 2015). Such behaviour has been demonstrated in particular for trees on mountain sites with frequent wind events (Bräcker
and Baumann 2006). However, even such adaptations cannot prevent damage to
forest stands in the case of very strong wind peaks, such as storm Lothar.

8.5.6 Stand and Landscape Structure

Air flows are attenuated in more structured stands by dissipating the energy into
smaller vortices (Hanewinkel et al. 2015). In poorly structured stands, the energy of
gusts strikes with no dampening effect. This effect has led to the hypothesis that
uneven-aged forests and selection cutting systems (plenter forests) are particularly
resistant to windthrow. According to Hanewinkel et al. (2015), however, this hypothesis cannot be unequivocally substantiated because different drivers interact in such
forests. Nonetheless, Hanewinkel et al. (2014) were able to show a weak, albeit

significant, negative effect of increasing structural diversity on the susceptibility to storm damage. Differences in susceptibility in structurally diverse stands can, however, be the result of differences in species composition (e.g. more shade tolerant trees such as silver fir, which also have a deep rooting system) or differences in tree height (as forests managed under a selection cutting system have many smaller trees that are generally less susceptible to wind). Since these effects are difficult to separate in practice, process-oriented simulation models can be used for further analysis of explanatory variables. For example, Seidl et al. (2014a) demonstrated that neglecting structural and spatial heterogeneity, even in forests with a relatively uniform age class distribution, leads to an underestimation of windthrow damage. While the susceptibility of structurally complex stands to wind is not yet fully resolved, it is largely undisputed that the recovery after a windthrow progresses more rapidly in structured stands because of the presence of young trees under the shelter of old trees. Even 10–20 years after a wind disturbance, the trees that were already present before the disturbance still have the highest top heights in windthrow areas, indicating a lasting advantage compared to trees regenerating only after the disturbance (Kramer et al. 2014). Therefore, the resilience of uneven-aged forest stands to storms may indeed be greater than that of even-aged forest stands.

In areas with clear-cutting as a silvicultural strategy, the size and distance to the nearest stand edge in the prevailing wind direction can be a significant factor influencing the susceptibility to windthrow. In Central Europe, attempts were made already more than 100 years ago to minimize storm damage by optimizing the 'spatial order' of forest stands (i.e. the spatial configuration of stands in terms of age and height in a landscape; Hanewinkel et al. 2010). The success of these measures is, however, relatively modest particularly in high-intensity wind events.

8.6 More Storms Under Climate Change?

The increasing frequency of winter storms in Europe towards the end of the twentieth century (Gardiner et al. 2010; Donat et al. 2011) led to major forest damage, a considerable part of which is attributable to the development trajectory of Europe's forests after World War II and the resulting increase in timber stocks (Schelhaas et al. 2003; Usbeck et al. 2010a). In most European countries, timber stocks have increased over the last century as a result of declining reduction in the use of wood as fuel and building material. A study of trends in disturbance damage in European forests showed that the observed increase in wind damage (+2.6% per year on average between 1958 and 2001) is in roughly equal proportions the result of changes in forests (growing stock, age, and tree species composition) and climate (wind speeds, ground frost, and other factors) (Seidl et al. 2011). Assuming that storm frequencies remain high or even increase, forest damage caused by storms will increase significantly compared to current levels, especially in Central European countries (Seidl et al. 2014b). On the other hand, a comprehensive modelling study, in which large temperature differences in the stratosphere were considered as the main factors driving strong winds, pointed to a slight decrease in wind peaks in winter in the twenty-first century (Bengtsson et al. 2009). The authors of this study also discuss the gradual shift of extratropical storms towards the North. However, storm frequencies are still extremely difficult to predict. This fact is reflected in the Intergovernmental Panel on Climate Change (IPCC) Technical Report on Climate Change 2013, which points to a tendency for cyclone tracks to shift towards the poles but does not infer any future decrease in winter storms in Europe (Stocker et al. 2013). In summary, it can be concluded that intensive winter storms are likely to continue to occur in Central Europe and that forest stocked with vulnerable species such as Norway spruce with tall trees must be considered to be particularly at risk from such future storms (Jakoby et al. 2016) (Box 8.1).

Box 8.1: Wind Force and Damage to Trees

Thomas Wohlgemuth [ID]
Forest Dynamics Research Unit, Swiss Federal Research Institute WSL, Birmensdorf, Switzerland

One might think that wind measurement is simple, given that information on wind speed is part of daily weather forecasts. However, storms do not occur in strictly ordered paths, but instead are spatially strongly heterogeneous, interspersed with vortices, and interact with the topography and vegetation of an area. To calculate gust peaks for larger regions, models based on wind measurements at meteorological stations are used. However, the authors of such models point out that caution is required when comparing modelled values with observed storm effects on forests (e.g. Etienne et al. 2013). Furthermore, wind events are not only spatially heterogeneous but also show considerable temporal variation. Therefore, average wind speeds over longer periods of time (hours to days) are less relevant for effects on vegetation than short-term wind peaks, which can last from a few seconds to a few minutes. Based on the correspondence of observed wind effects and local measurement series, it is assumed that wind speeds of at least 28 m/s (100 km/h), corresponding to a wind force on the Beaufort scale of 10 or more, must prevail over several seconds to cause uprooting or stem breakage, the two most conspicuous effects on trees. Films of hurricanes show that healthy trees incline up to 45° and only break or uproot after long and strong wind exposure. Hurricane-like storms with wind peaks above 35 m/s (126 km/h) can devastate both old and young stands and cause damage over large areas.

References

Albrecht A, Hanewinkel M, Bauhus J, Kohnle U (2012) How does silviculture affect storm damage in forests of south-western Germany? Results from empirical modeling based on long-term observations. Eur J For Res 131:229–247

Albrecht A, Kohnle U, Hanewinkel M, Bauhus J (2013) Storm damage of Douglas-fir unexpectedly high compared to Norway spruce. Ann For Sci 70:195–207

Albrecht A, Hanewinkel M, Bauhus J, Kohnle U (2015) Wie sturmstabil ist die Douglasie? AFZ-Der Wald 9/2015:30–34

Attiwill PM (1994) The disturbance of forest ecosystems – the ecological basis for conservative management. For Ecol Manag 63:247–300

Baur P, Bernath K, Holthausen N, Roschewitz A (2003) Lothar. Ökonomische Auswirkungen des Sturms Lothar im Schweizer Wald, Teil II. Verteilung der Auswirkungen auf bäuerliche und öffentliche WaldeigentümerInnen: Ergebnisse einer Befragung. Bundesamt für Umwelt, Wald und Landschaft, Bern, Umwelt-Materialien 158:1–204

Bengtsson L, Hodges KI, Keenlyside N (2009) Will extratropical storms intensify in a warmer climate? J Clim 22:2276–2301

Blennow K, Olofsson J (2008) The probability of wind damage in forestry under a changed wind climate. Clim Chang 87:347–360

Bouget C, Duelli P (2004) The effects of windthrow on forest insect communities: a literature review. Biol Cons 118:281–299

Bräcker O-U, Baumann E (2006) Growth reactions of sub-alpine Norway spruce (*Picea abies* (L.) Karst.) following one-sided light exposure (Case study at Davos "Lusiwald"). Tree-Ring Res 62:76–73

Brang P, Hilfiker S, Wasem U, Schwyzer A, Wohlgemuth T (2015) Langzeitforschung auf Sturmflächen zeigt Potenzial und Grenzen der Naturverjüngung. Schweiz Z Forstwes 166:147–158

Braun S, Schindler C, Volz R, Flückiger W (2003) Forest damages by the storm 'Lothar' in permanent observation plots in Switzerland: the significance of soil acidification and nitrogen deposition. Water Air Soil Poll 142:327–340

Busby PE, Motzkin G, Boose ER (2008) Landscape-level variation in forest response to hurricane disturbance across a storm track. Can J For Res 38:2942–2950

Čada V, Morrissey RC, Michalová Z, Bače R, Svoboda M (2016) Frequent severe natural disturbances and non-equilibrium landscape dynamics shaped the mountain spruce forest in central Europe. For Ecol Manag 363:169–178

de Langre E (2008) Effects of wind on plants. Annu Rev Fluid Mech 40:141–168

Dietz L, Collet C, Dupouey JL, Lacombe E, Laurent L, Gégout J-C (2020) Windstorm-induced canopy openings accelerate temperate forest adaptation to global warming. Glob Chang Biol 29:2067–2077

Dobbertin M, Seifert M, Schwyzer A (2002) Ausmass der Sturmschäden. Wald Holz 83:39–42

Donat MG, Renggli D, Wild S, Alexander LV, Leckebusch GC, Ulbrich U (2011) Reanalysis suggests long-term upward trends in European storminess since 1871. Geophys Res Lett 38(L14703):14701–14706

Etienne C, Goyette S, Kuszli CA (2013) Numerical investigations of extreme winds over Switzerland during 1990–2010 winter storms with the Canadian Regional Climate Model. Theor Appl Climatol 113:529–547

Everham EM, Brokaw NVL (1996) Forest damage and recovery from catastrophic wind. Bot Rev 62:113–185

Fischer A, Fischer HS (2012) Individual-based analysis of tree establishment and forest stand development within 25 years after wind throw. Eur J For Res 131:493–501

Fischer A, Marshall P, Camp A (2013) Disturbances in deciduous temperate forest ecosystems of the northern hemisphere: their effects on both recent and future forest development. Biodivers Conserv 22:1863–1893

Fournier M, Bailleres H, Chanson B (1994) Tree biomechanics: growth, cumulative prestresses and reorientations. Biometrics 2:229–252

Gardiner BA, Blennow K, Jean-Michel Carnus J-M, Fleischer P, Ingemarson F, Landmann G, Lindner M, Marzano M, Nicoll B, Orazio C, Peyron J-L, Reviron M-P, Schelhaas M-J, Schuck A, Spielmann M, Usbeck T (2010) Destructive storms in European forests: past and forthcoming impacts. Final report to DG Environment. European Forest Institute, Atlantic European Regional Office – EFIATLANTIC, 132 p

Hanewinkel M, Breidenbach J, Neeff T, Kublin E (2008) 77 years of natural disturbances in a mountain forest area – the influence of storm, snow and insect damage analysed with a long-term time-series. Can J For Res 38:2249–2261

Hanewinkel M, Hummel S, Albrecht A (2010) Assessing natural hazards in forestry for risk management: a review. Eur J For Res 130:329–351

Hanewinkel M, Kuhn T, Bugmann H, Lanz A, Brang P (2014) Vulnerability of uneven-aged forests to storm damage. Forestry 87:525–534

Hanewinkel M, Albrecht A, Schmidt M (2015) Können Windwurfschäden vermindert werden? Eine Analyse von Einflussgrössen. Schweiz Z Forstwes 166:118–128

Hofgaard A (1993) Structure and regeneration patterns in a virgin *Picea abies* forest in northern Sweden. J Veg Sci 4:601–608

Imbeck H, Ott E (1987) Verjüngungsökologische Untersuchungen in einem hochstaudenreichen subalpinen Fichtenwald, mit spezieller Berücksichtigung der Schneeablagerung und der Lawinenbildung. Eidg. Inst. Schnee- u. Lawinenforsch. Weissfluhjoch/Davos 42:1–202

Jaffe MJ (1973) Thigmomorphogenesis: the response of plant growth and development to mechanical stimulation. Planta 114:143–157

Jakoby O, Stadelmann G, Lischke H, Wermelinger B (2016) Borkenkäfer und Befallsdisposition der Fichte im Klimawandel. In: Pluess AR, Augustin S, Brang P (eds) Wald und Klimawandel. Grundlagen für Adaptationsstrategien, Haupt, Bern, pp 247–264

Johnson EA, Miyanishi K (2007) Plant disturbance ecology: the process and the response. Elsevier, Amsterdam, 698 p

Jonášová M, Vávrová E, Cudlín P (2010) Western Carpathian mountain spruce forest after a windthrow: natural regeneration in cleared and uncleared areas. For Ecol Manag 259:1127–1134

Jonsson BG, Esseen PA (1990) Treefall disturbance maintains high bryophyte diversity in boreal spruce forests. J Ecol 78:924–936

Kahl T, Arnstadt T, Baber K, Bassler C, Bauhus J, Borken W, Buscot F, Floren A, Heibl C, Hessenmoller D, Hofrichter M, Hoppe B, Kellner H, Kruger D, Linsenmair KE, Matzner E, Otto P, Purahong W, Seilwinder C, Schulze ED, Wende B, Weisser WW, Gossner MM (2017) Wood decay rates of 13 temperate tree species in relation to wood properties, enzyme activities and organismic diversities. For Ecol Manag 391:86–95

Kramer K, Brang P, Bachofen H, Bugmann H, Wohlgemuth T (2014) Site factors are more important than salvage logging for tree regeneration after wind disturbance in Central European forests. For Ecol Manag 331:116–128

Kulakowski D, Seidel R, Holeksa J, Kuuluvainen T, Nagel TA, Panayotov M, Svoboda M, Thorn S, Vacchiano G, Whitlock C, Wohlgemuth T, Bebi P (2017) A walk on the wild side: disturbance dynamics and the conservation and management of European mountain forest ecosystems. For Ecol Manag 388:120–131

Kuuluvainen T (1994) Gap disturbance, ground microtopography, and the regeneration dynamics of boreal coniferous forests in Finland: a review. Ann Zool Fenn 31:35–51

Kuuluvainen T, Aakala T (2011) Natural forest dynamics in boreal Fennoscandia: a review and classification. Silva Fenn 45:823–841

Lässig R, Egli S, Odermatt O, Schönenberger W, Stöckli B, Wohlgemuth T (1995) Beginn der Wiederbewaldung auf Windwurfflächen. Schweiz Z Forstwes 146:893–911

Matthek C, Breloer H (1995) The body language of trees: a handbook of failure analysis. H.M. Stationery Office, London, 240 p

Mayer P, Brang P, Dobbertin M, Hallenbarter D, Renaud JP, Walthert L, Zimmermann S (2005) Forest storm damage is more frequent on acidic soils. Ann For Sci 62:303–311

Mayer M, Matthews D, Schindlbacher A, Hauzenstelner R (2014) Soil CO_2 efflux from mountainous windthrow areas: dynamics over 12 years post-disturbance. Biogeosciences 11:6081–6093

Meyer FD, Paulsen J, Körner C (2008) Windthrow damage in *Picea abies* is associated with physical and chemical stem wood properties. Trees-Struct Funct 22:463–473

Michalová Z, Morrissey RC, Wohlgemuth T, Bače R, Fleischer P, Svoboda M (2017) Salvage-logging after windstorm leads to structural and functional homogenization of understory layer and delayed spruce tree recovery in Tatra Mts., Slovakia. Forests 8:88

Müller J, Bussler H, Gossner M, Rettelbach T, Duelli P (2008) The European spruce bark beetle *Ips typographus* in a national park: from pest to keystone species. Biodivers Conserv 17:2979–3001

Münchener Rück (1999) Naturkatastrophen in Deutschland: Schadenerfahrungen und Schadenpotentiale. Münchener Rückversicherungs-Gesellschaft, 98 p

Peltola H, Kellomäki S (1993) A mechanistic model for calculating windthrow and stem breakage at the stand edge. Silva Fenn 27:99–111

Peltola H, Kellomäki S, Hassinen A, Granander M (2000) Mechanical stability of Scots pine, Norway spruce, and birch: an analysis of tree-pulling experiments in Finland. For Snow Landsc Res 135:143–153

Pickett STA, White PS (1985) The ecology of natural disturbance and patch dynamics. Academic Press, Orlando, 472 p

Priewasser K, Brang P, Bachofen H, Bugmann H, Wohlgemuth T (2013) Impacts of salvage-logging on the status of deadwood after windthrow in Swiss forests. Eur J For Res 132:231–240

Quine CP, Gardiner BA (2007) Understanding how the interaction of wind and trees results in windthrow stem breakage, and canopy gap formation. In: Johnson EA, Miyanishi K (eds) Plant disturbance ecology; the process and the response. Elsevier, Amsterdam, pp 103–152

Schelhaas MJ, Nabuurs GJ, Schuck A (2003) Natural disturbances in the European forests in the 19th and 20th centuries. Glob Chang Biol 9:1620–1633

Schmidt M, Hanewinkel M, Kandler G, Kublin E, Kohnle U (2010) An inventory-based approach for modeling single-tree storm damage – experiences with the winter storm of 1999 in southwestern Germany. Can J For Res 40:1636–1652

Schmoeckel J, Kottmeier C (2008) Storm damage in the Black Forest caused by the winter storm "Lothar" – Part 1: Airborne damage assessment. Nat. Hazard Earth Sys 8:795–803

Schönenberger W, Fischer A, Innes JL (2002) Vivian's legacy in Switzerland – impact of windthrow on forest dynamics. For Snow Landsc Res 77:1–224

Seidl R, Blennow K (2012) Pervasive growth reduction in Norway spruce forests following wind disturbance. PLoS One 7(e33301):33301–33308

Seidl R, Schelhaas M-J, Lexer MJ (2011) Unraveling the drivers of intensifying forest disturbance regimes in Europe. Glob Chang Biol 17:2842–2852

Seidl R, Rammer W, Blennow K (2014a) Simulating wind disturbance impacts on forest landscapes: tree-level heterogeneity matters. Environ Modell Softw 51:1–11

Seidl R, Schelhaas MJ, Rammer W, Verkerk PJ (2014b) Increasing forest disturbances in Europe and their impact on carbon storage. Nat Clim Chang 4:806–810

Spatz H-C, Bruechert F (2000) Basic biomechanics of self-supporting plants: wind loads and gravitational loads on a Norway spruce tree. For Ecol Manag 135:33–44

Stadelmann G, Bugmann H, Wermelinger B, Bigler C (2014) Spatial interactions between storm damage and subsequent infestations by the European spruce bark beetle. For Ecol Manag 318:167–174

Stocker TF, Qin D, Plattner G-K, Alexander LV, Allen SK, Bindoff NL, Bréon F-M, Church JA, Cubasch U, Emori S, Forster P, Friedlingstein P, Gillett N, Gregory JM, Hartmann DL, Jansen E, Kirtman B, Knutti R, Krishna Kumar K, Lemke P, Marotzke J, Masson-Delmotte V, Meehl GA, Mokhov II, Piao S, Ramaswamy V, Randall D, Rhein M, Rojas M, Sabine C, Shindell D, Talley LD, Vaughan DG, Xie S-P (2013) Technical summary. In: Stocker TF et al (eds) Climate change 2013: the physical science basis. Contribution of Working Group I to the fifth assessment report of the Intergovernmental Panel on Climate Change. Cambridge University Press, Cambridge/New York, 84 p

Stokes A, Ghani MA, Salin F, Danjon F, Jeannin H, Berthier S, Kokutse AD, Frochot H (2007) Root morphology and strain distribution during tree failure on mountain slopes. In: Stokes A, Spanos I, Norris JE, Cammeraat E (eds) Eco-and ground bio-engineering: the use of vegetation to improve slope stability: proceedings of the first international conference on eco-engineering, 13–17 September 2004. Springer, Dordrecht, pp 165–173

Svoboda M, Pouska V (2008) Structure of a Central-European mountain spruce old-growth forest with respect to historical development. For Ecol Manag 255:2177–2188

Telewski FW (2006) A unified hypothesis of mechanoperception in plants. Am J Bot 93:1466–1476

Thom D, Seidl R (2016) Natural disturbance impacts on ecosystem services and biodiversity in temperate and boreal forests. Biol Rev 91:760–781

Thom D, Seidl R, Steyrer G, Krehan H, Formayer H (2013) Slow and fast drivers of the natural disturbance regime in Central European forest ecosystems. For Ecol Manag 307:293–302

Thorn S, Bässler C, Brandl R, Burton PJ, Cahall R, Campbell JL, Castro J, Choi CY, Cobb T, Donato DC, Durska E, Fontaine JB, Gautier S, Hebert C, Hothorn T, Hutto RL, Lee EJ, Leverkus A, Lindenmayer D, Obrist MK, Rost J, Seibold S, Seidl R, Thom D, Waldron K, Wermelinger B, Winter B, Zmihorski M, Müller J (2018) Impacts of salvage logging on biodiversity: a meta-analysis. J Appl Ecol 55:279–289

Ulanova NG (2000) The effects of windthrow on forests at different spatial scales: a review. For Ecol Manag 135:155–167

Usbeck T, Wohlgemuth T, Pfister C, Bürgi A, Dobbertin M (2010a) Increasing storm damage to forests in Switzerland from 1858 to 2007 Agr. For Meteorol 150:47–55

Usbeck T, Wohlgemuth T, Pfister C, Volz R, Beniston M, Dobbertin M (2010b) Wind speed measurements and forest damage in Canton Zurich (Central Europe) from 1891 to winter 2007. Int J Climatol 30:347–358

Van Couwenberghe R, Collet C, Lacombe E, Pierrat JC, Gegout JC (2010) Gap partitioning among temperate tree species across a regional soil gradient in windstorm-disturbed forests. For Ecol Manag 260:146–154

Vollsinger S, Mitchell SJ, Byrne KE, Novak MD, Rudnicki M (2005) Wind tunnel measurements of crown streamlining and drag relationships for several hardwood species. Can J For Res 35:1238–1249

Wagner C (1923) Die Grundlagen der räumlichen Ordnung im Walde, 4th edn. Verlag der H. Laupp'schen Buchhandlung, Tübingen, 387 p

White PS (1979) Pattern, process, and natural disturbance in vegetation. Bot Rev 45:229–299

White PS, Jentsch A (2001) The search for generality in studies of disturbance and ecosystem dynamics. Prog Bot 62:399–449

Wohlgemuth T, Kramer K (2015) Waldverjüngung und Totholz in Sturmflächen 10 und 20 Jahre nach Lothar (1999) und Vivian (1990). Schweiz Z Forstwes 166:135–146

Wohlgemuth T, Kull P, Wütrich H (2002) Disturbance of microsites and early tree regeneration after windthrow in Swiss mountain forests due to the winter storm Vivian 1990. For Snow Landsc Res 77:17–47

Wohlgemuth T, Bebi P, Schwitter R, Sutter F, Brang P (2017) Post-windthrow management in protection forests of the European Alps. Eur J For Res 136:1029–1040

Chapter 9
Avalanches and Other Snow Movements

Peter Bebi ⓘ, Perry Bartelt, and Christian Rixen ⓘ

Abstract Together with rockfall, landslides, and debris flows, snow avalanches are among the most important gravitational natural hazards in mountain regions. On the one hand, forests perform important protective functions against these natural hazards, and on the other hand, ecosystems are also strongly influenced by these disturbances. In this chapter we give an overview of the important interactions between gravitational natural events and mountain ecosystems using the example of avalanches and other snow movements. We also discuss the interactions with other natural disturbances and human land use.

Keywords Avalanche protection · Disturbance interaction · Forest management · Mountain forests · Snow avalanches · Snow gliding

9.1 Cause and Immediate Effect of Snow Movements

Wherever there is snow, there are also snow movements. Snow changes continuously, mostly in relation to the temperature and humidity of the snow cover. There are different types of snow movement, depending on the type of snow, the topographical conditions, and the weather.

P. Bebi (✉)
Alpine Environment and Natural Hazards Research Unit, WSL Institute for Snow and Avalanches Research SLF, Davos, Switzerland

Climate Change, Extremes and Natural Hazards in Alpine Regions Research Center CERC, Davos Dorf, Switzerland
e-mail: bebi@slf.ch

P. Bartelt · C. Rixen
Alpine Environment and Natural Hazards Research Unit, WSL Institute for Snow and Avalanches Research SLF, Davos, Switzerland

T. Wohlgemuth et al. (eds.), *Disturbance Ecology*, Landscape Series 32,
https://doi.org/10.1007/978-3-030-98756-5_9

9.1.1 Snow Settlement

Even on flat terrain, there are transformations in the snow cover, which lead to the settling of the snow cover. After a snowfall, the fresh snow crystals become more rounded; this results in more connections between the crystals being formed. This reduces the volume, which causes the snow cover to collapse or settle. This binds the layer of fresh snow. In the deeper snow layers, the weight of the new snow causes greater subsidence. Smaller trees and bushes can be pressed down and broken or split by snow settling, but if they do not break, then they usually straighten up again after the snow falls off (Frey 1977).

9.1.2 Snow Creeping

Snow creeping occurs during the settling of snow on a slope with a rough ground surface. In such cases, the snow movements are greatest in the uppermost layers of the snow cover, while the snow in the lowermost layer in contact with the ground does not move. Thus, the snow forces act like a lever on young trees and lead to characteristic changes in tree shape (Fig. 9.1). In addition to the sabre-shaped deformation of the trunk base, this also includes the increased occurrence of multiple stems and low-growing forms of growth (Schönenberger 1978).

9.1.3 Snow Gliding

Snow gliding is the movement of the total snow cover on an inclined ground surface in the trajectory of the fall line (In der Gand 1968, Höller et al. 2009). The creep movement is supplemented by a slow sliding of the snowpack on smooth and moist ground surfaces. Thus, the entire snowpack is usually sliding slowly, with glide rates of approximately less than 5 mm per day (according to Höller et al. 2009). This often happens on long grass or on low-lying dwarf shrubs (Newesly et al. 2000). Larger dwarf shrubs and trees act as obstacles, stabilizing the snow cover already to such an extent that the glide rates decrease significantly (Newesly et al. 2000). Snow gliding can also lead to the breaking and uprooting of young trees, whereby forces of about 1 kN are sufficient to uproot smaller or less flexible trees (Höller et al. 2009). Repeated snow gliding also leads to the typical sabre growth form and the distinct formation of compression wood (Fig. 9.1).

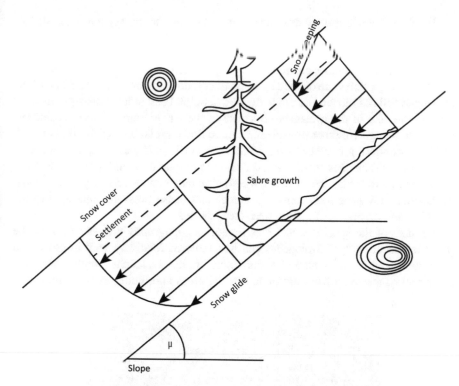

Fig. 9.1 Schematic representation of snow gliding and snow creeping. Snow movements cause the formation of compression wood (reinforced annual rings on the underside of the tree trunk) and adaptations in the growth form (sabre growth)

9.1.4   Avalanches

Avalanches are masses of snow that move down the slope relatively quickly. In addition to snow, avalanches can also contain rocks, soil, vegetation, and ice (Schweizer et al. 2003). Depending on the conditions under which they occur, avalanches are referred to as slab avalanches, loose-snow avalanches, or glide-snow avalanches (Harvey et al. 2012):

- Slab avalanches occur when a break occurs in a weak snow layer on a steep slope with a layered snowpack. The interconnected layers above the weak layer then slide off in compact form as a snow slab.
- Loose-snow avalanches start from a point and fan outwards. The moving snow then pulls more and more snow, creating a cone-shaped avalanche. Loose-snow avalanches occur either in dry and low-cohesion snow in steep terrain or in high solar radiation and wet snow.
- Glide-snow avalanches occur as a result of snow gliding when glide cracks open more and more, and the entire snow cover then quickly slides off the ground. This type of avalanche occurs on smooth and damp ground.

The flow behaviour of an avalanche, and thus its effect on vegetation, is significantly influenced by snow temperatures and snow moisture. The movement of dry, cold snow leads to the formation of dust avalanches or mixed flow and dust avalanches. These avalanches can reach very high speeds (>80 km/h) and consist of a core of snow particles surrounded by a powder cloud of ice and dust. The core of the avalanche follows the terrain, while the powder cloud can independently extend far beyond the core of the avalanche. Because of the air pressure of dust avalanches, trees with a large crown or impact area can be broken relatively easily (Feistl et al. 2015). In contrast, ground or base avalanches often form in wet and less cold conditions. These avalanches consist of a dense and granular core and move much more slowly than dust avalanches. Despite the significantly slower velocities, wet ground avalanches may have long runout distances and cause pressures that lead to the breaking and uprooting of mature trees (see Fig. 9.2).

Because of the specific conditions regarding topography and snow cover, the potential occurrence of avalanches can be predicted relatively well in space and time compared to other natural disturbances. Large avalanche events, therefore, often recur repeatedly from the same starting areas and, under extreme snow and

Fig. 9.2 Wet snow avalanche released slightly above the timberline. (Photo: T. Feistl, WSL/SLF)

weather conditions, often occur synchronously at least regionally (Schneebeli et al. 1997).

9.2 Influence of Avalanches on Ecosystems

Avalanches that start above the timberline are ecosystem-defining processes of mountain ecosystems. At high intensities or frequencies, avalanches define the composition and growth of the underlying plants and thus the ecosystem. Avalanches that strike far above the timberline usually generate such high pressures that they break even large adult trees (Fig. 9.2; Bartelt and Stöckli 2001; Feistl et al. 2015) and form characteristic avalanche tracks (Fig. 9.3). In the influence zone of larger avalanches, trees usually remain small and have a low and flexible growth. Typical representatives of such growth forms are the scrub mountain pine (*Pinus mugo* Turra subsp. *mugo*), green alder (*Alnus viridis* (Chaix) D.C.) (Rudolf-Miklau and Sauermoser 2011; Boscutti et al. 2014), and downy birch (*Betula pubescens* Ehrh.) in Scandinavia (Arbellay et al. 2013). Also, prostrate forms of beech (*Fagus sylvatica* L.) adapted to snow movements are known on exposed sites (Fanta 1981). With decreasing intensity and frequency of avalanche events, the survival of larger and more upright trees becomes possible. With the change in growth forms, both the

Fig. 9.3 Typical tracks, which were created by avalanches above the forest line (Dischma Valley, Davos). Where the avalanche disturbance is most severe for topographical reasons and in the absence of intensive grazing, a mosaic of different vegetation types with patches of green alder shrubs, young trees, and subalpine grassland is formed. Avalanche activity on the right has been suppressed 20 years before the picture was taken by snow supporting structures in the release area. (Photo: M. Schmidlin WSL/SLF)

importance of competition between trees and the influence of climatic factors on tree growth increase (Kulakowski et al. 2006).

Different avalanche regimes lead to characteristic forest structures and species compositions. In forests that are frequently disturbed by avalanches, trees typically have smaller diameters and shorter trunks than in undisturbed forests. The proportion of pioneer tree species is higher, and the proportion of shade-tolerant species is lower in such forests (Bebi et al. 2009; Rudolf-Miklau and Sauermoser 2011). The stem number in avalanche tracks is often low, but the density of small trees (diameter < ca. 10 cm) can also be significantly higher than in undisturbed areas, as avalanches regularly limit the size and growth of trees. These smaller trees are then also less at risk from subsequent avalanche events (Bebi et al. 2009; Rudolf-Miklau and Sauermoser 2011). Forest development after avalanche disturbances does not primarily take place in the sense of establishing tree regeneration but mainly in the form of a reorganization of surviving vegetation. Trees that survive an avalanche develop new leading shoots (Schönenberger 1978). When avalanches slide directly on the ground and thus expose mineral soil, tree regeneration via seed germination can follow (Kajimoto et al. 2004).

One of the most important ecological effects of avalanches is the increase in structural diversity (Kapayev 1978; Ellenberg 1988; Veblen et al. 1994). Where the forest is not influenced by avalanches, large areas of relatively poorly structured and species-poor coniferous forests often dominate in snow-rich mountainous areas. Early successional vegetation types characterized by small and light-demanding species would be much less frequent without avalanches or could only exist temporarily after other disturbance events. Where avalanches create open habitats in otherwise dense forests, they thus contribute significantly to the heterogeneity and diversity of the vegetation mosaic (Patten and Knight 1994; Rixen et al. 2007). For example, avalanche tracks are preferred habitats for brown bear (*Ursus arctos* subsp. *horribilis* Ord), caribou (*Rangifer tarandus* L.), and wolverine (*Gulo gulo* L.) in North America (Mace et al. 1996; Krajick 1998) and for chamois (*Rupicapra pyrenaica* Bonaparte) in Europe (Garcia-Gonzalez and Cuartas 1996). Various bird species, such as the hazel grouse (*Bonasa bonasia* L.) and the black grouse (*Lyrurus tetrix* L.), also live in open or semi-open avalanche tracks for feeding and breeding (Klaus et al. 1990; Schäublin and Bollmann 2011).

The structure and diversity of vegetation in avalanche tracks depend strongly on the avalanche frequency (Fig. 9.4). 'Active' avalanche tracks in the Swiss Alps, which are regularly influenced by avalanches, have a higher structural diversity than avalanche tracks which are suppressed by avalanche barriers (Kulakowski et al. 2006). The preservation of structure by regularly descending avalanches also has a positive effect on the diversity of plant and vegetation types (Rixen et al. 2007). The number of flowering plant species is larger in the centre of avalanche tracks, where disturbance frequency and intensity are higher and smaller in the peripheral areas (Brugger 2002).

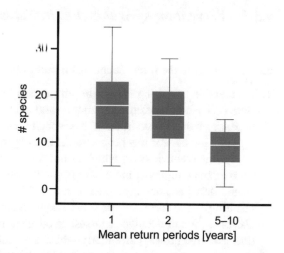

Fig. 9.4 Number of plant species in avalanche tracks with different return periods of avalanches. (Source: Rixen and Brugger 2004)

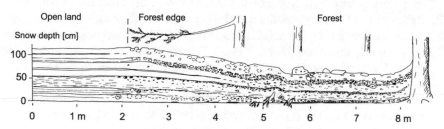

Fig. 9.5 Snow cover formation in the transition zone from an open field to coniferous forest (modified from Imbeck 1987). In the open field, the snow cover shows regular layers. In the forest, the snow depth is reduced due to interception, and the stratification is more structured than in the open field due to falling lumps of frozen snow

9.3   Avalanche Protection Function of Forests

Forests protect against avalanches because hardly any avalanches occur in sufficiently dense forest and because smaller avalanches may be slowed down in the forest (Bebi et al. 2009; Feistl et al. 2015). The effect of avalanche protection is not only influenced by forest structure but also by topographical factors and the structure and thickness of the snow cover (Fig. 9.5). Critical thresholds for the spontaneous release of avalanches may vary with changing forest cover. For example, avalanches in forested areas normally only occur on slopes with a gradient of more than 30° (Schneebeli and Bebi 2004), whereas in open areas they also occur in less steep terrain (McClung and Schaerer 2006). The surface roughness of the terrain is also a very important factor, at least as long as the snow cover thickness in the forest does not exceed the effective height of the dominant objects (de Quervain 1978; Leitinger et al. 2008).

9.3.1 Prevention of Avalanches in Forests

There are essentially four physical processes that contribute to the stabilization of the snow cover in the forest (Schneebeli and Bebi 2004):

1. Interception of falling snow: the proportion of the total snow quantity that is stopped by interception on branches and sublimated back into the atmosphere depends on the forest structure, tree species, and meteorological conditions during snowfall events; this proportion is usually in the range 5–60% (Moeser et al. 2015). Intercepted snow, which is not sublimated, enters the snowpack in the form of snow lumps or meltwater (Bründl et al. 1999). The snow cover in forest stands thus becomes more structured and less thick than that in open terrain, resulting in the formation of weak snow layers that are less continuous. Larch (*Larix* spp.) and deciduous broadleaved trees retain less snow in their crowns than evergreen trees, particularly when snowflakes are relatively small or when temperatures are cold during snowfall (Miller 1964; Pfister and Schneebeli 1999).
2. Balanced radiation regime: the duration of solar radiation as well as the longwave radiation during the night is reduced in the forest compared to that in the open land (Leonard and Eschner 1968; Tribeck et al. 2006) The formation of surface hoar, which can contribute significantly to the formation of weak snow layers in areas outside of the forest, is therefore rare within the forest (Höller 1998; McClung and Schaerer 2006). In addition, temperature differences within the snow cover in the forest are smaller, which counteracts the formation of unstable floating snow in layers close to the ground.
3. Reduced wind speeds: within the forest, near-surface wind speeds are lower than that on open land (Miller 1964). Large accumulations of drifting snow, which often create the conditions for avalanche formation, therefore occur less frequently with increasing stand density in the forest.
4. Direct mechanical support: the supporting effect of tree trunks, root plates, and lying logs helps to stabilize the snow cover and increase the surface roughness of the ground. However, the mechanical supporting effect of standing tree stems alone does normally not sufficiently control the snowpack. According to Salm (1978) at least 1000 trees per hectare would be necessary on a 40° steep slope if the avalanche protection function consisted solely of the mechanical effect of standing tree stems.

Because of these processes, the most important forest characteristics for the prevention of avalanche releases are canopy coverage, gap sizes, tree species, and terrain roughness (Schneebeli and Bebi 2004; Bebi et al. 2009). Critical thresholds that determine whether or not avalanches can occur in the forest can be derived from past events but must also be assessed in relation to topographical factors and snow properties (Bebi et al. 2009). The minimum gap width, which is sufficient for the formation of avalanches, is significantly smaller in deciduous broadleaved forests

Fig. 9.6 Critical gap widths depending on forest type: 1 = deciduous forests, 2 = deciduous coniferous mixed forests, 3 = evergreen coniferous forests, 4 = larch or larch–spruce forests. (Data from Meyer-Grass and Schneebeli 1992)

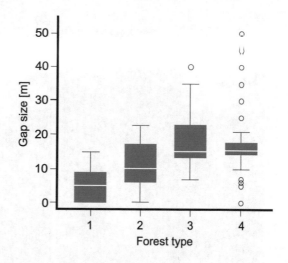

(approximately 5–10 m) than in evergreen forests (approximately 15–20 m), with considerable variation depending on steepness, ground roughness, and snow conditions (see Fig. 9.6). Critical lengths of forest gaps vertical to the slope are usually given in the range of 25–60 m, depending on the slope inclination (see Frehner et al. 2005), and are also relevant for whether a small forest avalanche may actually develop into a larger avalanche.

9.3.2 Braking Effect of the Forest

Smaller avalanches, which start in the forest or just above the timberline, may come to a halt again depending on the forest structure, topography, and snow characteristics in the forest. The braking effect of the forest is a consequence of various interactions between terrain, forest, snow, and avalanche characteristics. A more recent approach to quantifying the braking effect is based on the fact that avalanche snow is deposited behind trees and groups of trees, thus removing mass and the impulse from the avalanche (Feistl et al. 2014). This removal of avalanche snow mass is called 'detrainment'. By estimating this detrainment as a function of forest type and forest structure, the braking effect of the forest can be quantified (Teich et al. 2014). This allows prediction of runout distances for given avalanche incipient sizes, snow conditions, and forest structure parameters and for verification of the predictions with well-documented avalanche events (Fig. 9.7).

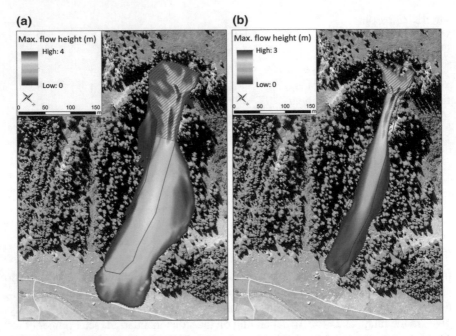

Fig. 9.7 Simulation of maximum flow heights of an avalanche in a forest area in comparison with the observed avalanche (red line) for two different avalanche release areas (hatched area). The runout distance, which was calculated taking into account the forest in the release area (B), corresponds well with the observed runout zone. (Simulation: N. Brozova, WSL/SLF)

9.4   Interaction Between Avalanches and Other Natural Disturbances

There are various interactions between avalanches and other natural disturbances in mountain forests. On the one hand, openings caused by natural disturbances in the forest can potentially create new avalanche starting zones and may suddenly reduce the protective function of the forest (Bebi et al. 2015; Wohlgemuth et al. 2017). On the other hand, avalanches can change a forest landscape to such an extent that the probability of occurrence and the potential extent of other natural disturbances are reduced.

Natural disturbances such as windthrow, snow breakage, fire, or insect outbreaks may open up the forest to such an extent that new avalanche starting zones are created or that the braking effect of the forest is reduced compared to smaller avalanches. This is particularly serious when an intense fire destroys not only the treetops but also the upper soil layer. Such an intensive fire would thus reduce not only the direct interception effect of the canopy but also the ground roughness and germination capacity, so that the protective function is reduced for a long time. Other natural disturbances usually leave a higher residual effect on living and dead biomass. The effect of this residual effect on avalanches depends not only on the intensity of the disturbance but also on the resilience of the original forest. Resilience is understood in this context as the ability of the forest to regenerate and adapt to change so that its protection effect against avalanches or other gravitational natural hazards is sustained (Bebi et al. 2016). Existing (advance) regeneration and residual effects of the remaining stand (including lying deadwood and root plates) contribute thus to an increased resilience. Investigations on former windthrow areas have shown that a time window with insufficient protection may develop as a result of a continuous decrease of the remaining stock in the case of unfavourable regeneration situations (lack of advance regeneration) (Wohlgemuth et al. 2017). Nevertheless, observations of avalanche events on uncleared windthrow areas are very rare (Bebi et al. 2015). This indicates that the effect of increased surface roughness on disturbed forest areas is likely to be more important for avalanche protection than previously assumed.

Distinct avalanche tracks can stop the expansion of other disturbances at their edge. For example, avalanche tracks can potentially serve as 'fire-breaks' (Malanson and Butler 1984; Veblen et al. 1994). Even more complex are interactions between bark beetle outbreaks and avalanches (see Fig. 9.8). On the one hand, avalanche tracks reduce connectivity between vulnerable pure Norway spruce (*Picea abies* (L.) H.Karst) stands. On the other hand, deadwood may accumulate directly after avalanche events, which in turn may create new breeding material for bark beetles (Forster and Meier 2010). Overall, interactions between avalanches and other

Fig. 9.8 Interactions between avalanches and bark beetle outbreaks in the Polish Carpathians. Avalanche tracks can reduce the spread of bark beetle outbreaks. However, bark beetle calamities and other large-scale disturbances may also lead to the formation of new avalanche starting zones. (Photo: A. Casteller WSL/SLF)

natural disturbances and their effects on natural hazard protection are difficult to predict; however, they should at least be taken into account to a greater extent as scenarios in future natural hazard assessments.

9.5   Avalanches and Forest Management

Interactions between forests and avalanches have been strongly influenced by humans for thousands of years. Slash-and-burn and deforestation of entire forests have created new avalanche zones, some with irreversible consequences for the forest (Küttel 1990). The importance of the forest as a protection against avalanches was also recognized early on. For example, various protective forests in Alpine regions were already protected in the fourteenth century by strict preservation regulations (Schuler 1988). However, until the first half of the nineteenth century, poverty and energy demand led to a further overuse of protective forests, resulting in a reduction of natural avalanche protection (Landolt 1860). In the second half of the nineteenth century, the reconstruction of the protective function began, which was initiated by stricter forest legislation, promoted by active reforestation, and favoured by passive re-growth of areas formerly used for agriculture (Bebi et al. 2017).

Simultaneously with the growth of avalanche protection forests in the Alps, structural avalanche protection measures have also been established to protect growing settlements and traffic routes. Above the forest line, permanent steel structures have been used since the 1950s. In contrast, temporary supporting structures and glide-snow protection measures made of wood in combination with reforestation are also used in the forest area (Margreth 2007). Compared to natural protection forests, the construction and maintenance of such technical structures are expensive (>€1 million per hectare for permanent structures). In addition, such structures may also affect the ecosystems in and below the area of the protected forest. In particular, the prevention of avalanches reduces positive effects on habitat structure and biodiversity (Rixen et al. 2007).

In recent decades, the conditions for protective forest management have changed significantly. Extensification in forestry and agriculture has greatly increased the area of avalanche protection forests in the Alps and elsewhere. However, the forests have also become denser and darker, which in turn have made sustainable forest regeneration more difficult. In addition, since the 1960s, the use of wood from steep mountain forests has become less and less profitable, so that the objectives of protection forest management have increasingly moved away from afforestation and thinning towards increasing resilience in recent decades (Bebi et al. 2016). Thereby, active management interventions bring enough light back into dense mountain forests, so that the sustainability of the protective effect is more likely to be guaranteed even in cases of disturbance events (Fig. 9.9a). Since resulting openings after management interventions should also not create new avalanche release areas, the management of steep avalanche protection forests is a challenging optimization task (Frehner et al. 2005). Artificial reforestation of avalanche starting zones is carried

Fig. 9.9 Avalanche protection forest management: (**a**) management intervention by cable crane technology to initiate regeneration (Photo: P. Bebi), (**b**) preservation of the avalanche protection function by a combination of remaining deadwood and supporting group afforestation in a windthrow area in Pfäfers, Switzerland. (Photo: R. Schwitter)

out less extensively today than in the past and is increasingly concentrated on group plantings (Fig. 9.9b; Schönenberger 2001).

9.6   Avalanches and Avalanche Protection in a Warmer Climate

Interactions between snow movements and ecosystems are not only influenced by land use and other natural disturbances but also by climate. A changing climate has a direct impact on the snow situation. In the future, the time period with critical snow thickness for avalanche events is likely to be shorter and the proportion of wet snow situations higher (Castebrunet et al. 2014). Climate-induced changes also have a long-term effect on the species composition and structure of mountain ecosystems.

In forests, avalanche releases occur mainly in snow depths of at least 50 cm (Teich et al. 2012). According to snow cover model calculations in various areas of the Alps, the mean residence time of such snow thickness at elevations of approximately 1000–1600 m a.s.l. will be massively reduced to a few days per year by the end of the twenty-first century (Schmucki et al. 2017). The influence of higher temperatures on the snow cover cannot be compensated by an increase in winter precipitation. Based on such models, we can assume that avalanches that break above the current timberline will continue to be relevant into the twenty-second century but that their significance will decrease in the longer term compared to other disturbance regimes and gravitational natural hazards (Bebi et al. 2016).

Among the most important climate-induced forest changes that are expected to affect snow movements are (i) an increased frequency and intensity of natural disturbances, (ii) forest expansions near cold-limited treelines, and (iii) changes in species composition (Bebi et al. 2016). Climate change will, together with land use-related changes, contribute to treeline shifts in many places. However, such shifts may be delayed by various other factors, including snow movements and avalanche releases above the treeline (Kulakowski et al. 2016). Changes in tree species composition may have a major impact on snow movements, especially when changing from evergreen to deciduous broadleaved tree species. As long as such ecosystem changes are not initiated on a large scale, direct climatic influences on the snow cover will likely have a greater effect on avalanche–forest interactions.

References

Arbellay E, Stoffel M, Decaulne A (2013) Dating of snow avalanches by means of wound-induced vessel anomalies in sub-arctic *Betula pubescens*. Boreas 42:568–574

Bartelt P, Stöckli V (2001) The influence of tree and branch fracture, overturning and debris entrainment on snow avalanche flow. Ann Glaciol 32:209–216

Bebi P, Kulakowski D, Rixen C (2009) Snow avalanche disturbances in forest ecosystems: state of research and implications for management. For Ecol Manag 257:1883–1892

Bebi P, Putallaz J-M, Fankhauser M, Schmid U, Schwitter R, Gerber W (2015) Die Schutzfunktion in Windwurfflächen. Schweiz Z Forstwes 166:168–176

Bebi P, Bugmann H, Lüscher P, Lange B, Brang P (2016) Auswirkungen des Klimawandels auf Schutzwald und Naturgefahren. In: Pluess AR, Augustin S, Brang P (eds) Wald im Klimawandel. Grundlagen für Adaptionsstrategien. Bundesamt für Umwelt BAFU, Bern; Eidg. Forschungsanstalt WSL, Birmensdorf, pp 269–287

Bebi P, Seidl R, Motta R, Fuhr M, Firm D, Krumm F, Conedera M, Ginzler C, Wohlgemuth T, Kulakowski D (2017) Changes of forest cover and disturbance regimes in the mountain forests of the Alps. For Ecol Manag 388:43–56

Boscutti F, Poldini L, Buccheri M (2014) Green alder communities in the Alps: phytosociological variability and ecological features. Plant Biosyst 148:917–934

Brugger S (2002) Auswirkungen von Lawinen auf die Vegetation: Eine Studie im Dischmatal. Diplomarbeit Geogr. Inst. Univ. Zürich, Zürich, 65 p

Bründl M, Schneebeli M, Flühler H (1999) Routing of canopy drip in the snowpack below a spruce crown. Hydrol Process 13:49–58

Castebrunet H, Eckert N, Giraud G, Durant Y, Morin S (2014) Projected changes of snow conditions and avalanche activity in a warming climate: the French Alps over the 2020–2050 and 2070–2100 periods. Cryosphere 8:1673–1697

de Quervain M (1978) Wald und Lawinen. In: Proceedings IUFRO Seminar "Mountain forests and avalanches". Davos, Switzerland, pp 219–231

Ellenberg H (1988) Vegetation ecology of Central Europe. Cambridge University Press, Cambridge, 731 p

Fanta J (1981) *Fagus sylvatica* L. und das Aceri-Fagetum an der alpinen Waldgrenze in Mitteleuropäischen Gebirgen. Vegetatio 44:13–24

Feistl T, Bebi P, Bühler Y, Christen M, Bartelt P (2014) Observations and modeling of the breaking effects of forests on small and medium avalanches. J Glaciol 60:124–138

Feistl T, Bebi P, Christen M, Margreth S, Diefenbach L, Bartelt P (2015) Forest damage and snow avalanche flow regime. Nat Hazard Earth Syst 15:1275–1288

Forster B, Meier F (2010) Sturm, Witterung und Borkenkäfer – Risikomanagement im Forstschutz. Merkbl Prax 44:1–8

Frehner M, Wasser B, Schwitter R (2005) Nachhaltigkeit und Erfolgskontrolle im Schutzwald. Wegleitung für Pflegemassnahmen in Wälder mit Schutzfunktion. Bundesamt für Umwelt, Wald und Landschaft, Bern, 564 p

Frey W (1977) Wechselseitige Beziehung zwischen Schnee und Pflanze – eine Zusammenstellung anhand von Literatur. Mitt. Eidg. Inst. Schnee und Lawinenforsch 34:1–233

Garcia-Gonzalez R, Cuartas P (1996) Trophic utilization of a montane/subalpine forest by chamois (*Rupicapra pyrenaica*) in the Central Pyrenees. For Ecol Manag 88:15–23

Harvey S, Rhyner H, Schweizer J (2012) Lawinenkunde. Bruckmann Verlag GmbH, München, 192 p

Höller P (1998) Tentative investigations on surface hoar in mountain forests. Ann Glaciol 26:31–34

Höller P, Fromm R, Leitinger G (2009) Snow forces on forest plants due to creep and glide. For Ecol Manag 257:546–552

Imbeck H (1987) Schneeprofile im Wald. Winterber. Eidg. Inst. Schnee und Lawinenforsch. 1985/86, 50:177–183

In der Gand H (1968) Aufforstungsversuche an einem Gleitschneehang. Ergebnisse der Winteruntersuchungen 1955/56 bis 1961/62. Mitt Eidg Forschungsanst 44:233–326

Kajimoto T, Daimaru H, Okamoto T, Otani T (2004) Effects of snow avalanche disturbance on regeneration of subalpine *Abies mariesii* forest, Northern Japan. Arct Antarct Alp Res 36:436–445

Kapayev SA (1978) Dynamics of avalanche natural complexes: an example from the high-mountain Teberda State Reserve, Caucasus Mountains. Arct Alp Res 3:215–224

Klaus S, Bergmann H-H, Marti C, Müller F, Vitovic OA, Wiesner J (1990) Die Birkhühner. *Tetrao tetrix* und *T. mlokosiewiczi*. Ziemsen Verlag, Wittenberg Lutherstadt, 288 p

Krajick K (1998) Animals thrive in an avalanche's wake. Science 279:1853–1853

Kulakowski D, Rixen C, Bebi P (2006) Changes in forest structure and in the relative importance of climatic stress as a result of suppression of avalanche disturbances. For Ecol Manag 223:66–74

Kulakowski D, Barbeito I, Casteller A, Kaczka RJ, Bebi P (2016) Not only climate: interacting drivers of treeline change in Europe. Geogr Pol 89:7–15

Küttel M (1990) The subalpine protection forest in the Urseren valley – an inelastic ecosystem. Bot Helv 100:183–197

Landolt E (1860) Bericht an den hohen Schweizerischen Bundesrath über die Untersuchung der Hochgebirgswaldungen in den Kantonen Tessin, Graubünden, St.Gallen und Appenzell. Orell Füssli und Compagnie Druck, Zürich, 183 p

Leitinger G, Höller P, Tasser E, Walde J, Tappeiner U (2008) Development and validation of a spatial snow-glide model. Ecol Model 211:363–374

Leonard R, Eschner AR (1968) Albedo of intercepted snow. Water Resour Res 4:931–935

Mace RD, Waller JS, Manley TL, Lyon LJ, Zuuring H (1996) Relationships among grizzly bears, roads and habitat in the Swan Mountains, Montana. J Appl Ecol 33:1395–1404

Malanson GP, Butler DR (1984) Avalanche paths as fuel breaks: implications for fire management. J Environ Manag 19:229–238

Margreth S (2007) Lawinenverbau im Anbruchgebiet. Technische Richtlinie als Vollzugshilfe. Umwelt-Vollzug 0704. Bundesamt für Umwelt, Bern, WSL-Institut für Schnee- und Lawinenforschung SLF, Davos, Davos, 136 p

McClung DM, Schaerer P (2006) The avalanche handbook, 3 edn. The Mountaineers Books, Seattle, 342 p

Meyer-Grass M, Schneebeli M (1992) Die Abhängigkeit der Waldlawinen vom Standorts, Bestandes- und Schneeverhältnissen. In: Schutz des Lebensraumes vor Hochwasser, Muren und Lawinen, vol 2. Interpraevent 92, pp 443–455

Miller DH (1964) Interception processes during snowstorms. U.S. Forest Service Research Paper PSW-18. Pacific Southwest Forest and Range Experiment Station Berkeley, USA, 24 p

Moeser D, Stähli M, Jonas T (2015) Improved snow interception modeling using canopy parameters derived from airborne LiDAR data. Water Resour Res 51:5041–5059

Newesly C, Tasser E, Spadinger P, Cernusca A (2000) Effects of land-use changes on snow gliding processes in alpine ecosystems. Basic Appl Ecol 1:61–67

Patten RS, Knight DH (1994) Snow avalanches and vegetation pattern in Cascade Canyon, Grand Teton National Park, Wyoming, USA. Arct Alp Res 26:35–41

Pfister R, Schneebeli M (1999) Snow accumulation on boards of different sizes and shapes. Hydrol Process 13:2345–2355

Rixen C, Brugger S (2004) Naturgefahren – ein Motor der Biodiversität. Forum Wissen 2004:67–71

Rixen C, Haag S, Kulakowski D, Bebi P (2007) Natural avalanche disturbance shapes plant diversity and species composition in the subalpine forest belt. J Veg Sci 18:735–742

Rudolf-Miklau F, Sauermoser S (2011) Handbuch Technischer Lawinenschutz. Ernst & Sohn. 464 p

Salm B (1978) Snow forces on forest plants. In: de Quervain M (ed) Proceedings of the IUFRO Seminar "Mountain forests and avalanches", September 1978. Eidgenössisches Schnee- und Lawinenforschungsinstitut Davos, pp 157–181

Schäublin S, Bollmann K (2011) Winter habitat selection and conservation of Hazel Grouse *(Bonasa bonasia)* in mountain forests. J Ornithol 152:179–192

Schmucki E, Marty C, Fierz C, Weingartner R, Lehning M (2017) Impact of climate change in Switzerland on socioeconomic snow indices. Theor Appl Climatol 127:875–889

Schneebeli M, Bebi P (2004) Snow and avalanche control. In: Burley J, Evans J, Youngquist JA (eds) Encyclopedia of forest sciences. Elsevier, Amsterdam, pp 397–402

Schneebeli M, Laternser M, Amman W (1997) Destructive snow avalanches and climate change in the Swiss Alps. Eclogae Geol Helv 90:457–461

Schönenberger W (1978) Ökologie der natürlichen Verjüngung von Fichte und Bergföhre in Lawinenzügen der nördlichen Voralpen. Mitt Eidg Anst For Versuchswes 54:215–361

Schönenberger W (2001) Cluster afforestation for creating diverse mountain forest structures – a review. For Ecol Manag 145:121–128

Schuler A (1988) Forest area and forest utilisation in the Swiss pre-alpine region. In: Salbitano F (ed) Human influences on forest ecosystems development in Europe. Pitagora Editrice, Bologna, pp 121–127

Schweizer J, Jamison B, Schneebeli M (2003) Snow avalanche formation. Rev Geophys 41:1016–1041

Teich M, Bartelt P, Grêt-Regamey A, Bebi P (2012) Snow avalanches in forested terrain: influence of forest parameters, topography, and avalanche characteristics on runout distance. Arct Alp Res 44:509–519

Teich M, Fischer J-T, Feistl T, Bebi P, Christen M, Grêt-Regamey A (2014) Computational snow avalanche simulation in forested terrain. Nat Hazard Earth Syst 14:2233–2248

Tribeck MJ, Gurney RJ, Morris EM (2006) The radiative effect of a fir canopy on a snowpack. J Hydrometeorol 7:808–895

Veblen T, Hadley KE, Nel EM, Kitzberger T, Reid M, Villalba R (1994) Disturbance regime and disturbance interaction in a Rocky Mountain subalpine forest. J Ecol 82:125–135

Wohlgemuth T, Schwitter R, Bebi P, Sutter F, Brang P (2017) Post-windthrow management in protection forests of the Swiss Alps. Eur J For Res 136:1029–1040

Part IV
Biotic Disturbances

Chapter 10
Tree Diseases

Marco Pautasso (iD)

Abstract Tree diseases are important agents of ecological disturbance in wooded ecosystems. Native tree diseases create forest gaps, thus diversifying forest structure and creating habitat for many organisms. Tree diseases are also part of biodiversity, but they have seldom been considered from a conservation biology perspective. More diverse forests are generally less susceptible to tree disease epidemics. Tree diseases can in turn modify the resilience of forests against other stresses. Because of increased plant trade and other global change drivers, exotic tree diseases are becoming more frequent. Interactions between tree diseases and other ecological disturbances need more public awareness and integration into forest simulation models.

Keywords Deadwood · Disease triangle · Ecosystem engineers · Emerging tree diseases · Forest health · Forest pathology · Interacting disturbances · Tree diversity · Tree fungal pathogens · Tree health

10.1 Introduction: Characterization of the System

Disturbances caused by tree diseases are variable in their extent, similarly to other types of disturbance – e.g. forest fires (Chap. 7), windthrow (Chap. 8), and human activity (Chaps. 14 and 15). Tree diseases can be caused by bacteria, fungi, parasitic plants (e.g. mistletoes), nematodes, oomycetes, phytoplasmas, viroids, viruses, and the combined action of these organisms (Tainter and Baker 1996). Traces of tree diseases are preserved in fossils, which document the long co-evolution between host trees and their pathogens (Labandeira and Prevec 2014). Some organisms, e.g.

The positions and opinions presented in this article are those of the authors alone and are not intended to represent the views or scientific works of EFSA.

M. Pautasso (✉)
Animal & Plants Health Unit, European Food Safety Authority (EFSA), Parma, Italy
e-mail: marpauta@gmail.com

T. Wohlgemuth et al. (eds.), *Disturbance Ecology*, Landscape Series 32,
https://doi.org/10.1007/978-3-030-98756-5_10

endophytes, are normally beneficial or neutral to trees, but they can cause disease under stressful environmental conditions (Sieber 2007). Tree disease symptoms may occur on the tree's foliage, flowers, fruits, branches, stems, and roots (Jarosz and Davelos 1995). Since microorganisms are often difficult to distinguish from each other under the microscope and the disease symptoms can resemble those of abiotic stress factors (such as drought), molecular methods are often necessary to reliably identify pathogens (Cooke et al. 2007). Trees of any age are often already infected (and can infect other trees) before symptoms become visible. Normally, most plant diseases have no substantial impact on the survival and fitness of their hosts (Roy et al. 2000). Given that ecological disturbances are defined as discrete events perturbing ecosystems, most tree diseases would thus not be classified as disturbance agents. Tree diseases become disturbance agents when they overcome the defence mechanisms of their host, eventually leading to their death.

Tree diseases act as agents of ecological disturbance at local to regional scales (Holdenrieder et al. 2004; Pautasso et al. 2015). On the one hand, native tree diseases typically create small gaps in forests, thus opening the canopy and leading to increased diversity in forest structure, safeguarding the habitat for a variety of organisms (Ostry and Laflamme 2008; Kõrkjas et al. 2021). On the other hand, exotic tree diseases may result in widespread forest disturbance, sometimes partly removing their naïve hosts from the landscape (Cobb et al. 2012; Brunet et al. 2014; Budde et al. 2016). When tree diseases result in the loss of a common and widespread tree species, fundamental ecosystem processes are disrupted, from decomposition rates to carbon sequestration and energy flows (Ellison et al. 2005; Loo 2009; Paseka et al. 2020). The distinct effects on ecosystems of native vs. exotic tree disease are a consequence of the adaptation of host trees to diseases with which they have co-evolved and the lack of defences against newly introduced tree diseases (Hansen 2008; Desprez-Loustau et al. 2016; Müller et al. 2016).

These differences in the extent of damage between native and exotic pathogens may dwindle as a result of environmental changes: when climatic conditions move outside of the range to which host trees are adapted, the ensuing tree stress can lead to increased incidence of native pathogens (Costanza et al. 2020). Conversely, invasive exotic tree diseases can sometimes result in increasing mortality rates of their host(s) without changing host abundance and the community composition (e.g. beech bark disease; Garnas et al. 2011). Shifts in climate can lead to natural range expansion of tree pathogens to new areas (Brodde et al. 2019). When tree species are planted outside of their native range, local and imported pathogens can affect them leading to new pathosystems (Schmid et al. 2014). Moreover, a long-term view of tree diseases as ecological disturbance might lead to reconsideration of the native–exotic disease dichotomy, as both native and exotic tree diseases undergo long-term cycles in abundance and virulence due to shifts in host availability, environmental conditions, and natural enemies. Such processes have also been documented, for example, for insect defoliators (Ferrenberg 2016, Chap. 11).

Naturally occurring tree pathogens typically create slowly expanding openings in the canopy of forests. Because of this mortality – which is often independent of tree vitality – pathogens are an important cause of disturbance during stand development, i.e. in the period between large-scale disturbances such as forest fires and clear-cuts. In the Swiss National Park, for example, fungal infections by *Armillaria ostoyae* and sometimes common root rot (*Heterobasidion annosum*) are the main causes of openings in the canopy in mountain pine (*Pinus mugo* L.) forests (Bendel et al. 2006a). The slow expansion of individual fungi can ultimately cover hundreds of hectares of forests (Ferguson et al. 2003). The distribution of common root rot extends from Europe to Russia; this naturally occurring tree disease leads to small-scale openings in the canopy of coniferous stands of larch (*Larix decidua* L.), Scots pine (*Pinus sylvestris* L.), Norway spruce (*Picea abies* L.), Italian stone pine (*Pinus pinea* L.), Sitka spruce (*Picea sitchensis* (Bong.) Carr.), and Douglas fir (*Pseudotsuga menziesii* (Mirb.) Franco) (Asiegbu et al. 2005; Garbelotto and Gonthier 2013). Its occurrence is promoted by thinnings (Rizzo et al. 2000). In addition, *Armillaria* spp. and bracket fungi (e.g. *Fomes fomentarius* on beech, *Phellinus hartigii* on silver fir (*Abies alba* L.) and *Climacocystis borealis* on spruce) are important for the dynamics of unmanaged forests in Central Europe. Other naturally occurring pathogens with an impact on forest dynamics in Central Europe include snow moulds (*Phacidium infestans*, *Herpotrichia juniperi*, and *Gremmeniella abietina*, the causative agent of scleroderris canker). These fungi are an obstacle to the regeneration of spruce and stone pine (*Pinus cembra* L.) in Alpine forests and afforestation at high elevations (Senn 1999; Barbeito et al. 2013).

An exotic tree disease causing widespread mortality of European ash (*Fraxinus excelsior* L.) trees throughout Europe is ash dieback. This emerging disease is caused by an introduced ascomycete, *Hymenoscyphus fraxineus*, which causes brown leaf spots, necrotic lesions along leaflet veins, rachises, stems, branches, and root collars, wilting of single shoots, and crown dieback. The fungus originates from East Asia and was introduced to Poland from where it started spreading at the beginning of the 1990s, through both wind-blown spores and trade of infected ash saplings (Gross et al. 2014). In the meantime, the disease has reached most of the distribution area of European ash (Pautasso et al. 2013). The pathogen has affected and killed European ash trees of all ages, but some mature trees tend to be able to survive for longer than younger trees (McKinney et al. 2014; Fig. 10.1). The disease is not just a threat to European ash but also for the associated biodiversity (Needham et al. 2016; Hultberg et al. 2020). Given the lack of co-evolution of host and pathogen, the observed tolerance of some ash tree individuals to the disease is an example of exaptation, i.e. the harnessing of a trait in a context for which it did not originally evolve (Landolt et al. 2016; Bartholomé et al. 2020).

Various factors affect epidemic development both for native and exotic tree diseases (Meentemeyer et al. 2012). These factors act at different levels: from individual trees (disease resistance, species identity, and tree size) to the community (microclimate, host tree density, and tree species richness) up to regions (landscape connectivity and seasonality) and over long distances (e.g. climate gradients along latitude and longitude and the long-distance spread of pathogens with the help of humans) (Dillon et al. 2014; Jules et al. 2014; Haas et al. 2016). At each spatial scale, these factors affect the three-way relationship between host, pathogen, and environment, the three aspects of the disease triangle (Fig. 10.2). For example, for

Fig. 10.1 Ash (*Fraxinus excelsior* L.) dieback as a consequence of infection by *Hymenoscyphus fraxineus* in Copenhagen, Denmark (Photo: M. Pautasso). Tree species diversity is also an insurance against emerging tree diseases in urban forests

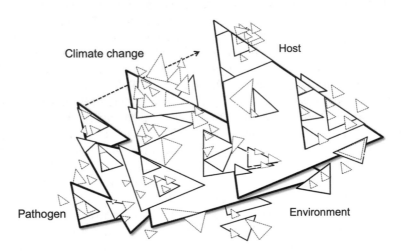

Fig. 10.2 The scale dependence of climate effects on the disease triangle of different tree diseases. The disease triangle describes the interplay between the host, pathogen, and environment. Its area is proportional to the severity of a disease. (From Jeger and Pautasso, 2008, redrawn)

tree diseases spreading along water courses (e.g. *Phytophthora alni*), the connectivity of the river networks is a key environmental factor (Pomportl et al. 2016). Long-
distance movement of susceptible plants for planting can lead to efficient dispersal
of pathogens that normally require the presence of water courses for dispersal (Jung
and Blaschke 2004). In addition, the genetic diversity of the tree population also
determines whether a disease leads to ecological disturbance or not, as it causes
local differences in the susceptibility to disease of individual stands or tree individuals (Nguyen et al. 2016).

10.2   Is Tree Diversity a Forest Adaptation to Tree Diseases?

Tree diseases have an impact on forest tree diversity, by reducing the abundance of
potentially dominating tree species, thereby creating opportunities for rare tree species to coexist in the forest (Hansen 1999; Gilbert 2002; Mordecai 2011; Bever et al.
2015). The diversity-promoting effect of tree diseases is observed when tree mortality risk increases with increasing density of neighbouring trees of the same species,
which promotes the activity of host-specific tree diseases (Janzen–Connell hypothesis) (Das et al. 2008).

In turn, the impact of tree diseases as ecological disturbance is mediated by forest tree diversity (Pautasso et al. 2005). More diverse forests are expected to reduce
the incidence and impact of tree diseases through various mechanisms (Felton et al.
2016). Compared with tree monocultures, in diverse forests the individuals of each
tree species grow at a lower population density, thus reducing host availability for
inoculum production of specialized pathogens (Hantsch et al. 2013). Moreover,
single individuals of different tree species tend to be separated from each other by
tree individuals of other species, thus reducing the chances of dispersal of pathogen
propagules (Gerlach et al. 1997; Hantsch et al. 2014). Forest trees often have considerable genetic diversity both within and among populations, thus increasing the
ability of tree species to adapt to new environmental challenges, including exotic
tree diseases (Elvira-Recuenco et al. 2014). Reduced diversity leads to increased
habitat connectivity for pathogens, which in combination with stress increases the
likelihood of disease-related tree mortality (Laćan and McBride 2008). Increased
host diversity within the landscape, on the other hand, increases the threshold at
which epidemics can develop (Papaïx et al. 2014; Rodewald and Arcese 2016).
There is evidence that dilution effects can provide protection not just against host-
specific tree diseases but also against generalist pathogens (e.g. *Phytophthora ramorum*) (Haas et al. 2011; Fig. 10.3). Tree diversity could thus be considered as an
adaptation at the forest community level to keep tree diseases in check.

The insurance hypothesis of tree diversity as a protection against tree diseases
still considers tree diseases as something a forest should be protected from. However,
tree diseases are part of biodiversity (Pusz and Urbaniak 2017). They play an important ecological role in forest ecosystems, and they therefore deserve to be considered by conservation biologists (Ingram 1999, 2002). As a rule, more diverse

Fig. 10.3 Tanoak (*Notholithocarpus densiflorus* (Hook. & Arn.) Manos, Cannon & S.H.Oh) mortality caused by the oomycete *Phytophthora ramorum* in Sonoma County, California, USA. (Source: Wikimedia Commons)

ecosystems also tend to harbour more diseases (as has been shown in grasslands by Rottstock et al. 2014). If every tree species in the world harboured ten species-specific pathogens, the diversity of tree pathogens would be an order of magnitude higher than that of trees. Many important tree diseases are not host-specific, e.g. *Erwinia amylovora*, the bacterium causing fire blight (Schroth et al. 1974); *Phytophthora cinnamomi*, the oomycete causing the jarrah forest dieback in southern Western Australia (Podger 1972); the rust fungus *Puccinia psidii* on a variety of species of the Myrtaceae family (Morin et al. 2012); and *Armillaria* root disease on many conifer and broadleaved tree species (Prospero et al. 2003; Edmonds 2013). However, there are also many tree diseases specific to a certain host, e.g. foliar diseases of broadleaved trees (Kowalski 2013). Nearly 60% of the tree fungal pathogens reported in French forests were found to affect one or two host taxa only (Vacher et al. 2008). In various cases, pathogens previously believed to be generalists were then recognized using molecular techniques to be a complex of different species, each specialized on a single or a limited group of host tree species, e.g. *Heterobasidion* root rot (Garbelotto and Gonthier 2013). This is a reminder of the still evolving knowledge about tree pathogens and that many cryptic pathogens are not yet identified. Answering the question of whether a plant serves as a host or not is also crucial for the development of plant protection strategies. For example, it was long believed that *Cronartium ribicola* completed its life cycle on two main hosts – (i) five-needled pines (aecial host) and (ii) currants (*Ribes* spp.) (telial host) – but various other plant species were then shown to be additional telial hosts of *C. ribicola* (Hamelin 2013).

Biogeographical research on environmental gradients in the richness of tree fungal pathogens is still in its infancy (Fukasawa and Matsuoka 2015; Khaliq et al 2021; Meyer et al. 2021; Prada-Salcedo et al. 2021). Indeed, most biodiversity conservation research has focused on high-profile, yet relatively species-poor groups such as vertebrates, plants, and some groups of insects, often neglecting fungi (Lonsdale et al. 2008; Heilmann-Clausen et al. 2015). Even the fungal conservation biology literature is more concerned with saproxylic fungi than with tree fungal pathogens, although tree diseases are an important provider of deadwood (Jönsson et al. 2017). A forest without a sufficient amount of tree diseases is thus today recognized as not entirely healthy (Ostry and Laflamme 2008; Holdenrieder and Pautasso 2014; Szwagrzyk 2020).

10.3   Tree Diseases and Biodiversity

As with human and animal diseases, tree death is not the inevitable outcome of tree diseases. Only a few tree diseases cause major harm to tree health (Boyd et al. 2013). Trees have developed a variety of ways to cope with many pathogens, e.g. by compartmentalizing stem infection, building new shoots after dieback, replacing lost leaves, and developing resistance over several generations. When there is little host resistance to tree diseases (whether they are exotic or not), they can lead to reduction in ecosystem service provision because of the mortality of a high proportion of the individuals of the susceptible tree species (e.g. white pine blister rust, Tomback et al. 2016; Dutch elm disease, Freer-Smith and Webber 2017; Fig. 10.4). In some cases, the human reaction to tree diseases causes more damage than the disease itself (Pezzi et al. 2011). Pesticides are no longer used in the forests of Central Europe for ecological and economic reasons. Because of the development of resistance of the pathogens and the side effects on other organisms, such use is not justifiable.

Traditionally, tree diseases have been negatively regarded by foresters because they lead to a reduction in the forest resources available for human needs (Anderson 2003; Manion 2003). Even recently, it was possible to read that "root diseases kill trees, decay wood, slow tree growth, predispose trees to other mortality agents, and cause trees to fail or fall over. In this manner, these diseases reduce timber volume, alter forest composition and structure, impair ecosystem function, and decrease carbon sequestration" (Lockman and Kearns 2016). This is a rather one-sided view of the role of tree diseases in forest ecosystems (Laflamme 2010). Similar to the role of top-down predators in food chains, tree diseases can be effective tools for diversification of uniform forests (Castello et al. 1995). Timber volume of tree plantations might be slightly reduced in the short term, but increased structural and genetic diversity caused by tree disease outbreaks can actually make (planted) forests more resilient against other stresses, e.g. environmental change (Kuparinen et al. 2010; Martin et al. 2015). Resilience and biodiversity – in addition to timber production – are frequently among the goals of forest management today. There is increasing

Fig. 10.4 Dutch elm disease as a consequence of infection by *Ophiostoma novo-ulmi* in Assisi, Italy (Photo: M. Pautasso). Elms have largely disappeared from European landscapes because of elm dieback; this has consequences for biodiversity and ecosystem services

recognition that tree death caused by tree diseases is an important ecological process that is a fundamental part of healthy and biodiverse forest ecosystems (Franklin et al. 1987; Haack and Byler 1993; Harmon 2001). Moreover, more diverse forests have been repeatedly shown to be more productive (Paquette and Messier 2011; Dymond et al. 2014), which also has a positive effect on the provision of forest services.

Tree diseases are ecosystem engineers, i.e. they modify their environment and create habitat for a variety of other organisms (Steeger and Hitchcock 1998). Together with storms and outbreaks of insects, tree diseases increase the amount of coarse and fine woody debris in managed forest ecosystems (Lundquist 2007). The deadwood provision of tree diseases has been demonstrated, for example, in unmanaged stands of the mountain pine (*Pinus mugo* Turra) in the Swiss National Park (Bendel et al. 2006b). Host-specific tree diseases differ from abiotic forest disturbances inasmuch as they selectively affect a single tree component of the forest – e.g. ash (*Fraxinus excelsior*) in the case of ash dieback (Broome et al. 2019), sweet chestnut (*Castanea sativa* Mill.) in the case of chestnut blight (causal organism *Cryphonectria parasitica*) (Rigling and Prospero 2018), and oriental plane (*Platanus orientalis* L.) in the case of plane wilt (causal organism *Ceratocystis platani*) (Tsopelas et al. 2017). Also forest fires (Chap. 7) and storms (Chap. 8) can have differing impacts on different tree species, but their consequences are not as species-specific as is the case for host-specific tree diseases. It follows that host-specific tree diseases, on their own, are not an ideal tool for deadwood provision, because they

will tend to provide deadwood only of the particular species they affect. For deadwood to provide habitat to the diversity of organisms dependent on it, there needs to
be spatial and temporal continuity in the availability of deadwood of a diversity of
tree species (Lassauce et al. 2011). Therefore, to increase deadwood resources in
forests, naturally occurring generalists, such as root pathogens of the genus
Armillaria, are more suitable, as they cover a broad host range. Tree diseases are
thus just one tool for deadwood restoration in managed forests, which needs to be
combined with other strategies, i.e. leaving deadwood in the forest after fires, logging, and storms (Hekkala et al. 2016).

10.4   Drivers of Emerging Tree Diseases

The risk of introducing exotic tree pathogens is on the rise, because of the increased
trade in plant material between regions and continents, shifts in climate, and other
global change drivers (Ayres and Lombardero 2000; Pautasso et al. 2010; Santini
et al. 2013). Key pathways leading to introduction of exotic tree pathogens include
wood, packing materials made from wood, soil and growing media, as well as plants
for planting (Brasier 2008; Tobin 2015; Fig. 10.5). If plants for planting, wood, and
wood packaging material were exclusively sourced locally, as proposed by the
Montesclaros Declaration (Klapwijk et al. 2016; Ayres and Lombardero 2018), the
risk of inadvertently introducing new tree diseases would be considerably reduced
(Liebhold et al. 2012; Hantula et al. 2014). The trend over recent decades has been

Fig. 10.5 Trees for planting are one key pathway for introduction of emerging tree diseases. This
photo shows trees recently planted in Copenhagen, Denmark. (Photo: M. Pautasso)

to obtain apparently cheaper material from faraway regions, with higher costs in the long term owing to the reduction in ecosystem services and the cost of eradication and containment activities following outbreaks of new tree diseases (Dehnen-Schmutz et al. 2010; Stenlid et al. 2011; Luvisi et al. 2017; Zadoks 2017).

The increase in new introductions of tree pathogens is caused not just by increased trade volumes and distances but also by changes in the structure of trade networks (Moslonka-Lefebvre et al. 2011). Research has shown that epidemics are more likely in the presence of hubs, i.e. trade players with many more connections than the average plant nursery or garden centre (Sutrave et al. 2012). Moreover, as plant trade connectivity is inherently directed (i.e. from nurseries to wholesalers, garden centres, and consumers), the risk of epidemics is magnified in the presence of hubs with a high number of both incoming and outgoing connections (Shaw and Pautasso 2014). The structure of plant trade networks is poorly characterized, owing to low availability of data because of commercial sensitivity (Eschen et al. 2015a, b). In the future, an improvement of the availability of plant trade data could contribute to reducing the introduction of pathogens despite the increasing trade volume (Pautasso and Jeger 2014). Such prevention is difficult when infestation levels are high. For instance, about 90% of over 700 surveyed European plant nurseries were found to harbour at least one species of *Phytophthora* (Jung et al. 2016), and new *Phytophthora* species are regularly discovered in nursery surveys (Bregant et al. 2021).

It is commonly assumed that introduced tree pathogens are limited to similar climatic conditions as in their areas of origin (Venette and Cohen 2006; Ireland et al. 2013). This may be a rather conservative assumption as it does not consider the evolutionary potential of microorganisms (La Porta et al. 2008; Santini and Ghelardini 2015; Lion and Gandon 2016). Indeed, it has been shown that introduced alien species can also thrive under climates not found in their home range (Camenen et al. 2016; Boiffin et al. 2017). This increases the uncertainty when assessing the risk posed by introduced tree diseases under current climates. An even more challenging task is to predict how known and potential forest pathosystems will develop under rapidly shifting climate conditions (Shaw and Osborne 2011; Sturrock 2012; Bebber 2015).

10.5 Climate Change and Tree Diseases

Climate is an essential factor shaping the distribution, abundance, and severity of tree diseases across forest ecosystems. Each tree disease requires a certain combination of humidity and temperature conditions to be able to infect their host tree(s). At the same time, each host tree is adapted to a certain range in climatic conditions. Stress caused by novel climatic conditions can lead to increased host susceptibility to tree diseases (Ghelardini et al. 2016). Because tree pathogens are able to adapt to novel climatic conditions more rapidly than their hosts, the overall expectation is for an increase in the incidence and severity of tree diseases under shifting climatic conditions (Sturrock et al. 2011). Many tree pathogens of temperate and boreal forests are currently limited by low winter temperatures (Desprez-Loustau et al. 2007). Climate warming could thus lead to a range expansion of tree pathogens into as yet unaffected areas (Haughian et al. 2012; Fig. 10.6).

Given the complex array of interactions between environmental factors, host susceptibility, and disease expression, it is often difficult to prove that observed climate shifts have caused a certain tree disease outbreak (Hennon et al. 2020). Long-term, standardized, regional data about tree disease incidence are thus needed (Jeger and Pautasso 2008; Barrett and Pattison 2017). Recent climate change has been shown to have facilitated increased outbreaks of *Diplodia* shoot blight on pine trees in France, a disease favoured by summer rain and mild winter temperatures (Fabre et al. 2011). Similarly, the recent increase in the severity of *Dothistroma* needle blight and its expansion into new areas have been linked to the warmer and wetter weather conditions associated with El Niño–Southern Oscillation (ENSO) events (Woods et al. 2016).

Predictions about the effects of future climate change on tree pathosystems are complicated by the still considerable uncertainty about how precipitation patterns will shift in particular regions (Thompson et al. 2014). For the many tree pathogens whose life cycle necessitates moist conditions, rates of reproduction, spread, and infection tend to be greater when conditions are moist rather than dry (Kolb et al. 2016). Yet, secondary tree diseases can be favoured by drought, when this makes trees more susceptible to insects and pathogens (Desprez-Loustau et al. 2006).

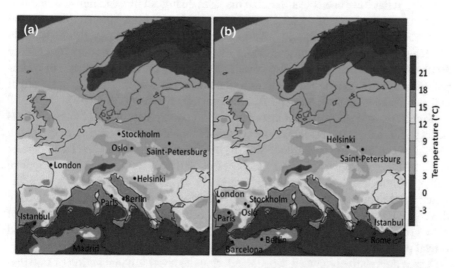

Fig. 10.6 Background mean temperature ranges of Europe (colours; 1961–1990) and location of cities in places that have their predicted temperature patterns for the end of the twenty-first century according to two climate models in an 'A2' global warming scenario (Source: West et al. 2012). The black dots on the map show the 'analogue cities' of some well-known European cities. An 'analogue' to a city A is a city B whose climate for the period 1961–1990 represents A's simulated future climate. (Kopf et al. 2008)

Weather extremes such as heavy rainfall and flooding can also influence the occurrence and severity of tree diseases (Jung 2009; Hubbart et al. 2016).

Regardless of future changes to the average climate and the frequency and degree of extreme events, most tree diseases only occur in part of their potential distribution. For example, Burgess et al. (2017) found that *Phytophthora cinnamomi* is only found in part of the area to which it is climatically suited. Given the observed and expected trends in plant trade and climate change across the world, it can be reasonably expected that many tree pathogens will be introduced into new areas over the coming decades (Ramsfield et al. 2016).

10.6   Interactions of Tree Diseases with Other Forest Disturbances

The combination of climate change and long-distance trade of live plants (including bonsai), wood, and other plant products poses a much more substantial threat to forest health through tree diseases than either of these two global change drivers by themselves (Stenlid and Oliva 2016). In addition, tree diseases interact with other forest disturbances, further complicating prediction and management (Keča et al. 2016; Cobb and Metz 2017).

In 1995, Castello et al. mentioned that the interaction between tree diseases and abiotic disturbances in the context of forest succession had received minimal attention. Yet, it has long been clear that tree diseases do not act in a vacuum and that tree health is the outcome of a complicated array of interactions between factors that may lead to trees succumbing to diseases or that may prevent tree death (Manion 1981; Pronos et al. 1999; Jönsson 2006; Fig. 10.7). It is thus important to consider feedback loops when studying tree diseases as ecological disturbances. For example, tree diseases can predispose trees to damage arising from drought, but drought can also make trees more susceptible to tree diseases (Oliva et al. 2014).

Similarly, tree diseases can also lead to the occurrence of other forest disturbances, e.g. increased availability of flammable material following tree and shrub mortality can lead to more frequent fires, as shown for Sudden Oak Death (Metz et al. 2011; Forrestel et al. 2015; Cobb et al. 2016). Forest fire can also induce pathogen infection and insect attack (Parker et al. 2006). Silviculture is an additional issue to be considered (see Chap. 14): in silver fir (*Abies alba* Mill.) stands in the central-western Spanish Pyrenees, it has been suggested that the retention of slow-growing trees might have made the forest more vulnerable to drought and fungal pathogens (Sangüesa-Barreda et al. 2015).

There are relatively few studies considering the relative contribution of tree diseases and of other forest disturbances to tree mortality rates (Jules et al. 2016; Kirschbaum et al. 2016; Kautz et al. 2017). In old-growth mountain forests in California's Sierra Nevada, insects and pathogens were the most important cause of mortality, particularly for large trees (Das et al. 2016). Often, given the interactions

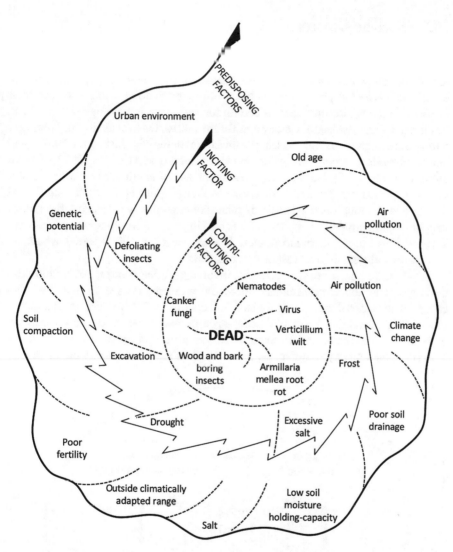

Fig. 10.7 Tree decline predisposing, inciting, and contributing factors. Pathogens rarely act alone but often interact in complex ways in combination with other disruptive factors. (Source: Manion 1981, redrawn)

between disturbances, it is difficult to attribute tree mortality to a single cause (Mulvey and Bisbing 2016). In Valais, Switzerland, blue-stain fungi contribute to pine decline, but this is primarily a consequence of drought in combination with stand competition and mistletoe, nematode, and insect infestations (Heiniger et al. 2011). Also, the role of various environmental factors interacting with a certain tree disease may vary among regions. Oak death, for example, is often attributed to a complex disease, with regional differences in the effect of individual factors of the complex (Thomas et al. 2002; Lynch et al. 2014).

10.7   Societal Aspects

Tree diseases are subtle but powerful agents of disturbance in forest ecosystems throughout the world. Once exotic tree diseases have been introduced and become established and widespread, there is often little we can do to avoid or reduce their impacts – although in some cases breeding for resistance and biological control can be effective (Pautasso 2013; Ganley and Bulman 2016; Oliva et al. 2016). Trees take a long time to grow, but new diseases can kill them rapidly. Just as for human and animal diseases, prevention is better than cure (Hepting and Cowling 1977; Howard 1996; Pautasso 2020). But how can new outbreaks of exotic tree diseases be prevented if people are generally not aware of them (Fuller et al. 2016) and do not realize that the long-distance trade of plant commodities is increasing the risk of new tree health problems (Fig. 10.8)? It is striking that even among tree professionals in various European countries, there is a relatively low self-reported awareness of a number of tree diseases (Marzano et al. 2016).

Information campaigns, for example, involving garden centres, could help raise awareness and educate consumers about the problems caused by paradoxically cheap, plant material from other regions and continents. Information availability is indeed a key condition for public acceptance of deadwood in forests (Gundersen et al. 2016; Thorn et al. 2020). Such initiatives would probably be more effective than strengthening phytosanitary regulations, given that these mostly target already

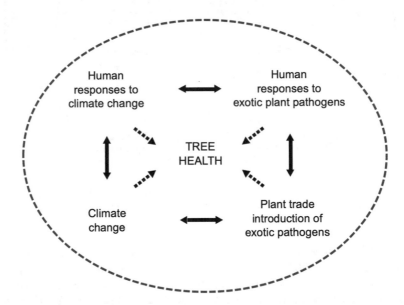

Fig. 10.8 Tree health in a changing world will be the outcome of the effects of climate change and emerging tree diseases on tree health, but it will also be influenced by how people will react to both climate change and emerging tree diseases. Socio-economic considerations must therefore be taken into account in the modelling and risk assessment of tree diseases. (Modified from Pautasso and Jeger 2014)

known threats, whereas the main danger is posed by unknown pathogens described only after their introduction (Roy et al. 2014). This is not to dismiss the importance of phytosanitary laws (Lovett et al. 2016). For example, the current EU regulations were estimated to reduce exposure of European oaks (*Quercus* spp.) to *Ceratocystis fagacearum* (a destructive pathogen which could be potentially introduced from North America through the wood trade) by a factor of 30,000 compared with a scenario without regulation (Robinet et al. 2016). Further use of pathway models to assess the effectiveness of control measures to reduce the risk posed by different pathways of entry for quarantine tree pathogens (Douma et al. 2016) needs to be accompanied by information campaigns to increase public awareness of such risks (Marzano et al. 2015).

An issue which requires attention by both phytosanitary regulators and the public is the need to maintain quarantine regulations also for exotic tree pathogens which are already widespread in the introduced range. This is because further introductions of strains of those pathogens likely increase their genetic diversity in the introduced range, thus potentially leading to enhanced virulence (Landolt et al. 2016). One problem with phytosanitary regulation is that administrative boundaries are often arbitrary and do not correspond to meaningful boundaries from an epidemiological point of view (Thompson et al. 2016). There is a need for cooperation across national borders, because early control measures are more effective than those that are delayed (Cunniffe et al. 2016; Evans et al. 2020).

10.8 Conclusions

Tree diseases are important, yet relatively neglected, agents of disturbance in forest ecosystems (Dinoor and Eshed 1984; Dobson and Crawley 1994; Alexander and Holt 1998). Naturally occurring tree diseases contribute to forest biodiversity by constantly generating deadwood and thus promoting variability in forest structures and habitat creation. Novel tree diseases can become problematic for entire ecosystems due to the lack of co-evolution of host and parasite (Heiniger 2003). Tree diversity (from the genetic to the community level) can in some cases also provide an insurance against exotic tree pathogens. The differences in impact between native and exotic tree pathogens are becoming increasingly blurred owing to ongoing climate change. In both cases, prediction and management of future tree pathosystems are made difficult by the interactions of tree diseases, other disturbances, and global change drivers.

Tree diversity experiments and surveys have delivered useful insights on the interconnectedness of forest diversity and tree diseases (Setiawan et al. 2014; Verheyen et al. 2016; von Gadow et al. 2016). A similarly fruitful approach is likely to be the use of botanic gardens as sentinel networks for early warning of potentially threatening exotic tree diseases (Sieber 2014; Vettraino et al. 2017, Noar et al. 2021). The increasing knowledge available on exotic tree diseases should be periodically synthesized by systematic reviews and made available for research and practice. Models of future forest development have become more realistic over the last decades, but the interactions of global change drivers, public awareness, and tree diseases are still rarely included in such endeavours (Whyte et al. 2016; Tonini et al. 2018; Honkaniemi et al. 2021). Together with the integration of evolutionary ecology considerations in the study and management of tree diseases (Jarosz and Davelos 1995; Alexander 2010; Burdon et al. 2013; Desprez-Loustau et al. 2016; Landolt et al. 2016), the modelling and communication of tree diseases as ecological disturbances are an important field of research for the future.

Acknowledgements Many thanks to O. Holdenrieder, A. Jentsch, R. Seidl, and T. Wohlgemuth for their helpful comments on a previous draft.

References

Alexander HM (2010) Disease in natural plant populations, communities, and ecosystems: insights into ecological and evolutionary processes. Plant Dis 94:492–503

Alexander HM, Holt RD (1998) The interaction between plant competition and disease. Perspect Plant Ecol 1:206–220

Anderson RL (2003) Changing forests and forest management policy in relation to dealing with forest diseases. Phytopathology 93:1041–1043

Asiegbu FO, Adomas A, Stenlid J (2005) Conifer root and butt rot caused by *Heterobasidion annosum* (Fr.) Bref. s.l. Mol Plant Pathol 6:395–409

Ayres MP, Lombardero MJ (2000) Assessing the consequences of global change for forest disturbance from herbivores and pathogens. Sci Total Environ 262:263–286

Ayres MP, Lombardero MJ (2018) Forest pests and their management in the Anthropocene. Can J For Res 48:292–301

Barbeito I, Brücker RL, Rixen C, Bebi P (2013) Snow fungi-induced mortality of *Pinus cembra* at the alpine treeline: evidence from plantations. Arct Antarct Alp Res 45:455–470

Barrett TM, Pattison RR (2017) No evidence of recent (1995–2013) decrease of yellow-cedar in Alaska. Can J For Res 47:97–105

Bartholomé J, Brachi B, Marçais B, Mougou-Hamdane A, Bodénès C, Plomion C, Robin C, Desprez-Loustau ML (2020) The genetics of exapted resistance to two exotic pathogens in pedunculate oak. New Phytol 226:1088–1103

Bebber DP (2015) Range-expanding pests and pathogens in a warming world. Annu Rev Phytopathol 53:335–356

Bendel M, Kienast F, Bugmann H, Rigling D (2006a) Incidence and distribution of *Heterobasidion* and *Armillaria* and their influence on canopy gap formation in unmanaged mountain pine forests in the Swiss Alps. Eur J Plant Pathol 116:85–93

Bendel M, Kienast F, Rigling D, Bugmann H (2006b) Impact of root-rot pathogens on forest succession in unmanaged *Pinus mugo* stands in the Central Alps. Can J For Res 36:2666–2674

Bever JD, Mangan SA, Alexander HM (2015) Maintenance of plant species diversity by pathogens. Ann Rev Ecol Syst 46:305–325

Boiffin J, Badeau V, Breda N (2017) Species distribution models may misdirect assisted migration: insights from the introduction of Douglas-fir to Europe. Ecol Appl 27:446–457

Boyd IL, Freer-Smith PH, Gilligan CA, Godfray HCJ (2013) The consequence of tree pests and diseases for ecosystem services. Science 342:823–831

Brasier CM (2008) The biosecurity threat to the UK and global environment from international trade in plants. Plant Pathol 57:792–808

Bregant C, Mulas AA, Rossetto G, Deidda A, Maddau L, Piras G, Linaldeddu BT (2021) *Phytophthora mediterranea* sp. nov., a new species closely related to *Phytophthora cinnamomi* from nursery plants of Myrtus communis in Italy. Forests 12(6),682

Brodde L, Adamson K, Camarero JJ, Castaño C, Drenkhan R, Lehtijärvi A, Luchi N, Migliorini D, Sánchez-Miranda Á, Stenlid J, Özdağ Ş, Oliva J (2019) Diplodia tip blight on its way to the north: drivers of disease emergence in northern Europe. Front Plant Sci 9:1818

Broome A, Ray D, Mitchell R, Harmer R (2019) Responding to ash dieback (*Hymenoscyphus fraxineus*) in the UK: woodland composition and replacement tree species. Forestry 92(1):108–119

Brunet J, Bukina Y, Hedwall PO, Holmstrom E, von Oheimb G (2014) Pathogen induced disturbance and succession in temperate forests: evidence from a 100-year data set in Southern Sweden. Basic Appl Ecol 15:114–121

Budde KB, Nielsen LR, Ravn HP, Kjr ED (2016) The natural evolutionary potential of tree populations to cope with newly introduced pests and pathogens: lessons learned from forest health catastrophes in recent decades. Curr For Rep 2:18–29

Burdon JJ, Thrall PH, Ericson L (2013) Genes, communities & invasive species: understanding the ecological and evolutionary dynamics of host-pathogen interactions. Curr Opin Plant Biol 16:400–405

Burgess TI, Scott JK, Mcdougall KL, Stukely MJC, Crane C, Dunstan WA, Brigg F, Andjic V, White D, Rudman T, Arentz F, Ota N, Hardy GES (2017) Current and projected global distribution of *Phytophthora cinnamomi*, one of the world's worst plant pathogens. Glob Change Biol 23:1661–1674

Camenen E, Porté AJ, Benito Garzón M (2016) American trees shift their niches when invading Western Europe: evaluating invasion risks in a changing climate. Ecol Evol 6:7263–7275

Castello JD, Leopold DJ, Smallidge PJ (1995) Pathogens, patterns, and processes in forest ecosystems. Bioscience 45:16–24

Cobb RC, Metz MR (2017) Tree diseases as a cause and consequence of interacting forest disturbances. Forests 8:147

Cobb RC, Filipe JAN, Meentemeyer RK, Gilligan CA, Rizzo DM (2012) Ecosystem transformation by emerging infectious disease: loss of large tanoak from California forests. J Ecol 100:712–722

Cobb RC, Meentemeyer RK, Rizzo DM (2016) Wildfire and forest disease interaction lead to greater loss of soil nutrients and carbon. Oecologia 182:265–276

Cooke DEL, Schena L, Cacciola SO (2007) Tools to detect, identify and monitor *Phytophthora* species in natural ecosystems. J Plant Pathol 89:13–28

Costanza KK, Livingston WH, Fraver S, Munck IA (2020) Dendrochronological analyses and whole-tree dissections reveal Caliciopsis canker (*Caliciopsis pinea*) damage associated with the declining growth and climatic stressors of eastern white pine (*Pinus strobus*). Forests 11:347

Cunniffe NJ, Cobb RC, Meentemeyer RK, Rizzo DM, Gilligan CA (2016) Modeling when, where, and how to manage a forest epidemic, motivated by sudden oak death in California. Proc Natl Acad Sci USA 113:5640–5645

Das A, Battles J, van Mantgem PJ, Stephenson NL (2008) Spatial elements of mortality risk in old-growth forests. Ecology 89:1744–1756

Das AJ, Stephenson NL, Davis KP (2016) Why do trees die? Characterizing the drivers of background tree mortality. Ecology 97:2616–2627

Dehnen-Schmutz K, Holdenrieder O, Jeger MJ, Pautasso M (2010) Structural change in the international horticultural industry: some implications for plant health. Sci Hortic 125:1–15

Desprez-Loustau ML, Marçais B, Nageleisen LM, Piou D, Vannini A (2006) Interactive effects of drought and pathogens in forest trees. Ann For Sci 63:597–612

Desprez-Loustau ML, Robin C, Reynaud G, Déqué M, Badeau V, Piou D, Husson C, Marçais B (2007) Simulating the effects of a climate-change scenario on the geographical range and activity of forest-pathogenic fungi. Can J Plant Pathol 29:101–120

Desprez-Loustau ML, Aguayo J, Dutech C, Hayden KJ, Husson C, Jakushkin B, Marçais B, Piou D, Robin C, Vacher C (2016) An evolutionary ecology perspective to address forest pathology challenges of today and tomorrow. Ann For Sci 73:45–67

Dillon WW, Haas SE, Rizzo DM, Meentemeyer RK (2014) Perspectives of spatial scale in a wildland forest epidemic. Eur J Plant Pathol 138:449–465

Dinoor A, Eshed N (1984) The role and importance of pathogens in natural plant-communities. Annu Rev Phytopathol 22:443–466

Dobson A, Crawley W (1994) Pathogens and the structure of plant-communities. Trends Ecol Evol 9:393–398

Douma JC, Pautasso M, Venette RC, Robinet C, Hemerik L, Mourits MCM, Schans J, van der Werf W (2016) Pathway models for analysing and managing the introduction of alien plant pests: an overview and categorization. Ecol Model 339:58–67

Dymond CC, Tedder S, Spittlehouse DL, Raymer B, Hopkins K, McCallion K, Sandland J (2014) Diversifying managed forests to increase resilience. Can J For Res 44:1196–1205

Edmonds RL (2013) General strategies of forest disease management. In: Gonthier P, Nicolotti G (eds) Infectious forest diseases. CABI, Wallingford, pp 29–49

Ellison AM, Bank MS, Clinton BD, Colburn EA, Elliott K, Ford CR, Foster DR, Kloeppel BD, Knoepp JD, Lovett GM, Mohan J, Orwig DA, Rodenhouse NL, Sobczak WV, Stinson KA, Stone JK, Swan CM, Thompson J, Von Holle B, Webster JR (2005) Loss of foundation species: consequences for the structure and dynamics of forested ecosystems. Front Ecol Environ 3:479–486

Elvira-Recuenco M, Iturritxa E, Majada J, Alia R, Raposo R (2014) Adaptive potential of maritime pine (*Pinus pinaster*) populations to the emerging pitch canker pathogen, *Fusarium circinatum*. PLoS One 9:e114971

Eschen R, Britton K, Brockerhoff E, Burgess T, Dalley V, Epanchin-Niell RS, Gupta K, Hardy G, Huang Y, Kenis M, Kimani E, Li HM, Olsen S, Ormrod R, Otieno W, Sadof C, Tadeu E, Theyse M (2015a) International variation in phytosanitary legislation and regulations governing importation of plants for planting. Environ Sci Pol 51:228–237

Eschen R, Grégoire JC, Hengeveld GM, de Hoop BM, Rigaux L, Potting RPJ (2015b) Trade patterns of the tree nursery industry in Europe and changes following findings of citrus longhorn beetle, *Anoplophora chinensis* Forster. NeoBiota 26:1–20

Evans KJ, Scott JB, Barry KM (2020) Pathogen incursions – integrating technical expertise in a socio-political context. Plant Dis 104:3097–3109

Fabre B, Piou D, Desprez-Loustau ML, Marçais B (2011) Can the emergence of pine *Diplodia* shoot blight in France be explained by changes in pathogen pressure linked to climate change? Glob Change Biol 17:3218–3227

Felton A, Nilsson U, Sonesson J, Felton AM, Roberge JM, Ranius T, Ahlström M, Bergh J, Björkman C, Boberg J, Drössler L, Fahlvik N, Gong P, Holmstrom E, Keskitalo ECH, Klapwijk MJ, Laudon H, Lundmark T, Niklasson M, Nordin A, Pettersson M, Stenlid J, Stens A, Wallertz K (2016) Replacing monocultures with mixed-species stands: ecosystem service implications of two production forest alternatives in Sweden. Ambio 45:S124–S139

Ferguson BA, Dreisbach TA, Parks CG, Filip GM, Schmitt CL (2003) Coarse-scale population structure of pathogenic *Armillaria* species in a mixed-conifer forest in the Blue Mountains of Northeast Oregon. Can J For Res 33:612–623

Ferrenberg S (2016) Landscape features and processes influencing forest pest dynamics. Curr Landsc Ecol Rep 1:19–29

Forrestel AB, Ramage BS, Moody T, Moritz MA, Stephens SL (2015) Disease, fuels and potential fire behavior: impacts of Sudden Oak Death in two coastal California forest types. For Ecol Manag 348:23–30

Franklin JF, Shugart HH, Harmon ME (1987) Tree death as an ecological process. Bioscience 37:550–556

Freer-Smith PH, Webber JF (2017) Tree pests and diseases: the threat to biodiversity and the delivery of ecosystem services. Biodivers Conserv 26:3167–3181

Fukasawa Y, Matsuoka S (2015) Communities of wood-inhabiting fungi in dead pine logs along a geographical gradient in Japan. Fungal Ecol 18:75–82

Fuller L, Marzano M, Peace A, Quine CP, Dandy N (2016) Public acceptance of tree health management: results of a national survey in the UK. Environ Sci Pol 59:18–25

Ga...... M, Duli...... L.... (2016) D........ ....
....s. Plant Pathol 65:1047–1055

Garbelotto M, Gonthier P (2013) Biology, epidemiology, and control of *Heterobasidion* species worldwide. Annu Rev Phytopathol 51:39–59

Garnas JR, Ayres MP, Liebhold AM, Evans C (2011) Subcontinental impacts of an invasive tree disease on forest structure and dynamics. J Ecol 99:532–541

Gerlach JP, Reich PB, Puettmann K, Baker T (1997) Species, diversity, and density affect tree seedling mortality from *Armillaria* root rot. Can J For Res 27:1509–1512

Ghelardini L, Pepori AL, Luchi N, Capretti P, Santini A (2016) Drivers of emerging fungal diseases of forest trees. For Ecol Manag 381:235–246

Gilbert GS (2002) Evolutionary ecology of plant diseases in natural ecosystems. Annu Rev Phytopathol 40:13–43

Gross A, Holdenrieder O, Pautasso M, Queloz V, Sieber TN (2014) *Hymenoscyphus pseudoalbidus*, the causal agent of European ash dieback. Mol Plant Pathol 15:5–21

Gundersen V, Stange EE, Kaltenborn BP (2016) Public visual preferences for dead wood in natural boreal forests: the effects of added information. Landsc Urban Plan 158:12–24

Haack RA, Byler JW (1993) Insects and pathogens. Regulators of forest ecosystems. J For 91:32–37

Haas SE, Hooten MB, Rizzo DM, Meentemeyer RK (2011) Forest species diversity reduces disease risk in a generalist plant pathogen invasion. Ecol Lett 14:1108–1116

Haas SE, Cushman JH, Dillon WW, Rank NE, Rizzo DM, Meentemeyer RK (2016) Effects of individual, community, and landscape drivers on the dynamics of a wildland forest epidemic. Ecology 97:649–660

Hamelin RC (2013) Tree rusts. In: Gonthier P, Nicolotti G (eds) Infectious forest diseases. CABI, Wallingford, pp 547–566

Hansen EM (1999) Disease and diversity in forest ecosystems. Aust Plant Path 28:313–319

Hansen EM (2008) Alien forest pathogens: *Phytophthora* species are changing world forests. Boreal Environ Res 13:33–41

Hantsch L, Braun U, Scherer-Lorenzen M, Bruelheide H (2013) Species richness and species identity effects on occurrence of foliar fungal pathogens in a tree diversity experiment. Ecosphere 4:81

Hantsch L, Bien S, Radatz S, Braun U, Auge H, Bruelheide H (2014) Tree diversity and the role of non-host neighbour tree species in reducing fungal pathogen infestation. J Ecol 102:1673–1687

Hantula J, Müller MM, Uusivuori J (2014) International plant trade associated risks: laissez-faire or novel solutions. Environ Sci Pol 37:158–160

Harmon ME (2001) Moving towards a new paradigm for woody detritus management. Ecol Bull 49:269–278

Haughian SR, Burton PJ, Taylor SW, Curry C (2012) Expected effects of climate change on forest disturbance regimes in British Columbia. J Ecosyst Manag 13:1–24

Heilmann-Clausen J, Barron ES, Boddy L, Dahlberg A, Griffith GW, Norden J, Ovaskainen O, Perini C, Senn-Irlet B, Halme P (2015) A fungal perspective on conservation biology. Conserv Biol 29:61–68

Heiniger U (2003) Das Risiko eingeschleppter Krankheiten für die Waldbäume. Schweiz Z Forstwes 154:410–414

Heiniger U, Theile F, Rigling A, Rigling D (2011) Blue-stain infections in roots, stems and branches of declining *Pinus sylvestris* trees in a dry inner alpine valley in Switzerland. For Pathol 41:501–509

Hekkala A-M, Ahtikoski A, Päätalo M-L, Tarvainen O, Siipilehto J, Tolvanen A (2016) Restoring volume, diversity and continuity of deadwood in boreal forests. Biodivers Conserv 25:1107–1132

Hennon PE, Frankel SJ, Woods AJ, Worrall JJ, Norlander D, Zambino PJ, Warwell MV, Shaw CG (2020) A framework to evaluate climate effects on forest tree diseases. For Pathol 50:e12649

Hepting GH, Cowling EB (1977) Forest pathology: unique features and prospects. Annu Rev Phytopathol 15:431–450

Holdenrieder O, Pautasso M (2014) Wie viel Krankheit braucht der Wald? Bündner Wald 2014:5–10

Holdenrieder O, Pautasso M, Weisberg PJ, Lonsdale D (2004) Tree diseases and landscape processes: the challenge of landscape pathology. Trends Ecol Evol 19:446–452

Honkaniemi J, Rammer W, Seidl R (2021) From mycelia to mastodons – a general approach for simulating biotic disturbances in forest ecosystems. Environ Model Soft 138:104977

Howard RJ (1996) Cultural control of plant diseases: a historical perspective. Can J Plant Path 18:145–150

Hubbart JA, Guyette R, Muzika R-M (2016) More than drought: precipitation variance, excessive wetness, pathogens and the future of the western edge of the eastern deciduous forest. Sci Total Environ 566:463–467

Hultberg T, Sandström J, Felton A, Öhman K, Rönnberg J, Witzell J, Cleary M (2020) Ash dieback risks an extinction cascade. Biol Conserv 244:108516

Ingram D (1999) Biodiversity, plant pathogens and conservation. Plant Pathol 48:433–442

Ingram D (2002) The diversity of plant pathogens and conservation: bacteria and fungi *sensu lato*. In: Sivasithamparama K, Dixon KW, Barrett RL (eds) Microorganisms in plant conservation and biodiversity. Springer, Berlin, pp 241–267

Ireland KB, Hardy GESJ, Kriticos DJ (2013) Combining inferential and deductive approaches to estimate the potential geographical range of the invasive plant pathogen, *Phytophthora ramorum*. PLoS One 8:e63508

Jarosz AM, Davelos AL (1995) Effects of disease in wild plant populations and the evolution of pathogen aggressiveness. New Phytol 129:371–387

Jeger MJ, Pautasso M (2008) Plant disease and global change–the importance of long-term data sets. New Phytol 177:8–11

Jönsson MT, Ruete A, Kellner O, Gunnarsson U, Snäll T (2017) Will forest conservation areas protect functionally important diversity of fungi and lichens over time? Biodivers Conserv 26:2547–2567

Jönsson U (2006) A conceptual model for the development of *Phytophthora* disease in *Quercus robur*. New Phytol 171:55–68

Jules ES, Carroll AL, Garcia AM, Steenbock CM, Kauffman MJ (2014) Host heterogeneity influences the impact of a non-native disease invasion on populations of a foundation tree species. Ecosphere 5:1–17

Jules ES, Jackson JI, van Mantgem PJ, Beck JS, Murray MP, Sahara EA (2016) The relative contributions of disease and insects in the decline of a long-lived tree: a stochastic demographic model of whitebark pine (*Pinus albicaulis*). For Ecol Manag 381:144–156

Jung T (2009) Beech decline in Central Europe driven by the interaction between *Phytophthora* infections and climatic extremes. For Pathol 39:73–94

Jung T, Blaschke M (2004) Phytophthora root and collar rot of alders in Bavaria: distribution, modes of spread and possible management strategies. Plant Pathol 53:197–208

Jung T, Orlikowski L, Henricot B, Abad-Campos P, Aday A, Aguín Casal O et al (2016) Widespread *Phytophthora* infestations in European nurseries put forest, semi-natural and horticultural ecosystems at high risk of Phytophthora diseases. For Pathol. 46:134–163

Kautz M, Meddens AJ, Hall RJ, Arneth A (2017) Biotic disturbances in Northern Hemisphere forests – a synthesis of recent data, uncertainties and implications for forest monitoring and modelling. Glob Ecol Biogeogr 26:533–552

Keča N, Koufakis I, Dietershagen J, Nowakowska JA, Oszako T (2016) European oak decline phenomenon in relation to climatic changes. Folia For Polon 58:170–177

Khaliq I, Burgess TI, Hardy GE, White D, McDougall KL (2021) *Phytophthora* and vascular plant species distributions along a steep elevation gradient. Biol Invasions 23:1443–1459

Kirschbaum AA, Pfaff E, Gafvert UB (2016) Are US national parks in the Upper Midwest acting as refugia? Inside vs. outside park disturbance regimes. Ecosphere 7:e01467

Klapwijk MJ, Hopkins AJ, Eriksson L, Pettersson M, Schroeder M, Lindelöw Å, Rönnberg J, Keskitalo ECH, Kenis M (2016) Reducing the risk of invasive forest pests and pathogens: combining legislation, targeted management and public awareness. Ambio 45:223–234

Kolb TE, Fettig CJ, Ayres MP, Bentz BJ, Hicke JA, Mathiasen R, Stewart JE, Weed AS (2016) Observed and anticipated impacts of drought on forest insects and diseases in the United States. For Ecol Manag 380:321–334

Kopf S, Ha-Duong M, Hallegatte S (2008) Using maps of city analogues to display and interpret climate change scenarios and their uncertainty. Nat Hazards Earth Syst Sci 8:905–918

Kõrkjas M, Remm L, Lõhmus A (2021) Development rates and persistence of the microhabitats initiated by disease and injuries in live trees: a review. For Ecol Manag 482:118833

Kowalski T (2013) Foliar diseases of broadleaved trees. In: Gonthier P, Nicolotti G (eds) Infectious forest diseases. CABI, Wallingford, pp 488–518

Kuparinen A, Savolainen O, Schurr FM (2010) Increased mortality can promote evolutionary adaptation of forest trees to climate change. For Ecol Manag 259:1003–1008

La Porta N, Capretti P, Thomsen IM, Kasanen R, Hietala AM, Von Weissenberg K (2008) Forest pathogens with higher damage potential due to climate change in Europe. Can J Plant Pathol 30:177–195

Labandeira CC, Prevec R (2014) Plant paleopathology and the roles of pathogens and insects. Int J Paleopathol 4:1–16

Laćan I, McBride JR (2008) Pest vulnerability matrix (PVM): a graphic model for assessing the interaction between tree species diversity and urban forest susceptibility to insects and diseases. Urban For Urban Green 7:291–300

Laflamme G (2010) Root diseases in forest ecosystems. Can J Plant Pathol 32:68–76

Landolt J, Gross A, Holdenrieder O, Pautasso M (2016) Ash dieback due to *Hymenoscyphus fraxineus*: what can be learnt from evolutionary ecology? Plant Pathol 65:1056–1070

Lassauce A, Paillet Y, Jactel H, Bouget C (2011) Deadwood as a surrogate for forest biodiversity: meta-analysis of correlations between deadwood volume and species richness of saproxylic organisms. Ecol Indic 11:1027–1039

Liebhold AM, Brockerhoff EG, Garrett LJ, Parke JL, Britton KO (2012) Live plant imports: the major pathway for forest insect and pathogen invasions of the US. Front Ecol Environ 10:135–143

Lion S, Gandon S (2016) Spatial evolutionary epidemiology of spreading epidemics. Proc R Soc B 283:20161170

Lockman IB, Kearns HS (2016) Forest root diseases across the United States, RMRS-GTR-342. USDA, Forest Service, Rocky Mountain Research Station, Ogden, 55 p

Lonsdale D, Pautasso M, Holdenrieder O (2008) Wood-decaying fungi in the forest: conservation needs and management options. Eur J. For Res 127:1–22

Loo JA (2009) Ecological impacts of non-indigenous invasive fungi as forest pathogens. Biol Invasions 11:81–96

Lovett GM, Weiss M, Liebhold AM, Holmes TP, Leung B, Lambert KF, Orwig DA, Campbell FT, Rosenthal J, McCullough DG (2016) Nonnative forest insects and pathogens in the United States: impacts and policy options. Ecol Appl 26:1437–1455

Lundquist J (2007) The relative influence of diseases and other small-scale disturbances on fuel loading in the Black Hills. Plant Dis 91:147–152

Luvisi A, Nicolì F, De Bellis L (2017) Sustainable management of plant quarantine pests: the case of olive quick decline syndrome. Sustainability 9:659

Lynch S, Zambino P, Scott T, Eskalen A (2014) Occurrence, incidence and associations among fungal pathogens and *Agrilus auroguttatus*, and their roles in *Quercus agrifolia* decline in California. For Pathol 44:62–74

Manion PD (1981) Tree disease concepts. Prentice-Hall, Cornell University, 399 p

Manion PD (2003) Evolution of concepts in forest pathology. Phytopathology 93:1052–1055

Martin PA, Newton AC, Cantarello E, Evans P (2015) Stand dieback and collapse in a temperate forest and its impact on forest structure and biodiversity. For Ecol Manag 358:130–138

Marzano M, Dandy N, Bayliss HR, Porth E, Potter C (2015) Part of the solution? Stakeholder awareness, information and engagement in tree health issues. Biol Invasions 17:1961–1977

Marzano M, Dandy N, Papazova-Anakieva I, Avtzis D, Connolly T, Eschen R, Glavendekić M, Hurley B, Lindelöw Å, Matošević D (2016) Assessing awareness of tree pests and pathogens amongst tree professionals: a pan-European perspective. Forest Policy Econ 70:164–171

McKinney LV, Nielsen LR, Collinge DB, Thomsen IM, Hansen JK, Kjær ED (2014) The ash dieback crisis: genetic variation in resistance can prove a long-term solution. Plant Pathol 63:485–499

Meentemeyer RK, Haas SE, Václavík T (2012) Landscape epidemiology of emerging infectious diseases in natural and human-altered ecosystems. Annu Rev Phytopathol 50:379–402

Metz MR, Frangioso KM, Meentemeyer RK, Rizzo DM (2011) Interacting disturbances: wildfire severity affected by stage of forest disease invasion. Ecol Appl 21:313–320

Meyer S, Rusterholz HP, Baur B (2021) Saproxylic insects and fungi in deciduous forests along a rural–urban gradient. Ecol Evol 11:1634–1652

Mordecai EA (2011) Pathogen impacts on plant communities: unifying theory, concepts, and empirical work. Ecol Monogr 81:429–441

Morin L, Aveyard R, Lidbetter JR, Wilson PG (2012) Investigating the host-range of the rust fungus *Puccinia psidii* sensu lato across tribes of the family Myrtaceae present in Australia. PLoS One 7:e35434

Moslonka-Lefebvre M, Finley A, Dorigatti I, Dehnen-Schmutz K, Harwood T, Jeger MJ, Xu X, Holdenrieder O, Pautasso M (2011) Networks in plant epidemiology: from genes to landscapes, countries, and continents. Phytopathology 101:392–403

Müller MM, Hamberg L, Hantula J (2016) The susceptibility of European tree species to invasive Asian pathogens: a literature based analysis. Biol Invasions 18:2841–2851

Mulvey RL, Bisbing SM (2016) Complex interactions among agents affect shore pine health in Southeast Alaska. Northwest Sci 90:176–194

Needham J, Merow C, Butt N, Malhi Y, Marthews TR, Morecroft M, McMahon SM (2016) Forest community response to invasive pathogens: the case of ash dieback in a British woodland. J Ecol 104:315–330

Nguyen D, Castagneyrol B, Bruelheide H, Bussotti F, Guyot V, Jactel H, Jaroszewicz B, Valladares F, Stenlid J, Boberg J (2016) Fungal disease incidence along tree diversity gradients depends on latitude in European forests. Ecol Evol 6:2426–2438

Noar RD, Jahant-Miller CJ, Emerine S, Hallberg R (2021) Early warning systems as a component of integrated pest management to prevent the introduction of exotic pests. J Integr Pest Manag 12:16

Oliva J, Stenlid J, Martínez-Vilalta J (2014) The effect of fungal pathogens on the water and carbon economy of trees: implications for drought-induced mortality. New Phytol 203:1028–1035

Oliva J, Castaño C, Baulenas E, Domínguez G, González-Olabarria J, Oliach D (2016) The impact of the socioeconomic environment on the implementation of control measures against an invasive forest pathogen. For Ecol Manag 380:118–127

Ostry M, Laflamme G (2008) Fungi and diseases—natural components of healthy forests. Botany 87:22–25

Papaïx J, Touzeau S, Monod H, Lannou C (2014) Can epidemic control be achieved by altering landscape connectivity in agricultural systems? Ecol Model 284:35–47

Paquette A, Messier C (2011) The effect of biodiversity on tree productivity: from temperate to boreal forests. Glob Ecol Biogeogr 20:170–180

Parker TJ, Clancy KM, Mathiasen RL (2006) Interactions among fire, insects and pathogens in coniferous forests of the interior western United States and Canada. Agric For Entomol 8:167–189

Paseka RE, White LA, Van de Waal DB, Strauss AT, González AL, Everett RA, Peace A, Seabloom EW, Frenken T, Borer ET (2020) Disease-mediated ecosystem services: pathogens, plants, and people. Trends Ecol Evol 75:731–743

Pautasso M (2013) Responding to diseases caused by exotic tree pathogens. In: Gonthier P, Nicolotti G (eds) Infectious forest diseases. CABI, Wallingford, pp 592–612

Pautasso M, Jeger MJ (2014) Impacts of climate change on plant diseases: new scenarios for the future. In: Choffnes ER, Mack A (eds) The influence of global environmental change on infectious disease dynamics: workshop summary. Institute of Medicine, US National Academy of Science, Washington, DC, pp 359–374

Pautasso M, Holdenrieder O, Stenlid J (2005) Susceptibility to fungal pathogens of forests differing in tree diversity. In: Scherer-Lorenzen M, Körner C, Schulze E-D (eds) Forest diversity and function. Springer, Berlin, pp 263–289

Pautasso M, Dehnen-Schmutz K, Holdenrieder O, Pietravalle S, Salama N, Jeger MJ, Lange E, Hehl-Lange S (2010) Plant health and global change–some implications for landscape management. Biol Rev 85:729–755

Pautasso M, Aas G, Queloz V, Holdenrieder O (2013) European ash (*Fraxinus excelsior*) dieback–a conservation biology challenge. Biol Conserv 158:37–49

Pautasso M, Schlegel M, Holdenrieder O (2015) Forest health in a changing world. Microb Ecol 69:826–842

Pezzi G, Maresi G, Conedera M, Ferrari C (2011) Woody species composition of chestnut stands in the Northern Apennines: the result of 200 years of changes in land use. Landsc Ecol 26:1463–1476

Podger FD (1972) *Phytophthora cinnamomi*, a cause of lethal disease in indigenous plant communities in Western Australia. Phytopathology 62:972–981

Prada-Salcedo LD, Goldmann K, Heintz-Buschart A, Reitz T, Wambsganss J, Bauhus J, Buscot F (2021) Fungal guilds and soil functionality respond to tree community traits rather than to tree diversity in European forests. Mol Ecol 30:572–591

Pronos J, Merrill L, Dahlsten D (1999) Insects and pathogens in a pollution-stressed forest. In: Miller PR, McBride JR (eds) Oxidant air pollution impacts in the montane forests of Southern California. Springer, New York, pp 317–337

Prospero S, Rigling D, Holdenrieder O (2003) Population structure of *Armillaria* species in managed Norway spruce stands in the Alps. New Phytol 158:365–373

Pusz W, Urbaniak J (2017) Foliar diseases of willows (*Salix* spp.) in selected locations of the Karkonosze Mts. (the Giant Mts). Eur J Plant Pathol 148:45–51

Ramsfield TD, Bentz BJ, Faccoli M, Jactel H, Brockerhoff EG (2016) Forest health in a changing world: effects of globalization and climate change on forest insect and pathogen impacts. Forestry 89:245–252

Rigling D, Prospero S (2018) *Cryphonectria parasitica*, the causal agent of chestnut blight: invasion history, population biology and disease control. Mol Plant Pathol 19:7–20

Rizzo DM, Slaughter GW, Parmeter JR Jr (2000) Enlargement of canopy gaps associated with a fungal pathogen in Yosemite Valley, California. Can J For Res 30:1501–1510

Robinet C, Douma JC, Piou D, van der Werf W (2016) Application of a wood pathway model to assess the effectiveness of options for reducing risk of entry of oak wilt into Europe. Forestry 89:456–472

Rodewald AD, Arcese P (2016) Direct and indirect interactions between landscape structure and invasive or overabundant species. Curr Landsc Ecol Rep 1:30–39

Romportl D, Chumanová E, Havrdová L, Pešková V, Černý K (2016) Potential risk of occurrence of *Phytophthora alni* in forests of the Czech Republic. J Maps 12:280–284

Rottstock T, Joshi J, Kummer V, Fischer M (2014) Higher plant diversity promotes higher diversity of fungal pathogens, while it decreases pathogen infection per plant. Ecology 95:1907–1917

Roy B, Kirchner J, Christian CE, Rose L (2000) High disease incidence and apparent disease tolerance in a North American Great Basin plant community. Evol Ecol 14:421

Roy BA, Alexander HM, Davidson J, Campbell FT, Burdon JJ, Sniezko R, Brasier C (2014) Increasing forest loss worldwide from invasive pests requires new trade regulations. Front Ecol Environ 12:457–465

Sangüesa-Barreda G, Camarero JJ, Oliva J, Montes F, Gazol A (2015) Past logging, drought and pathogens interact and contribute to forest dieback. Agric For Meteorol 208:85–94

Santini A, Ghelardini L (2015) Plant pathogen evolution and climate change. CAB Rev 10:1–8

Santini A, Ghelardini L, Pace CD, Desprez-Loustau M-L, Capretti P, Chandelier A, Cech T, Chira D, Diamandis S, Gaitniekis T (2013) Biogeographical patterns and determinants of invasion by forest pathogens in Europe. New Phytol 197:238–250

Schmid M, Pautasso M, Holdenrieder O (2014) Ecological consequences of Douglas fir (*Pseudotsuga menziesii*) cultivation in Europe. Eur J For Res 133:13–29

Schroth M, Thomson S, Hildebrand D, Moller W (1974) Epidemiology and control of fire blight. Annu Rev Phytopathol 12:389–412

Senn J (1999) Tree mortality caused by *Gremmeniella abietina* in a subalpine afforestation in the Central Alps and its relationship with duration of snow cover. For Pathol 29:65–74

Setiawan NN, Vanhellemont M, Baeten L, Dillen M, Verheyen K (2014) The effects of local neighbourhood diversity on pest and disease damage of trees in a young experimental forest. For Ecol Manag 334:1–9

Shaw M, Pautasso M (2014) Networks and plant disease management: concepts and applications. Annu Rev Phytopathol 52:477–493

Shaw MW, Osborne TM (2011) Geographic distribution of plant pathogens in response to climate change. Plant Pathol 60:31–43

Sieber TN (2007) Endophytic fungi in forest trees: are they mutualists? Fungal Biol Rev 21:75–89

Sieber TN (2014) Neomyzeten–eine anhaltende Bedrohung für den Schweizer Wald. Schweiz Z Forstwes 165:173–182

Steeger C, Hitchcock CL (1998) Influence of forest structure and diseases on nest-site selection by red-breasted nuthatches. J Wildl Manag 62:1349–1358

Stenlid J, Oliva J (2016) Phenotypic interactions between tree hosts and invasive forest pathogens in the light of globalization and climate change. Philos Trans R Soc B 371:20150455

Stenlid J, Oliva J, Boberg JB, Hopkins AJ (2011) Emerging diseases in European forest ecosystems and responses in society. Forests 2:486–504

Sturrock R (2012) Climate change and forest diseases: using today's knowledge to address future challenges. For Syst 21:329–336

Sturrock R, Frankel S, Brown A, Hennon P, Kliejunas J, Lewis K, Worrall J, Woods A (2011) Climate change and forest diseases. Plant Pathol 60:133–149

Sutrave S, Scoglio C, Isard SA, Hutchinson JS, Garrett KA (2012) Identifying highly connected counties compensates for resource limitations when evaluating national spread of an invasive pathogen. PLoS One 7:e37793

Szwagrzyk J (2020) A healthy forest needs diseased trees. Fragm Florist Geobot Polon 27:5–15

Tainter FH, Baker FA (1996) Principles of forest pathology. Wiley, New York, 832 S

Thomas FM, Blank R, Hartmann G (2002) Abiotic and biotic factors and their interactions as causes of oak decline in Central Europe. For Pathol 32:277–307

Thompson SE, Levin S, Rodriguez-Iturbe I (2014) Rainfall and temperatures changes have confounding impacts on *Phytophthora cinnamomi* occurrence risk in the southwestern USA under climate change scenarios. Glob Change Biol 20:1299–1312

Thompson RN, Cobb RC, Gilligan CA, Cunniffe NJ (2016) Management of invading pathogens should be informed by epidemiology rather than administrative boundaries. Ecol Model 324:28–32

Thorn S, Seibold S, Leverkus AB, Michler T, Müller J, Noss RF, Stork N, Vogel S, Lindenmayer DB (2020) The living dead: acknowledging life after tree death to stop forest degradation. Front Ecol Environ 18:505–512

Tobin PC (2015) Ecological consequences of pathogen and insect invasions. Curr For Rep 1:25–32

Tomback DF, Resler LM, Keane RE, Pansing ER, Andrade AJ, Wagner AC (2016) Community structure, biodiversity, and ecosystem services in treeline whitebark pine communities: potential impacts from a non-native pathogen. Forests 7:21

Tonini F, Jones C, Miranda BR, Cobb RC, Sturtevant BR, Meentemeyer RK (2018) Modeling epidemiological disturbances in LANDIS-II. Ecography 41:2038–2044

Tsopelas P, Santini A, Wingfield MJ, Wilhelm de Beer Z (2017) Canker stain: a lethal disease destroying iconic plane trees. Plant Dis 101:645–658

Vialle C, Plan Ti Ti qpi int iiiti iii iii t i iitii j ii iiiiiiiiiiiiiii ii All AAlagiiiiihi lioiilaiiguu iici work: the asymmetric influence of past evolutionary history. PLoS One 3:e1740

Venette RC, Cohen SD (2006) Potential climatic suitability for establishment of *Phytophthora ramorum* within the contiguous United States. For Ecol Manag 231:18–26

Verheyen K, Vanhellemont M, Auge H, Baeten L, Baraloto C, Barsoum N, Bilodeau-Gauthier S, Bruelheide H, Castagneyrol B, Godbold D (2016) Contributions of a global network of tree diversity experiments to sustainable forest plantations. Ambio 45:29–41

Vettraino AM, Li H-M, Eschen R, Morales-Rodriguez C, Vannini A (2017) The sentinel tree nursery as an early warning system for pathway risk assessment: fungal pathogens associated with Chinese woody plants commonly shipped to Europe. PLoS One 12:e0188800

von Gadow K, Zhao XH, Tewari V, Zhang CY, Kumar A, Rivas JJC, Kumar R (2016) Forest observational studies: an alternative to designed experiments. Euro J For Res 135:417–431

West JS, Townsend JA, Stevens M, Fitt BD (2012) Comparative biology of different plant pathogens to estimate effects of climate change on crop diseases in Europe. Eur J Plant Pathol 133:315–331

Whyte G, Howard K, Hardy GSJ, Burgess T (2016) The Tree Decline Recovery seesaw; a conceptual model of the decline and recovery of drought stressed plantation trees. For Ecol Manag 370:102–113

Woods A, Martín-García J, Bulman L, Vasconcelos MW, Boberg J, La Porta N, Peredo H, Vergara G, Ahumada R, Brown A (2016) Dothistroma needle blight, weather and possible climatic triggers for the disease's recent emergence. For Pathol 46:443–452

Zadoks JC (2017) On social and political effects of plant pest and disease epidemics. Phytopathology 107:1144–1148

Chapter 11
Insect Defoliators

Christa Schafellner ⓘ **and Katrin Möller**

Abstract The US Department of Agriculture considers insect outbreaks as the most expensive type of natural disturbance. Outbreaks of phyllophagous and xylophagous forest pests have significant ecological and economic impacts. Severe defoliation in combination with specific weather conditions is able to induce widespread forest dieback. Since most insect pests that undergo regular outbreaks are thermophilic, the frequency and magnitude of outbreaks are likely to increase with the predicted climate changes, specifically with rising temperatures. Thus, reliable monitoring and forecast tools as well as containment strategies for insect pests are urgently needed in order to maintain the multi-functionality of our forests in the future.

Keywords Broadleaved forests · Climate change · Coniferous forests · Host plants · Insect population dynamics · Natural enemies · Pest management · Predisposition · Tree mortality

11.1 Population Dynamics and Outbreaks

In ecology, a disturbance is a discrete event limited in space and time that disrupts the function of an ecosystem either by loss of living biomass or changes in availability of resources in biotic communities (see Chap. 2). Tree mortality resulting from phyllophagous (i.e. leaf and needle feeding) insects is one type of disturbance.

C. Schafellner (✉)
Institute of Forest Entomology, Forest Pathology and Forest Protection (IFFF),
University of Natural Resources and Life Sciences, Vienna, Austria
e-mail: christa.schafellner@boku.ac.at

K. Möller
Department of Forest Conservation and Wildlife Ecology, Landeskompetenzzentrum Forst
Eberswalde, Eberswalde, Germany

T. Wohlgemuth et al. (eds.), *Disturbance Ecology*, Landscape Series 32,
https://doi.org/10.1007/978-3-030-98756-5_11

Worldwide, insects consume annually about 3–10% of the plant biomass in forest ecosystems. However, in years with high population densities, pest-induced biomass loss on a local or regional scale can be substantially higher. Single or multiple defoliation events may trigger large-scale tree mortality with both ecological consequences (e.g. water balance, erosion, CO_2 storage capacity) and economic losses (e.g. timber production, tourism) (Niesar et al. 2015).

It is only when the adverse effects of insect outbreaks intersect unfavourably with resources valued by humans that such insects are considered as pests. In German-speaking countries, the term *Schädlinge* [English, 'pest'] was first used in 1880 for the grapevine phylloxera (*Viteus vitifoliae* Fitch) (Jansen 2003), an aphid-like pest of commercial grapevines worldwide. Originally native to eastern North America, this insect was accidentally introduced to Europe in the mid-nineteenth century. Typically, the insect forest pests that result in tree mortality or loss of timber value are found in a number of insect orders, particularly in the beetles (Coleoptera), moths (Lepidoptera), and sawflies (Hymenoptera). While beetles (e.g. the European spruce bark beetle, *Ips typographus* L.) and sawflies (e.g. the common pine sawfly, *Diprion pini* (L.)) are important forest pests, the most important defoliators of forest trees in Europe belong to the order Lepidoptera. However, these insects only become pests at certain times or under special environmental conditions that allow them to realize their incredibly high reproductive potential. Insect outbreaks consist of several qualitatively different phases (Fig. 11.1): progradation (building phase, i.e. increase of population density), culmination (peak, i.e. highest population density), retrogradation (collapse, i.e. decline of population density), and latent phase (i.e. non-outbreak, low population density) (Schwerdtfeger 1981).

Insect outbreaks result from a complex interplay between host plant, phytophagous insect, and the natural enemies (i.e. predators, parasites, pathogens). Insect populations explode because of the coincidence of several factors and collapse

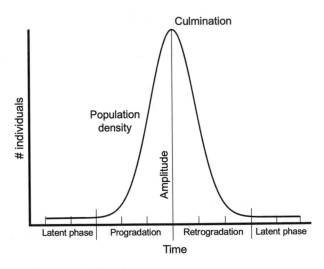

Fig. 11.1 Phases of insect outbreak cycles

Table 11.1 Factors influencing insect population dynamics

Abiotic	Biotic
Temperature	Offspring number
Precipitation	Vitality
Photoperiod	Intra- and interspecific competition
Fire	Pathogens
Storm	Parasitoids
Snow	Predators
Flooding	Food quantity
Drought	Food quality

because of other factors (Table 11.1). These factors may be density independent (e.g. site and weather conditions or suitable host plants) or density dependent (e.g. quantity and quality of food resources or presence/absence of natural enemies). Insect population dynamics is one of the fundamental research areas of ecology that addresses temporal patterns of insect abundance and factors causing or regulating fluctuations in populations (Dettner and Peters 2003).

Favourable weather, especially during the early stages of development (e.g. newly hatched larvae), and abundance of high-quality food usually shorten development times, increase insect survival, and promote fecundity. Insect populations grow when progeny production exceeds mortality rates, eventually reaching outbreak levels if conditions remain beneficial for a prolonged period. Outbreaks collapse as a consequence of one or several of the following factors: destruction of resources, food depletion, host resistance (Benz 1974; Schultz and Baldwin 1982; Schopf 1986; Clancy et al. 1988; Quiring and McKinnon 1999; Mumm and Hilker 2006), natural enemies (Murray and Elkinton 1989; Elkinton et al. 2004; Moreau and Lucarotti 2007; Hilker and McNeil 2008; Möller and Bemmann 2009; Mirchev et al. 2013), or adverse weather (Pernek et al. 2008; Netherer and Schopf 2010; Möller et al. 2017; Hentschel et al. 2018). Population dynamics of important forest pests are discussed in detail in Berryman (1988).

The first and most obvious kind of outbreak occurs when the environment changes in some way to favour the pest or to disadvantage its natural enemies. For example, extensive monocultures of host plants provide a huge amount of highly susceptible food for insects that feed on these plants. This is especially true of boreal forests, but even more so of the intensively managed, and often even-aged and single-species secondary/plantation forests. A high impact and degree of human interventions on forest ecosystems (hemeroby) are seen as a major reason for pest problems, not only because of a low(er) level of biodiversity but also because of missing structures and functions of the forest ecosystems, lack of resilience to disturbances, and impaired regeneration processes. The trees either adapt to periodically occurring insect outbreaks and defoliation, or they become locally extinct. Today, forest management strategies often aim to create a dynamic forest structure consisting of stands with unevenly aged trees and species composition close to the natural vegetation that tolerate defoliation episodes without high tree mortality and still have adequate timber production.

11.2   Forest Insect Defoliators

Both spatial and temporal fluctuations in the population density of insects are common in nature (Schwenke 1978). Forest Macrolepidoptera such as larch budmoth (*Zeiraphera griseana* Hübner) (Esper et al. 2007), gypsy moth (*Lymantria dispar* L.) (Pernek et al. 2008; Bjørnstad et al. 2010), or autumnal moth (*Epirrita autumnata* Borkhausen) (Ruohomäki et al. 2000) undergo regular (cyclic) population fluctuations. The time intervals from the latent phase to culmination (peak) and back to the latent phase remain fairly constant for a given species within specific geographic areas and in similar climate conditions. Larch budmoth populations, for example, can grow by a factor of more than 10,000 within four to five generations (progradation); the phase of population decline (retrogradation) lasts just as long, resulting in cycles of 8–9 years (Wermelinger et al. 2018).

Outbreaks of other forest Lepidoptera, such as pine-tree lappet moth (*Dendrolimus pini* L.), pine beauty moth (*Panolis flammea* Denis and Schiffermüller), or pine hawk-moth (*Sphinx pinastri* L.), are more irregular. These species exhibit population fluctuations with extreme oscillations between lower and upper boundaries. Their populations remain in latent phase for years or even decades (Varley 1949); however, at irregular and (mostly) unpredictable intervals, the insect density increases dramatically within a few generations (progradation). These outbreaks sometimes affect hundreds of square kilometres of forests and may last for several years, inducing tree mortality as the consequence of repeated defoliation or even forest dieback, especially in managed forests with certain dominant tree species (monocultures). Even without human intervention, the resulting openings in the canopy (small-scale gaps, large-scale patches) are rapidly (re)colonized by numerous plants, microbes, and animals; new communities emerge, and after various stages of succession, the forest ecosystem will arrive eventually at a climax community.

In the remaining sections of this chapter, some phyllophagous moths that are strongly affected by weather are discussed as representative examples of insect defoliators in broadleaved and coniferous forests with significant implications for the ecology and economy of temperate forest ecosystems in Central Europe.

11.2.1   Oak Forests

The gypsy moth and the oak processionary moth (*Thaumetopoea processionea* L.) are serious forest pests in Europe that are capable of causing total defoliation of broadleaved trees in deciduous forests (Delb and Block 1999; Roques 2015). The polyphagous gypsy moth larvae prefer foliage of oaks (*Quercus* spp.), hornbeam (*Carpinus betulus* L.), and beech (*Fagus sylvatica* L.). The monophagous oak processionary moth, on the other hand, feeds exclusively on oak leaves. Within Europe, these thermophilic moths have their core distribution area in the south, where outbreaks occur frequently and regularly. In temperate regions, high population densities occur only at warm and xeric sites (Table 11.2). Gypsy moth and pine

Table 11.2 Pest species profiles: gypsy moth and oak processionary moth

Defoliation of oak	Gypsy moth *Lymantria dispar* L.	Oak processionary moth *Thaumetopoea processionea* L.
Classification	Order Lepidoptera, family Erebidae	Order Lepidoptera, family Notodontidae
Distribution	Native to Europe, northern Africa, central (Siberia) and eastern Asia, and Japan; in Europe, main distribution in southern and southeastern areas; rapid spread across the northeastern USA after its introduction 150 years ago	Native to Europe, main distribution in Central and southern Europe, significant range expansion northwards, possibly as a result of global warming
Habitat	Warm and dry regions with pure oak and mixed oak forests, parks, orchards	Sunny, open oak forests, sunlit forest edges, avenues, single oaks in urban areas
Hosts	Polyphagous: In Central Europe, mainly sessile oak (*Quercus petraea* Liebl.) and English oak (*Quercus robur* L.), hornbeam (*Carpinus betulus* L.), European beech (*Fagus sylvatica* L.), also apple (*Malus domestica* Borkh.) and plum (*Prunus domestica* L.) In southern Europe, cork oak (*Quercus suber* L.). In the Balkans, pubescent oak (*Quercus pubescens* Willd.); during outbreaks as result of food shortage other tree species, e.g. poplar (*Populus* spp.), willow (*Salix* spp.), larch (*Larix decidua* Mill.)	Monophagous (oak leaves): In Central Europe, mainly sessile oak (*Quercus petraea* Liebl.) and English oak (*Quercus robur* L.) In southern Europe, cork oak (*Quercus suber* L.) and pubescent oak (*Quercus pubescens* Willd.)
Life cycle	One generation per year	One generation per year
Adult moths	July–September	July–August
Larvae	April/May–June	March/April–June

(continued)

Table 11.2 (continued)

	Gypsy moth	Oak processionary moth
Overwintering	Eggs in egg masses in tree bark crevices on stems and lower branches	Eggs in flat batches on twigs and small branches in the upper canopy
Natural enemies	Egg parasitoids (chalcidid wasps), larval parasitoids (braconid and ichneumonid wasps, tachinids), pupal parasitoids (chalcidid wasps, ichneumonid wasps), predatory beetles, birds, bats, mice, shrews, pathogens (viruses, bacteria, microsporidia, entomopathogenic fungi)	Egg parasitoids (chalcidid wasps), larval and pupal parasitoids (braconid and ichneumonid wasps, tachinids), wood ants, predatory beetles, bats, some birds, pathogens (viruses, bacteria, microsporidia, entomopathogenic fungi)
Population dynamics	Cyclic (8–12 years or longer), time intervals between successive outbreaks longer in western Europe, shorter in eastern Europe; mainly in oak and mixed oak forests in Hungary, Slovakia, Croatia, Serbia, Rumania, Bulgaria, France, Germany, Switzerland, Austria; Asia; USA after introduction; population collapse after 1–3 years	Non-cyclic, mainly in oak forests in southern Europe Since the early 1990s rapid spread and new outbreaks in Germany, the Netherlands, Belgium, France, and the UK
Outbreak factors	Warm and dry springtime, close coincidence of budburst and larval hatch, high food quality of preferred host plant (oaks), conditions leading to absence of predators and parasitoids (e.g. dry soils, poor vegetation cover, low diversity of flora and fauna)	Warm and dry springtime, highly predisposed oaks, conditions leading to absence of predators and parasitoids (e.g. lack of habitats suitable for alternative hosts of the predators and parasitoids)
Population collapse	Food shortage, low food quality (lack of preferred host plant), susceptibility of larvae to diseases (viral, fungal, bacterial), low fecundity (small egg numbers), late frost events following larval hatch, increasing densities of parasitoids	Natural enemies (predators and parasitoids), especially increasing densities of parasitoids, long and heavy rainfall during early larval development

Photos: K. Möller (left, right), B. Wermelinger (centre)

processionary moth are important representatives for a vast number of early-season defoliators that essentially increase the vulnerability of oaks to other factors that contribute to the overall oak decline syndrome (Patočka et al. 1999).

In Central Europe, the frequency and magnitude of gypsy moth and oak processionary moth outbreaks have increased in recent decades. Dry and warm weather, especially in spring and early summer, promotes insect population growth (Schröder et al. 2016) and the probability of heavy defoliation. During severe outbreaks, the hungry larvae completely defoliate the tree crown by the end of May or mid-June. Typically, oaks compensate for leaf loss by reflushing (producing new leaves and

Fig. 11.2 Remnants of an oak forest near Genthin (Saxony-Anhalt, Germany) after repeated oak processionary moth defoliation. The surviving trees show severe symptoms of decline. (Photo: K. Möller)

so-called Lammas shoots) later in the season. However, if the new foliage is destroyed again by insects, powdery mildew, or drought, tree health declines fast (Fig. 11.2). Tree mortality, individual, in groups, or extended, usually increases with more defoliation events due to outbreaks of gypsy moths and oak processionary moths (Block et al. 1995; Delb and Block 1999; Lobinger 2006; Strasser et al. 2013; Niesar et al. 2015).

In Europe, oaks are major components of temperate and Mediterranean forest ecosystems. Of all native tree species, oaks are associated with a greater number of insect species than any other trees; Bussler (2014) reported that more than 500 wood-inhabiting (xylobiont) beetles (Coleoptera) and almost 180 butterfly species (Macrolepidoptera) were found on sessile oak (*Q. petraea*) with many of them being monophagous (i.e. feeding only on oaks). Thus, canopy defoliation is not limited to the short-term adverse effects caused by the removal of leaves for the individual trees affected by the defoliation, but there may also be long-term consequences associated with tree death (e.g. food and habitat loss) for the many species associated with oaks.

The specific wood anatomy and physiology of oaks predispose them to drought- or defoliation-induced dieback. One factor contributing to this predisposition is that oaks form ring-porous wood (Matyssek et al. 2010), meaning that water is efficiently transported in the wide, current-year earlywood xylem vessels; however, these vessels normally become dysfunctional within 1 year. The effects of total defoliation shortly after bud opening in spring can be dramatic because the tree is growing rapidly and food reserves in the shoots and roots are at their lowest. With such low food supplies, the tree cannot support all its branches and roots until new leaves are formed; the tree must use its stored reserves for the new flush. How badly the reserves are depleted depends on how quickly new leaves are formed (regeneration shoots) and the amount of carbohydrates they produce until the end of the vegetation period. Partial or total defoliation over 2–3 consecutive years eventually results in a pronounced depletion of reserves (sugar, starch) and formation of xylem vessels with narrow diameters (Blank 1997). Thus, less water is transported in small diameter vessels, and fewer fine roots are produced, which in turn reduces water and mineral absorption from the soil. Such changes in earlywood-vessel sizes are reliable indicators of repeated defoliation events (Bréda and Granier 1996; Hansen 1999).

Throughout Europe, oak forests have regularly undergone declines resulting from complex interacting factors, both biotic and abiotic. Although causal factors may differ among the oak species and the regions affected, repeated early-season defoliation by insects in combination with weather extremes (late frost, drought) is a predisposing, inciting, or contributing factor of the most recent oak decline [German, *Eichensterben*] in Central Europe since the early 1990s (Führer 1998; Gaertig et al. 2005; Lobinger 2006). Once the decline has been initiated, trees undergo a continuous loss of vigour, making them more vulnerable to attack by secondary pests such as jewel beetles (Buprestidae, e.g. *Agrilus* spp. and *Coraebus* spp.) and wood-decaying fungi (e.g. *Armillaria* spp. and *Phytophthora* spp.), which accelerate the death of the weakened trees (Delb 2012; Kätzel et al. 2013). Small or large openings in the canopy resulting from individual or small-group tree mortality change the habitat characteristics (e.g. light, water regime) considerably for current plant and animal assemblages, with sometimes harmful effects on rare or endangered species. Species that prefer open habitats generally benefit from gaps and patches, while species adapted to closed canopies become less competitive.

11.2.2 Coniferous Forests

In general, Central European forests are dominated by broadleaved deciduous tree species; however, conifers form an essential part of the forest composition under certain edaphic (sandy, poor-nutrient soils) and environmental conditions (low precipitation) or at higher elevations and latitudes (short vegetation period, snow cover, low temperatures). Firs (*Abies* spp.), spruces (*Picea* spp.), pines (*Pinus* spp.), and larches (*Larix* spp.) are able to grow in almost every habitat that supports broadleaved trees, but these conifers can only dominate when protected from competition from broadleaved species. Because conifers grow quickly and with straight stems, they reach timber age quickly and have high timber yields and are considerably more productive in broadleaved forest habitats than in natural conifer communities. Thus, they have been widely planted both in the lowlands and on mountain ranges. However, these pure and often single-species coniferous forests are highly susceptible to insect herbivore attacks from bark beetles (see Chap. 12) and defoliators including several Macrolepidoptera (*Lymantria monacha* L., *Panolis flammea* Den. & Schiff., *Bupalus piniaria* L., *Dendrolimus pini* L.) and sawfly species (*Diprion pini* L., *Gilpinia hercyniae* Hartig, *Pachynematus montanus* Zadd., *Pristiphora abietina* Christ, *Cephalcia abietis* L.) (Altenkirch et al. 2002).

Pest populations in secondary coniferous forests and conifer plantations in Central European lowlands specifically benefit from favourable stand and climate conditions and rich food resources, especially when the pest insect is monophagous. Conditions for their natural enemies, however, are often unfavourable (Jäkel and Roth 2004). For example, many parasitic ichneumonid and chalcidid wasps depend on alternative hosts at times when the main host (the pest insect) is scarce (latent phase) or not available in the suitable stage (e.g. for overwintering). Additionally, monoculture coniferous forests do not have rich understories that provide food (e.g. nectar, pollen, honeydew) and shelter for adult parasitoids.

However, even coniferous forests at middle and high elevations suffer from recurrent insect defoliations. Pure larch and mixed larch–stone pine forests in the subalpine area across the arc of the European Alps are affected by outbreaks of the larch budmoth (*Zeiraphera griseana* Hübner) (Nola et al. 2006). The European larch (*Larix decidua* Mill.) and the Swiss stone pine (*Pinus cembra* L.) form the uppermost subalpine forest communities in the Alps; these mountain forests play a crucial role in stabilizing slopes, thus providing protection for people (e.g. settlements, infrastructure) against natural hazards (e.g. rockfall, snow avalanches, landslides, soil erosion). The importance of the protective function has even increased in recent decades because remote mountain regions that were not accessible in the past are now used year-round for tourism (Wehrli et al. 2007).

We selected two lepidopteran species as representatives of the vast number of foliage-feeding insects on conifers with distinctive outbreaks in Europe: the pine-tree lappet moth (*Dendrolimus pini* L.) and the larch budmoth (*Zeiraphera griseana* Hübner). The factsheets presented in Table 11.3 provide information on their biology and ecology.

The pine-tree lappet moth is one of the most important insect pests of pines across Europe. Outbreaks of this insect sometimes occur over large geographical areas and, if left uncontrolled, may result in widespread tree death caused by severe defoliation. After hatching in late summer, the early instar larvae feed on needles until November when they descend from the canopy to the forest floor to hibernate in the needle litter or topsoil. As the temperatures rise in spring, the larvae become active again, crawl back up to the canopy, and continue feeding throughout early summer until they pupate. A single larva eats 600–1000 needles, and female larvae are up to 8 cm long when fully grown (Björkman et al. 2013). During outbreaks, the larvae consume both old and young needles, buds, and even the bark of young shoots. In cases of food shortage, mature larvae feed on the needle stumps and destroy the needle base (sheath). Since needles cannot regrow without a sheath, the tree develops small needle rosettes by mobilizing reserves (Weckwerth 1952). However, the new shoots often do not produce enough photosynthates to meet the energy demand of the tree. Depending on local weather conditions and the degree of defoliation, pine mortality is 50–60%, but can reach 100% in totally defoliated trees (Wenk and Möller 2013). Drought periods after defoliation increase tree mortality rates markedly (Fig. 11.3), but individual trees with few needles left can still

Table 11.3 Pest species profiles: pine-tree lappet moth and larch budmoth

	Pine-tree lappet moth	Larch budmoth
Defoliation of larch	*Dendrolimus pini* L.	*Zeiraphera griseana* Hübner
Classification	Order Lepidoptera, family Lasiocampidae	Order Lepidoptera, family Tortricidae
Distribution	Eurasia, serious pest in Europe; from Spain to Russia and from Scandinavia to Romania	Palearctic; pest in Siberia and in the European Alps; in the Alps, genetically and morphologically distinct, host-associated biotypes on European larch and stone pine
Habitat	Pure pine stands on warm, xeric, well-drained, sandy sites	In Europe: subalpine conifer forests (1700–2000 m), outbreaks mainly on south-facing slopes
Hosts	Monophagous on pine, primary host plant Scots pine (*Pinus sylvestris* L.), during outbreaks also on Douglas fir (*Pseudotsuga menziesii* (Mirb.) Franco) and Norway spruce (*Picea abies* (L.) H. Karst.)	Exclusively conifers, host plants vary with region In the Alps, mainly European larch (*Larix decidua* Mill.), stone pine (*Pinus cembra* L.) In the Ore Mountains, Giant Mountains, and Tatra Mountains, Norway spruce (*Picea abies* (L.) H. Karst.), less frequently mountain pine (*Pinus mugo* Turra)
Life cycle	One, partly 2-year life cycle	One generation per year
Adult moths	July–(August), active at dawn	Depending on elevation mid-June to early October, active at dawn
Larvae	August–November (December), overwintering, (February) March–June	Depending on elevation (end of March) April–August
Overwintering	Larvae in leaf litter or upper surface soil layers at tree base	Eggs, under lichen or scales of cones or bark
Natural enemies	Parasitoids (eggs: platygastroid wasps; larvae and pupae: ichneumonid and braconid wasps, tachinids), predators (wood ants, birds, bats; during overwintering wild boars, mice), pathogens (viruses, bacteria, entomopathogenic fungi)	Parasitoids (chalcidid, ichneumonid and braconid wasps, tachinids), predators (mites, true bugs, wood ants, birds), pathogens (viruses)

(continued)

Table 11.3 (continued)

	Pine-tree lappet moth	Larch budmoth
Population dynamics	Monoculture pine stands in warm and xeric areas with low precipitation (<650 mm), non-cyclic, 2–3 years in progradation, followed by mostly 2 years in culmination	Pure and mixed larch stands, regular cycles (8–10 years) In central Alps, favoured by warm, dry spring (coincidence hatching/budburst in larch) Irregular in larch plantations (Pyrenees, England, Japan) Extended outbreaks in central Siberia; population collapse after 2–3 years
Outbreak factors	Warm, dry late summer, improved food quality due to changes in stand properties	High-quality needles, constant local climate conditions, close coincidence of larch budburst and larval hatch
Population collapse	Food shortage, natural enemies, specifically nuclear polyhedrosis viruses, tachinids as larval and pupal parasitoids, egg parasitoids	Food competition and food shortage, low food plant quality, low larval vitality, predisposition to viruses, reduced female fecundity, increasing abundance of parasitoids and predators

Photos: K. Möller (centre), B. Wermelinger (left, right)

Fig. 11.3 Dead pine stand in Brandenburg (Germany). The trees were defoliated by the pine-tree lappet moth in 2005 and died after the extensive summer drought of 2006. Living pines in the background had been sprayed with insecticides. (Photo: K. Möller)

survive under favourable weather conditions (adequate water supply, moderate temperatures) (Möller 2014).

In recent years, a significant dieback of pine forests associated with outbreaks of the pine-tree lappet moth has been observed in the lowlands of Germany (Möller 2015; Habermann 2017) and Poland (Sierpinska 1988), sometimes associated with previous attack by the nun moth (*Lymantria monacha* L.) (Wenk 2016). When the tree is completely defoliated, the hungry larvae migrate to the forest understorey and feed on pine seedlings and saplings. Consequently, pine regeneration mortality in outbreak areas is very high (Menge and Pastowski 2016).

Defoliation reduces tree vitality and vigour and facilitates attack by secondary pests such as bark beetles and wood-boring insects, which ultimately accelerate tree death. During an outbreak of the pine-tree lappet moth in Brandenburg, the trees were also infested by the six-toothed bark beetle (*Ips sexdentatus* Boerner), resulting in a considerably higher forest dieback than expected from either of the two insects alone (Fig. 11.4). Thus, tree mortality leads to ecological (e.g. habitat loss), economic (e.g. timber loss), and social (e.g. recreation, tourism) impacts and increases the risk of fire as the dead material dries. To evaluate a disturbance event from a scientific perspective, it is necessary to consider the time it takes the ecosystem to recover or adapt to new conditions and also to calculate economic loss or ecological consequences. Plant and animal communities under dead or dying trees change, specifically as more light is able to reach the soil. For example, plant and arthropod assemblages typical of pine forests disappear in favour of thermophilic and xerophilic species that prefer open habitats (Möller 2002). The magnitude of change depends, among other things, on how fast various grass species invade the area.

Fig. 11.4 Dead pine trees in south Brandenburg (Germany). The trees were defoliated by the nun moth in 2013 and by the pine-tree lappet moth in 2014, followed by bark beetle attack, mainly the six-toothed bark beetle. For nature conservation reasons, no plant protection measures were implemented. (Photo: K. Möller)

The larch budmoth finds optimal conditions for growth and reproduction in the subalpine conifer forests of the central Alps. At elevations of 1600 to 2200 m, peri- ᴉᴉ ᴉᴉᴉ ᴀᴜᴛᴮᴦᴏᴀᴋᴜ ᴏᴦ ᴛᴴᴄ ᴉᴉᴤᴄᴄᴛ ᴏᴄᴄᴜᴦ ᴀᴛ ᴦᴄᴦᴜᴦ ᴉᴉᴛᴄᴦᴠᴀᴉ ᴉᴉᴇᴦᴠᴀᴉᴤ ᴏᴦ 8–10 years, ᴡᴉᴛᴴ ᴀᴄᴦᴏᴦᴉᴀᴛᴉᴏᴦᴨ levels ranging from visible needle loss on individual trees up to full defoliation of larch and pine stands (Baltensweiler et al. 2008). On larch, defoliation is character- ized by a colour change of the tree crown from green to reddish brown in early summer and yellow brown later in the season. Accordingly, outbreak waves, which move along the Alpine arc from west to east at about 200 km per year, can be fol- lowed visually through the discoloration of the larch needles (Bjørnstad et al. 2002). Larch budmoth population sizes can fluctuate by four to five orders of magnitude at the highest and lowest densities, respectively. During heavy outbreaks up to 30,000 larvae feed on a single tree. By contrast, budmoth density fluctuations on stone pine are less pronounced (Wermelinger et al. 2018).

Periodic outbreaks of the larch budmoth in the subalpine areas of the European Alps are driven by several factors. Host plant quality seems to play a significant role in the population dynamics of the insect since needle quality declines after defolia- tion. The trees usually produce a second flush of needles following early-season defoliation, but these cannot compensate for the loss of foliage. Thus, the lower photosynthetic capacity results in significantly reduced radial growth and xylem cell wall development so that larch budmoth outbreaks leave distinct signatures in tree rings (Fig. 11.5). In the following spring, budbreak is delayed, and the needles remain short and contain fewer nutrients (e.g. nitrogen) but more indigestible (e.g. fibre) or toxic compounds (e.g. resins). The newly emerged budmoth larvae com- pete for a now quantitatively and qualitatively poorer food which in turn affects larval survival and female fecundity, resulting in fewer offspring (negative feed- back). Concomitantly, a growing number of natural enemies (i.e. parasitoids, patho- gens, predators) decimate the larch budmoth population. Lower survival and fecundity of the budmoths after defoliation are drivers of the population cycles. Female moths exhibit an increased flight activity in years of heavy defoliation and prefer to oviposit on green foliage rather than defoliated trees. Hence populations are redistributed to less attacked areas (migration effects), producing the typical outbreak waves that move from valley to valley. In the population growth phase (progradation), high survival and fecundity allow a ten-fold increase from one gen- eration to the next, until the number of insects reaches defoliation threshold densi- ties, and the cycle starts all over again (Benz 1974; Wermelinger et al. 2018).

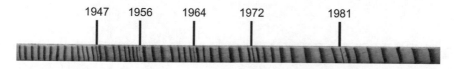

1947   1956   1964   1972   1981

Fig. 11.5 Larch budmoth outbreak years (numbers above tree rings) occur as narrow tree ring widths in larch stems. (Photo: W. Oberhuber; in Oberhuber 2009). Dendrochronological recon- structions of wood core samples revealed regular larch budmoth outbreaks since 800 AD. (Esper et al. 2007)

Temporarily, excessive feeding during larch budmoth outbreaks results in a more open canopy, sunlight reaches the forest floor, and the soil warms up. Additionally, a fertilizing effect has been observed from the accumulation of frass excreted by larvae. This source of extra nutrients (nitrogen) becomes easily accessible for the tree roots. It takes 2–7 years for the larch to recover from defoliation and produce high-quality needles (Benz 1974).

Healthy larches usually survive total defoliation. Even during heavy outbreaks, less than 1% of the trees die. However, tree mortality has been observed in combination with harmful weather conditions, such as summer drought (Esper et al. 2007). Often, defoliation results in growth reductions and narrow tree rings for up to 3 years following outbreaks. Individual trees with low vitality are predisposed to attack from secondary pests such as bark and longhorn beetles. Compared to larch, pines are generally more sensitive to defoliation; their chances of survival are low when needle loss exceeds 60% (Baltensweiler and Fischlin 1988).

Interestingly, the subalpine mixed larch–stone pine forests are not endangered by regular, intensive outbreaks of the larch budmoth. In fact, quite the opposite is true with the population cycles of the larch budmoth being of great ecological importance for the inner-Alpine larch forests. Historical outbreaks reconstructed from tree ring analyses back to 800 AD confirmed the long-lasting, successful coexistence of larch and larch budmoth in these forests (Esper et al. 2007). Larch trees are highly dominant in mountain areas today because of human activities in the past. Pine and spruce trees were deliberately and selectively removed to use the wood. The remaining open spaces turned into wood pastures; larch was favoured for livestock herding because its light canopy permitted suitable forage plants to grow beneath the canopy. Positive selection of larch was achieved by cutting down the competing tree species. In a natural forest succession, the shade-tolerant pines and spruces would outcompete larch and – with no other disturbances occurring – ultimately form the climax forest. Recurrent outbreaks of the larch budmoth, however, counteract the dominance of pine and spruce. While larch recovers from defoliation, pines can do so only to a very limited extent; thus, when they lose their needles, parts of the crown die. Defoliated pines often suffer from subsequent attack by weevils, bark beetles, and mealybugs, leading to poor growth or even tree death. The larch trees, on the other hand, may even take advantage of the fertilization effect from the larval faeces that accelerate soil carbon and nitrogen turnover following insect feeding. As such, the subalpine larch–stone pine forest with its regular budmoth outbreaks has developed into a surprisingly stable ecosystem, from which both larch and larch budmoth mutually benefit (Holtmeier 2002).

11.3   Insect Disturbances and Society

Insects have formed plagues since prehistory. There are reports of terrible insect plagues devastating crops, contributing to famines, and inducing human migrations from the ancient civilizations of Egypt and China. Also in Central Europe, locust invasions were a serious threat to agriculture and even forestry until the mid-eighteenth century. Successful oviposition and development of locusts generally depend on specific landscape structures and habitat conditions. These include suitable vegetation cover, optimal soil temperature, moisture, and water sources. Most outbreaks started in the deltas of rivers (e.g. Danube) flowing into the Black and Caspian seas. When the large European waterways were regulated in the past 100–150 years, locust outbreaks became rare, most likely because their breeding grounds were destroyed (Herrmann 2014).

In the nineteenth and twentieth centuries, crop and livestock production in Europe increased enormously, as innovative technologies were developed by farmers (e.g. novel crops, fertilization methods, rotation patterns, selective breeding, drainage, irrigation). Intensive farming and industrial agriculture, however, have had and continue to have significant negative effects on the environment, including recurrent insect disturbances in agricultural and forest ecosystems, favoured by the introduction of non-indigenous species via global trade. As mentioned in the first section of this chapter, the grapevine phylloxera, an aphid-like insect pest feeding on vine roots, was accidentally introduced to France with American grapevines in the mid-nineteenth century. Eurasian grapevine species have evolved in the absence of this aphid, and phylloxera quickly spread through much of the European wine-growing areas, leaving some areas devastated within a few decades (Goode 2005). Another example is the gypsy moth that was introduced in the mid-nineteenth century from Europe to North America as a potential silk producer, but which subsequently escaped and became established. To date this polyphagous species has defoliated more than 34 million ha of oak forests, with the largest outbreaks occurring from 1980 to 1983 covering over 11 million ha (Tobin and Blackburn 2007).

European forests experienced some of the largest and most severe insect outbreaks in the twentieth century (Table 11.4). The economic and ecological impacts of such vast disturbances are manifold and affect forest owners, the wood-processing industry, and tourism, as well as forest biodiversity and the carbon storage capacity of forest ecosystems (Radermacher 2011; Schulze et al. 2021). While until the 1970s tree-feeding insects were controlled using highly toxic substances such as mercury and arsenic compounds or DDT (dichloro-diphenyl-trichloroethane), today's concept of control is integrated pest management, including biological (e.g. *Bacillus thuringiensis*-based insecticides) and chemical control agents such as tebufenozide that triggers the moulting process in larvae (e.g. Mimic®). Both chemical and biological insecticides are applied to protect mature, high-value forest stands or to slow the spread of invasive species across the landscape and thus mitigate the economic and ecological impacts. In cases where no effective chemical or biological controls are applied or possible, devastating losses of forest ecosystems may occur (see Fig. 11.4).

Although the ecotoxicological risks associated with most forest-use insecticides are quite unlike those of historic compounds, the potential for negative effects still exists. Such risks often dominate the public and political debate associated with the use of pesticides as a forest management practice, as well as with forest certification schemes. In relation to agricultural pesticide use, the few insecticides employed in

Table 11.4 Major forest defoliations by lepidopteran pests in Central Europe from 1920 to 2020

Insect	Country	Years	Infested area (ha)
Pine beauty moth (*Panolis flammea*)	Germany	1922–1924	200,000
Gypsy moth (*Lymantria dispar*)	Bulgaria	1953–1960	1,023,000
Nun moth (*Lymantria monacha*)	Poland	1978–1984	3,000,000
Nun moth	Germany	1978–1984	500,000
Gypsy moth	Romania	1987–1989	700,000
Gypsy moth	Bulgaria	1989–2000	985,000
Gypsy moth	France	1992–1994	104,000
Gypsy moth	Germany	1992–1994	129,000
Pine-tree lappet moth (*Dendrolimus pini*)	Poland	1992–1994	58,000
Pine-tree lappet moth	Germany	1993–1998	259,000
Gypsy moth	Austria	1993	4000
Gypsy moth	Serbia	1997	500,000
Gypsy moth	Slovakia	2002–2006	42,000
Gypsy moth	South-Tyrol	2003	2000
Oak processionary moth (*Thaumetopoea processionea*)	Germany	2009–2018	46,000
Pine-tree lappet moth	Poland	2012–2014	131,000
Pine-tree lappet moth	Germany	2013–2014	25,000
Gypsy moth	Bavaria	2018–2019	10,000
Gypsy moth	Austria	2018	4000

forest management, their relatively low application frequency, the minor proportion of total forest land treated, and the resulting lower environmental impact, public concern raised over forest-use pesticides seems disproportionately high. Decisions about forest protection measures are usually based on solid forest monitoring data, which consider the degree of infestation, susceptibility to the pest, and estimated costs.

All things considered, the divergent ideas about nature conservation mean that there is a huge potential of conflicts between different stakeholder groups at all levels. Many countries have adopted specific laws that strictly prohibit human intervention in protected areas (e.g. national park core areas, wilderness areas). Ecosystems are dynamic, vary across space and time, and are modulated by disturbances (e.g. storms, fires, flooding, insect pest outbreaks) that may change the habitat conditions dramatically in chaotic and unforeseen ways. A disturbance event allows space and resources to be used by new individuals or species. The result, however, often does not correspond to our ideas of 'forest' or 'nature'. For example, large-scale disturbances may result in a landscape-scale extinction of a tree species or create long-term open spaces without trees. In this concept of 'nature conservation', dynamic processes are willingly accepted and considered as key factors to achieve – sooner or later – the highest possible degree of naturalness of an ecosystem (i.e. the presumed natural state before it was affected by man). The other important concept of nature conservation was developed initially in terms of species preservation; later the concept of 'habitat' has been adopted as a more powerful tool integrating functionality, the species within the habitat, their mutual relationships, and interactions with the environment. Here, the goal is to preserve specific stages of development with their respective biodiversity (habitats, organisms) by means of management measures (conservative nature conservation). The preference of one or the other depends on how we define 'nature' (Scherzinger 1996).

11.3.1 Monitoring and Control

Tree-feeding insects are ubiquitous in forest ecosystems. While relatively few species cause widespread mortality, insect outbreaks are important ecological disturbances that can have devastating economic effects. To develop a good pest control strategy, identification of the pest is essential; it allows determination of basic information about the insect life cycle and the time when it is most susceptible to being controlled. Insect outbreaks have been studied in the context of forest disturbances for a long time. However, the interest of forest managers and policy makers seems to change in cycles that follow the population dynamics of the insect itself, although the area affected by outbreaks is often much greater than that destroyed by fire or logging. Thus, to reduce the impact of outbreaks in forest ecosystems, a great focus nowadays is put on the insect species during the endemic (latent) phase (long-term monitoring) when forest management decisions have a strong effect on future outbreak severity. Additionally, modelling approaches that link outbreak intensity with forest conditions are increasingly used (forecasting). Decision support systems are useful during outbreaks to calculate the potential damage with or without interventions (e.g. pesticide applications, salvage cutting, adaptation in harvest schedule) to minimize negative short-term and long-term consequences. In contrast to other land-use types such as agriculture, fruit, and viniculture, significantly higher damage thresholds (i.e. stand-level dieback) apply to the use of insecticides in forest ecosystems. Even though forest insect outbreaks can cause widespread economic loss, they can also be seen as natural processes.

In Europe, oaks are major components of temperate and Mediterranean forests with significant ecological, economic, and cultural importance. All oak species are long-lived trees, and as such they are exposed to varying environmental conditions throughout their lifetime. During the past centuries, mature oak forests have undergone several declines (Wulf and Kehr 1996). Chronic oak decline results from complex interactions of various damaging abiotic (e.g. drought, frost) and biotic factors (e.g. insects, fungi) that occur either simultaneously or sequentially and bring about a serious, long-term decline in tree health and condition. Phyllophagous insects play an essential role in this syndrome, and they are the only factor that can be influenced in the short term. Insect defoliation has been identified as a specific risk factor for oak, with repeated defoliation in spring causing a shortage of carbohydrates and reduced earlywood production resulting in failure of latewood formation (Blank 1997). Once the decline has been initiated (e.g. by repeated defoliation), oaks undergo a loss of vigour resulting in an increased susceptibility to secondary pests and pathogens. Contributing factors (e.g. drought, heat, pathogens) exacerbate the deleterious effects that may ultimately kill the tree.

A key element of any forest pest management programme is a careful evaluation of the forest stand to determine whether pest population levels indicate that intervention is appropriate. For example, the economic threshold level for insecticide application in oak stands is the forecast of repeated defoliation by early-feeding Lepidoptera, including gypsy moth, oak processionary moth (Fig. 11.6), green oak leaf roller (*Tortrix viridana* L.), winter moth (*Operophtera brumata* L.), and mottled umber moth (*Erannis defoliaria* Clerck). On the other hand, insecticide use to combat oak processionary moth infestations in non-outbreak areas where people are living, working, or recreating is justified because of the poisonous hairs (setae) of the larvae that pose a serious risk to human and animal health (Lamy 1990; Maier 2013).

Gypsy moth population densities vary by several orders of magnitude. Commonly, populations persist for many years at densities that are so low that it may be difficult to detect egg masses or larvae at all. Occasionally, and often for unknown reasons, populations grow, and when reaching epidemic (outbreak) levels within a few generations, they have spectacular effects on their habitat (i.e. total defoliation of the host trees). The two most commonly used survey practices of gypsy moth densities are counts of male moths in pheromone traps and pre-season counts of overwintering egg masses on trunks or branches (Delb 2016). In the oak processionary moth, risk analysis and forecasting infestation levels have proven difficult. Monitoring relies on the labour-intensive, time-consuming counting of egg clutches on 1- to 2-year-old twigs in the tree crown or resting/pupation nests on trunks and larger branches or visual assessment of leaf biomass loss (Gößwein and Lobinger 2014).

In the USA, after decades of unsuccessful attempts to eradicate the gypsy moth, an integrated pest management programme was launched to slow the spread of the

Fig. 11.6 Oak defoliation by larvae of the oak processionary moth. (Photo: A. Reichling)

insect (Liebhold et al. 2007). The programme focuses on early detection and suppression of low-level populations along the advancing front, disrupting population
ⅰⅰⅰⅰⅰⅰⅰ ⅰⅰⅰⅰ ⅰⅰⅰⅰⅰⅰ ⅰⅰⅰⅰⅰⅰ ⅰⅰⅰⅰⅰⅰⅰⅰ ⅰⅰⅰⅰⅰⅰⅰ ⅰ ⅰⅰⅰⅰ ⅰⅰⅰⅰⅰⅰⅰ ⅰⅰⅰ ⅰⅰⅰⅰⅰⅰⅰ ⅰⅰⅰⅰⅰⅰⅰ-
cides (i.a. *Bacillus thuringiensis* 'Kurstaki'). Since implementation of the national
programme, the average rate of natural spread of the gypsy moth has been reduced
from 21 km per year (average during the period from 1966 to 1989) to well below
10 km per year (average in the period following implementation of the national
programme in 1999).

A new means to control the gypsy moth, the entomopathogenic fungus
Entomophaga maimaiga (Humber, Shimazu & R.S. Soper), appeared in the last
decades. Native to east Asia, the fungus has become ubiquitous in gypsy moth populations in the northeastern USA, closely following the spread of gypsy moth populations and causing widespread epizootics, eventually yielding mortalities of gypsy
moth caterpillars of up to 99%. In 1999, fungal spores were imported from the USA
and released in Bulgaria and Serbia (Mirchev et al. 2013). The fungus quickly
spread throughout the Balkan Peninsula to Central Europe and is expected to spread
further in the coming years, favoured by trade, tourism, and suitable weather conditions (Zúbrik et al. 2016). Different to viral and bacterial-induced infections, larval
mortality caused by the fungus is largely density independent (i.e. occurring both at
high and low host densities), but fungal infection rates relate strongly to spring
moisture. Rainfall, soil moisture, and air humidity promote fungal infections, while
cool and dry weather conditions decrease fungal infections. Although *E. maimaiga*
is probably the most effective natural mortality agent leading to suppression of outbreak populations, as yet it is unclear how the gypsy moth and the fungus will
interact under altered weather conditions as the climate changes.

Increased surveillance of forest stands increases the costs of forest protection
measures. For example, monitoring of growing populations of the pine-tree lappet
moth is achieved through collecting overwintering larvae from litter and mineral
soil. In spring, sticky bands on tree trunks are used to capture the larvae migrating
from soil to the tree crown. These control methods allow prediction of the upcoming
feeding damage in infested areas (Möller et al. 2007). Near real-time pest incidence
data coupled with remote sensing (e.g. satellite) and geographic information system
(GIS) tools facilitate early warning of impending population build-up and assessment of defoliation levels from a temporal and spatial perspective (Fig. 11.7). The
information is used as a decision support tool to determine where the use of insecticides is needed, as well as for assessment of plant protection measures in retrospect
and restoration/regeneration of infested stands after defoliation (Marx et al. 2015;
Möller and Heinitz 2016).

After World War II, the newly developed pesticides were such successful poisons
against virtually every insect pest that their use became a common procedure in just
about every agricultural crop and, subsequently, in urban and recreational areas as
well. During the post-war years, DDT was applied to control the spectacular larch
budmoth outbreaks in the Engadin valley in southeastern Switzerland and so to
counteract the expected negative consequences of defoliated ('dead') forests on
tourism (Meyer 1946). Today, owing to the detrimental environmental effects and

Fig. 11.7 Satellite grid damage map showing defoliation levels of pine forests in Brandenburg, Germany (green, 0–25%; yellow, >25–50%; red, >50–70%; blue, >70–100%; black, no pine). Pixel size is 5 m × 5 m (RapidEye). The black area in the centre (pixel no. 8065, designated as open space) is the dead pine stand in Fig. 11.4. Total defoliation by the pine-tree lappet moth had been predicted; however, for conservation reasons no pesticide treatment was applied. Data: Black Bridge®, Landesbetrieb Forst Brandenburg

health hazards of DDT, the insecticide is banned for use in agriculture and forestry in almost all countries around the world. For larch budmoth outbreaks, there is consensus that no direct control measures are required, and no pesticides are applied in mountain forests in the European Alps.

11.3.2   Carbon Cycles

Forests represent a prominent part of the global carbon cycle, both storing and releasing carbon dioxide (CO_2) in a dynamic process of growth, decay, disturbance, and renewal. Trees act both as carbon sinks and carbon sources. Natural disturbances may change the forest structure rapidly, causing fast changes in the magnitude and direction of carbon fluxes (i.e. transfers between carbon pools, including those in the ecosystem and the atmosphere) and carbon stocks (i.e. carbon reservoirs in trees and soil). Wildfires, outbreaks of insects, diseases, and windstorms are among the major natural disturbances in the northern hemisphere that have profound effects on forest carbon cycling. During outbreaks, defoliators consume the carbon from leaves and needles and release nitrogen, carbon, and other nutrients into the ecosystem through leaf fragments, frass, and exuviae (the remains of insect ꞮꞮꞮꞮꞮꞮꞮꞮ ꞀꞮ ꞮꞮꞮꞮꞮ ꞮꞮꞮ ꞮꞮꞮꞮꞮꞮ ꞮꞮꞮꞮꞮꞮꞮꞮ ꞮꞮꞮꞮ ꞮꞮꞮꞮꞮꞮ). ꞮꞮ ꞮꞮꞮꞮꞮꞮ, ꞮꞮꞮꞮꞮ ꞮꞮꞮꞮ ꞮꞮ the soil and carbon to nitrogen ratios in frass increased significantly within only a few months in pine forests that were heavily attacked by the pine-tree lappet moth (le Mellec et al. 2010).

Insect outbreaks have the capacity to reduce productivity or even kill trees across extensive areas, and thus they have a large influence on the carbon budget of forests. The most common effects of tree defoliation are reduced gross primary production (i.e. total amount of atmospheric carbon fixed by plants), growth reductions, or even mortality after repeated, severe defoliation, resulting in the reallocation of carbon. Dead trees generate a large amount of dead organic matter, which begins to decay instantly. Soil organisms release much of the stored carbon to the atmosphere (i.e. CO_2 from heterotrophic respiration) during mineralization in the humus layer and mineral soil. Additionally, carbon from easily degradable structures (such as fallen leaves and needles) is released to the atmosphere, while the decomposition of needles that remain on dead coniferous trees for 1 or 2 years is delayed. As a result, carbon release through decomposition following disturbances occurs over an extended period.

Moderate to severe outbreaks may change some of the functional characteristics of the ecosystem (e.g. light regime, water availability, nutrient cycling) in a way that affects the rate of succession and recovery. Severe outbreaks that modify soil moisture and nutrients, for example, alter tree density or species composition so that long-term carbon storage or the rate of carbon cycling is affected. The amount of change of individual carbon fluxes and, therefore, of the net carbon flux is determined by the intensity of the disturbance on stands. For severe outbreaks, the magnitude of the effect may be large enough to transform a forest from a carbon sink to a carbon source.

Following outbreaks, stand productivity can increase if growth of surviving trees or understorey accelerates after release from competition. However, recovery rates depend critically on the number of surviving trees, severity of outbreak, and seedling establishment. Accordingly, several factors are responsible for the variability in carbon cycle, including the type of insect, the time since disturbance, the number of trees affected, and the capacity of the remaining vegetation to increase growth rates following outbreaks. Because of the long time it takes a stand to develop a closed canopy, substantial carbon removal from the atmosphere (CO_2 uptake through photosynthesis) can only be expected after some decades following the disturbance event (Fig. 11.8).

Insects are strongly influenced by climate and weather, and future warming will likely increase the severity and extent of outbreaks, at least in some species. Alterations in forest composition and structure and, therefore, carbon sequestration have implications for atmospheric CO_2 concentrations and thus for future climates.

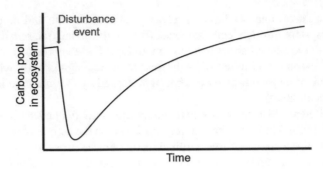

Fig. 11.8 Carbon fluxes and storage in forest ecosystems following disturbance. (After Körner, 2003)

11.4 Outbreaks of Forest-Defoliating Insects and Forest Resilience

Primary (i.e. undisturbed by man) and (most) semi-natural forests are forests where the natural structure, composition, and function have been shaped by natural dynamics (succession) with no or little human interventions over a long time. These forests are characterized by a mosaic of different stages of succession (e.g. open areas with pioneer tree species, dense areas with dominant tree species, areas with dying or dead trees), which coexist and usually have a high diversity of species. Forest succession is a continual process, and the forest ecosystem is thus in a constant state of flux. Natural disturbances like storms, fires, and insect attacks will clear an area, often combined with an increase in structural diversity, and the successional process will start all over again.

The capacity of forests to maintain a high level of stability, resilience (i.e. recover from or cope with disturbances), and biodiversity is measured by their naturalness which relates to the natural ('original') state of the forest. Near-natural forests with a species composition that corresponds to (or is close to) the natural forest community are a prerequisite for ecosystems that are stable over a longer period. Mixed species stands based on the native tree species composition largely offer favourable conditions for the natural enemies of phyllophagous insect defoliators. For example, the number of secondary hosts for parasitic wasps or food plants for adult wasps increases with plant diversity in the overstorey (i.e. trees) and understorey (i.e. shrubs, herbs) layers of the forest (Hawkins 1994; Kratochwil and Schwabe 2001). In general, species richness and diversity increase the complexity of the ecosystem and thus the number of ecological niches.

In Central Europe, outbreaks of forest-defoliating insects (e.g. gypsy moth, pine-tree lappet moth, larch budmoth) collapse after a few years owing to local ecological and environmental factors (e.g. weather, natural enemies, food shortage). Collapse of gypsy moth outbreaks, for example, is mainly driven by some species of parasitic flies (tachinids) and a specific nucleopolyhedrosis virus that is persistent within the population but tends to spread widely at high population densities (Schwenke 1978).

A tiny wasp (*Telenomus laeviusculus* Ratzeburg) that parasitizes the eggs causes outbreak populations of the pine-tree lappet moth to collapse naturally (Möller and Dernmann 2009). The cyclic behaviour of the larch budmoth has been attributed to foliage quality, with the amplitude of the outbreak cycle set by the insect–plant relationship (Benz 1974). Reductions in larch needle quality (lower nitrogen, higher fibre content) after heavy defoliation induce a dramatic decline in female fecundity. The delayed recovery in foliage quality explains the continued decline in insect numbers and vitality that is realized in subsequent generations and is typical for phytophagous forest insect populations undergoing multi-annual cycles and outbreaks.

11.5 Disturbance Interactions

Over the past decades, ecologists have increasingly focused not only on understanding individual disturbance agents but also on understanding how multiple disturbances interact. Much research is done by looking at how the occurrence of one disturbance affects the occurrence of subsequent disturbances, at how the occurrence of two or more disturbances in short succession cumulatively affects forest ecosystems, and at how the magnitude of disturbances determines interactive effects.

Abiotic factors can clearly promote insect population build-up to outbreak levels within a short time. Thus, warm and dry weather favours development of thermophilic butterflies and moths. Abiotic disturbances such as windthrow, snow break, ice damage, and avalanches produce an abundance of downed trees or trees physiologically weakened by drought. These trees are highly suitable hosts for successful brood production of bark beetles. In addition, abiotic disturbances also affect the extent and impact of insect outbreaks. For example, defoliation followed by drought events reduces the regenerative capacity of trees dramatically (see Sect. 11.2.2, Fig. 11.3). Interacting disturbance effects have been observed for the six-toothed bark beetle and the pine-tree lappet moth. Pine trees, which were first defoliated by moth larvae and subsequently infested by bark beetles, experienced significantly higher mortality rates than defoliated trees alone (see Fig. 11.4).

In continuous canopies, the crowns may shelter each other from wind, frost, or heat, but mortality of individual trees creates gaps in the canopy that predispose the stand to greater risk of damage (Kallweit and Mayer 2008). Usually, biotic agents (e.g. bark beetles, defoliators, fungal diseases) initiate the formation of gaps, while windthrows are important agents of gap expansion (Fig. 11.9; see Chaps. 8 and 14). Insect disturbances affecting long-lived, dominant trees may be sporadic, but their impacts can produce long-term and complex changes in forest structure and species composition.

Fig. 11.9 Effects of winter storm Kyrill (2007) on a pine stand in Schorfheide, Brandenburg (Germany). Tree mortality in the thinned stand, which had been defoliated by nun moth larvae 2 years earlier, was significantly higher than in a non-attacked stand nearby. (Photo: E. Hafemann)

11.6   Outbreaks of Forest-Defoliating Insects and Climate Change

In Europe, the projected increase in weather extremes (e.g. heat waves, storms, heavy precipitation) anticipated by the Intergovernmental Panel on Climate Change (IPCC) in the upcoming decades will have significant consequences for forest ecosystems (Kovats et al. 2014). Extreme events such as drought, heavy rainfall, frost, or hail generally influence the course and effect of so-called complex forest disorders, which result from a combination of primary adverse stress factors and attack by secondary organisms. For example, oak dieback is clearly associated with recurrent drought spells in Central Europe (Thomas et al. 2002).

In terms of insect outbreaks, climate change will influence population dynamics by modifying insect reproduction, survival, and ranges. Changes can also be expected between host plants, insect pests, and the natural enemies (predators, parasitoids, pathogens) of the pests. Warm weather usually favours survival and rapid development of larvae, thus accelerating pest population growth and eventually intensifying outbreak severity (Hentschel et al. 2018). To compensate for loss of foliage from insect feeding, host trees activate or develop defence mechanisms (e.g. tough leaves/needles, and resin or tannin production) and allocate carbon from stored reserves for regrowth. These depletion of reserves and consumption of extra energy often make the trees more vulnerable to other stressors (e.g. drought) than non-defoliated trees. Thus, the different stress levels and stand conditions need to be considered in future risk assessments (Möller 2015).

Given the significantly shorter generation cycles and usually higher mobility of phytophagous insects compared to their host plants, we can expect that the insects will adapt quickly to any changes in their environment. Polyphagous insects like gypsy moth or nun moth larvae compensate for poor nutritional food qualities by consuming more of the poor-quality food or by switching to suitable alternative plants (Hättenschwiler and Schafellner 1999, 2004). Temperature-related transitions from 2-year (or more) life cycles to 1-year life cycles have been observed in some insects. For example, spruce sawflies with extended life cycles as a consequence of slow growth and repeated or long periods of dormancy shorten their development significantly as ambient temperatures increase (Battisti 2004; Antonitsch and Schafellner 2017). In a similar way, warm and dry weather in early summer induces the common pine sawfly (*Diprion pini*) to switch from a univoltine (one generation per year) to a bivoltine (two generations per year) cycle (Möller et al. 2017). With the increase in temperatures in the past decades, range shifts and outbreaks beyond traditional limits have been observed in several insect species such as oak processionary moth, pine processionary moth (*Thaumetopoea pityocampa* Denis and Schiffermüller) (Roques 2015), gypsy moth, nun moth, or pine-tree lappet moth (Vanhanen et al. 2007; Ray et al. 2016).

Long-term observation data on population dynamics and damage levels under changing environments and climate conditions are limited to only a few pest species (Netherer and Schopf 2010; Hentschel et al. 2018). From 1990 to 2005, large gypsy moth outbreaks were recorded in Central Europe, specifically in the lowland areas of Germany, Switzerland, Austria, Hungary, the Czech Republic, and Slovakia. Gypsy moth population levels increased even in countries with less favourable weather conditions like England. Although the causes for the outbreaks are not fully understood, the strongest population eruptions correlated with years with high spring temperatures and little rainfall (Wulf and Graser 1996; Bub 2006; Csóka et al. 2015; Hlásny et al. 2016). In recent years, beech forests at higher elevations in Hungary and Croatia have increasingly been subjected to defoliation, which is also seen as an effect of the changing climate (Pernek et al. 2008; Csóka et al. 2015).

Similarly, the rapid increase in population densities of the oak processionary moth and the corresponding damage levels coincide with early spring onset and well above-average temperatures. Such weather conditions induce early budburst in oaks and promote high survival of newly emerged larvae (Wagenhoff et al. 2014; Schröder et al. 2016). Recently, the span between two outbreak cycles of the pine-tree lappet moth has shortened in northeast Germany (Gräber et al. 2012), with high temperatures and little rainfall as reliable indicators of impending outbreaks (Ray et al. 2016). In particular, 'favourable climate windows' during certain life stages promote population growth, and specifically, warm and dry weather enhances feeding and growth of old larvae in May; moth flight, dispersal, and oviposition in late summer; survival of newly hatched larvae in September; and fitness of the overwintering larvae in October (Hentschel et al. 2018). Also, warm and dry weather intensifies large-scale defoliation caused by pine sawflies (e.g. *Diprion pini*) and spruce sawflies (e.g. *Pristiphora abietina* Christ, *Pachynematus montanus* Zadd.) (Fig. 11.10). In all cases, reliable predictions of population dynamics require

Fig. 11.10 Large-scale defoliation of pines in the northeast lowlands of South Brandenburg (Germany) by the common pine sawfly (*Diprion pini*) in 2016. A second sawfly generation developed in the same year, increasing feeding damage and tree mortality. (Photo: M. Kopka)

appropriate input data, including climate/weather parameters, stand conditions, and temperature-based insect phenology data. Complexity and difficulty for modelling arise from the fact that the input data must be determined for each insect species and location (Schafellner and Schopf 2014; Möller et al. 2017; Hentschel et al. 2018).

Climate change, however, does not always promote insect population growth. For example, the occurrence of the Alpine-wide spectacular outbreaks of the larch budmoth at elevations between 1600 and 2000 m ceased after 1990 (Iyengar et al. 2016). This period coincides with the global increase in temperature observed in recent decades, but the temperature rise in the European Alps was almost twice the global average. Mild winters seem to be a main driver for the disruption of larch budmoth outbreaks. Higher temperatures lead to increased egg mortality during the winter diapause (because energy reserves are exhausted), a time lag between larval and leaf development in spring (phenological mismatch), and a warming-induced upward shift to higher elevations where sufficient food resources are lacking. As a result, populations remain at sub-outbreak levels, and defoliation capacity of the larvae remains low (Wermelinger et al. 2018).

It is likely that temperate forests will be more affected by insect pests in the future than has been the case in the past, both due to climate change and inadequate forest management practices. In the case of severe ecosystem disturbances, such as a landscape-level forest dieback, carbon that has entered the forests over decades or centuries is rapidly released into the atmosphere ('slow in, fast out'; Körner 2003). This loss of carbon sinks in terrestrial ecosystems, in turn, amplifies climate change.

References

Altenkirch W, Majunke C, Ohnesorge B (2002) Waldschutz auf ökologischer Grundlage. Eugen Ulmer, Stuttgart, 434 S

Antonitsch A, Schafellner C (2017) Auswirkungen einer Klimaänderung auf die Populationsdynamik der Fichten-Gespinstblattwespe im Waldviertel. In: Bednar-Friedl B, Formayer H, Brand U, Kienberger S, Auer I (eds) 18. Klimatag, Aktuelle Klimaforschung in Österreich, Wien. S. 20–21

Baltensweiler W, Fischlin A (1988) The larch budmoth in the Alps. In: Berryman AA (ed) Dynamics of forest insect populations: patterns, causes, implications. Plenum Press, New York, pp 331–351

Baltensweiler W, Weber UM, Cherubini P (2008) Tracing the influence of larch-bud-moth insect outbreaks and weather conditions on larch tree-ring growth in Engadine (Switzerland). Oikos 117:161–172

Battisti A (2004) Forests and climate change – lessons from insects. Forest 1:17–24

Benz G (1974) Negative Rückkoppelung durch Raum- und Nahrungskonkurrenz sowie zyklische Veränderung der Nahrungsgrundlage als Regelprinzip in der Populationsdynamik des Grauen Lärchenwicklers, Zeiraphera diniana (Guenée) (Lep., Tortricidae). J Appl Entomol 76:196–228

Berryman AA (1988) Dynamics of forest insect populations: patterns, causes, implications. Plenum Press, New York, 603 p

Björkman C, Lindelöw Å, Eklund K, Kyrk S, Klapwijk MJ, Fedderwitz F, Nordlander G (2013) A rare event – an isolated outbreak of the pine-tree lappet moth (Dendrolimus pini) in the Stockholm archipelago. [En ovanlig händelse – ett isolerat utbrott av tall-spinnare (Dendrolimus pini) i Stockholms skärgård.]. Entomologisk Tidskrift 134:1–9. Uppsala, Sweden 2013. ISSN 0013-886x

Bjørnstad ON, Peltonen M, Liebhold AM, Baltensweiler W (2002) Waves of larch budmoth outbreaks in the European Alps. Science 298:1020–1023

Bjørnstad ON, Robinet C, Liebhold AM (2010) Geographic variation in North American gypsy moth cycles: subharmonics, generalist predators, and spatial coupling. Ecology 91:106–118

Blank R (1997) Ringporigkeit des Holzes und häufige Entlaubung durch Insekten als spezifische Risikofaktoren der Eichen. For Holz 52:235–242

Block J, Delb H, Hartmann G, Seemann D, Schröck H (1995) Schwere Folgeschäden nach Kahlfraß durch Schwammspinner im Bienwald. AFZ/Wald 50:1278–1281

Bréda N, Granier A (1996) Intra-and interannual variations of transpiration, leaf area index and radial growth of a sessile oak stand (Quercus petraea). Ann Sci For 53:521–536

Bub G (2006) Massenvermehrung von Schwammspinner und Eichenprozessionsspinner. FVA-Einblick 10(1):1–5

Bussler H (2014) Käfer und Großschmetterlinge an der Traubeneiche. LWF Wissen 75:89–93

Clancy KM, Wagner MR, Tinus RW (1988) Variations in nutrient levels as a defense: identifying key nutritional traits of host plants of the western spruce budworm. In: Mattson WJ, Levieux J, Bernard-Dagan C (eds) Mechanisms of woody plant defenses against insects: search for pattern. Springer, New York, pp 203–214

Csóka G, Pödör Z, Nagy G, Hirka A (2015) Canopy recovery of pedunculate oak, Turkey oak and beech trees after severe defoliation by gypsy moth (Lymantria dispar): case study from western Hungary. For J 61:143–148

Delb H (2012) Eichenschädlinge im Klimawandel in Südwestdeutschland. FVA-Einblick 16(2):11–14

Delb H (2016) Monitoring und Prognose der Schadorganismen im Wald: eine Kernaufgabe des Waldschutzes. FVA-Einblick 20(1):4–9

Delb H, Block J (1999) Untersuchungen zur Schwammspinner-Kalamität von 1992 bis 1994 in Rheinland-Pfalz. Mitt Forstl Vers Anst Rheinland-Pfalz 45:1–241

Dettner K, Peters W (2003) Lehrbuch der Entomologie. Spektrum Akademischer Verlag, Heidelberg/Berlin, 921 p

Elkinton JS, Liebhold AM, Muzika R-M (2004) Effects of alternative prey on predation by small mammals on gypsy moth pupae. Popul Ecol 46:171–178

Esper J, Büntgen U, Frank DC, Nievergelt D, Liebhold A (2007) 1200 years of regular outbreaks in alpine insects. Proc R Soc Lond B Biol Sci 274:671–679

Führer E (1998) Oak decline in Central Europe: a synopsis of hypotheses. In: Proceedings of population dynamics, impacts, and integrated management of forest defoliating insects, USDA forest service general technical report NE-247, pp 7–24

Gaertig T, Wilpert K, Seemann D (2005) Differentialdiagnostische Untersuchungen zu Eichenschäden in Baden-Württemberg. Ber Freibg Forstl Forsch 61:1–161

Goode J (2005) The science of wine: from vine to glass. University of California Press, Berkeley/Los Angeles, 216 p

Gößwein S, Lobinger G (2014) Waldschutzrelevante Organismen an der Traubeneiche. LWF Wissen 75:80–88

Gräber J, Ziesche T, Möller K, Kätzel R (2012) Gradationsverlauf der Kiefernschadinsekten im Norddeutschen Tiefland. AFZ/Wald 9:35–38

Habermann M (2017) Auswirkungen der Anwendungsbestimmungen für die Ausbringung von Pflanzenschutzmitteln mit Luftfahrzeugen im Wald. J Kult 69:249–254

Hansen J (1999) Radialzuwachsverlauf und Gefäßstruktur der Jahrringe von Eichen in ausgewählten Beständen des Bienwaldes im Rahmen der Schwammspinner-Kalamität 1993/94. Mitt Forstl Vers anst Rheinland-Pfalz 45:151–175

Hättenschwiler S, Schafellner C (1999) Opposing effects of elevated CO_2 and N deposition on *Lymantria monacha* larvae feeding on spruce trees. Oecologia 118:210–217

Hättenschwiler S, Schafellner C (2004) Gypsy moth feeding in the canopy of a CO_2-enriched mature forest. Glob Chang Biol 10:1899–1908

Hawkins BA (1994) Pattern and process in host-parasitoid interactions. Cambridge University Press, Cambridge, 190 p

Hentschel R, Möller K, Wenning A, Degenhardt A, Schröder J (2018) Importance of ecological variables in modeling population dynamics of pine pest insects. Front Plant Sci 9:1667

Herrmann I (2014) Räumliche Optimierung der Bestandesstruktur unter Berücksichtigung von Einzelbaumeffekten. Universität Dresden; Fakultät Umweltwissenschaften, Dresden, 233 p

Hilker M, McNeil J (2008) Chemical and behavioral ecology in insect parasitoids: how to behave optimally in a complex odorous environment. In: Wajnberg E, Bernstein C, van Alphen J (eds) Behavioral ecology of insect parasitoids: from theoretical approaches to field applications. Blackwell, Malden, pp 92–112

Hlásny T, Trombik J, Holuša J, Lukášová K, Grendár M, Turčáni M, Zúbrik M, Tabaković-Tošić M, Hirka A, Buksha I, Modlinger R, Kacprzyk M, Csóka G (2016) Multi-decade patterns of gypsy moth fluctuations in the Carpathian Mountains and options for outbreak forecasting. J Pest Sci 89:413–425

Holtmeier FK (2002) Tiere in der Landschaft – Einfluss und ökologische Bedeutung (2. Aufl.). Ulmer, Stuttgart, 367 p

Iyengar SV, Balakrishnan J, Kurths J (2016) Impact of climate change on larch budmoth cyclic outbreaks. Sci Rep 6:27845

Jäkel A, Roth M (2004) Umwandlung einschichtiger Kiefernmonokulturen in strukturierte (Misch) bestände: Auswirkungen auf parasitoide Hymenoptera als Schädlingsantagonisten. Mitt Dtsch Ges allg angew Entomol 14:265–268

Jansen S (2003) Schädlinge. Geschichte eines wissenschaftlichen und politischen Konstrukts 1880–1920. Campus Verlag, Frankfurt/New York, 437 p

Kallweit R, Mayer U (2008) § 10 LWaldG (Kahlschlag) – Was sind freilandähnliche Verhältnisse nach Holznutzungsmaßnahmen? Eberswalder Forstl. Schrreihe 35:26–34

Kätzel R, Löffler S, Schröder J (2013) Sterben vor der Zeit – Das Eichensterben als Komplexkrankheit. Eberswalder Forstl Schrreihe 53:21–34

Körner C (2003) Slow in, rapid out – carbon flux studies and Kyoto targets. Science 300:1242–1243

Kovats RS, Valentini R, Bouwer LM, Georgopoulou E, Jacob D, Martin E, Rounsevell M, Soussana J-F (2014) Europe. In: Climate change 2014: impacts, adaptation, and vulnerability. Part B: regional aspects. Contribution of working group II to the fifth assessment report of the intergovernmental panel on climate change. Cambridge University Press, Cambridge/New York, pp 1267–1326

Kratochwil A, Schwabe A (2001) Ökologie der Lebensgemeinschaften. Biozönologie. UTB. Ulmer, Stuttgart, 756 p

Lamy M (1990) Contact dermatitis (erucism) produced by processionary caterpillars (genus *Thaumetopoea*). J Appl Entomol 110:425–437

le Mellec A, Krummel T, Korczynsk I, Reinhardt A, Vogt-Altena H, Slowik J, Erasmi S, Thies C, Gerold G, Rohe W, Roloff A, Rust S, Lasch P, Möller K, Kätzel R, Zeller B, Karg J, Mazur A, Bernacki Z, Rennenberg H (2010) From carbon sinks to carbon sources – insect mass outbreaks and the increased risk of carbon loss in forest ecosystems. In: Endlicher W, Gerstengarbe F-W (eds) Continents under climate change, Report number 115. Potsdam Institute for Climate Impact Research (PIK), Berlin, 115 p

Liebhold AM, Sharov AA, Tobin PC (2007) Population biology of gypsy moth spread. In: Tobin PC, Blackburn LM (eds) Slow the spread: a national program to manage the gypsy moth, General Technical Reports NRS-6. U.S. Department of Agriculture, Forest Service, Northern Research Station, Newtown Square, pp 15–32

Lobinger G (2006) Entwicklung neuer Strategien im Borkenkäfermanagement. Forstschutz Aktuell 37:11–13

Maier H (2013) The Pussy Caterpillar; Gesundheitliche Gefahren durch die Brennhaare des Eichenprozessionsspinners (*Thaumetopoea processionea* Linné). Julius-Kühn-Archiv 440:33–34

Marx A, Möller K, Wenk M (2015) RapidEye-Waldschutzmonitoring in Brandenburg. AFZ/Wald 11:40–42

Matyssek R, Fromm J, Rennenberg H, Roloff A (2010) Biologie der Bäume. Von der Zelle zur globalen Ebene. UTB. Ulmer, Stuttgart, 439 p

Menge A, Pastowski F (2016) Kahlfraß der Naturverjüngung – wenig Chancen auf Erholung. AFZ/Wald 15:35–37

Meyer A (1946) Untersuchungen über die Bekämpfung des grauen Lärchenwicklers (*Semasia diniana* Gn.) in den Wäldern des Ober-Engadins. Mitt Schweiz Entomol Ges 20:452–474

Mirchev P, Linde A, Pilarska D, Pilarski P, Georgieva M, Georgiev G (2013) Impact of *Entomophaga maimaiga* on gypsy moth populations in Bulgaria. IOBC-WPRS Bull 90:359–363

Möller K (2002) Der Einfluss von Störungen auf die Arthropodenfauna in Kiefernforsten Brandenburgs. Beitr Forstwirtsch Landsch ökol 36:77–80

Möller K (2014) Klimawandel und integrierter Waldschutz – Risikomanagement mit mehr Unbekannten und weniger Möglichkeiten. Eberswalder Forstl Schrreihe 55:59–65

Möller K (2015) Nur ein toter Baum ist ein guter Baum – Das Ende der Multifunktionalität des Waldes? Eberswalder Forstl. Schrreihe 59:70–78

Möller K, Bemmann M (2009) Eiparasitoide als Gegenspieler von Kiefernschädlingen. AFZ/Wald 8:396–399

Möller K, Heinitz M (2016) Waldschutz in Brandenburg – das Risikomanagement erfordert die Zusammenarbeit von Forst- und Naturschutzbehörden. Natschutz Landschpfl Brandenburg 1(2):30–39

Möller K, Apel K-H, Engelmann A, Hielscher K, Walter C (2007) Die Überwachung der Waldschutzsituation in den Kiefernwäldern Brandenburgs – Weiterentwicklung bewährter Methoden. Eberswalder Forstl Schrreihe 32:288–296

Möller K, Hentschel R, Wenning A, Schröder J (2017) Improved outbreak prediction for common pine sawfly (*Diprion pini* L.) by analyzing floating «climatic windows» as keys for changes in voltinism. Forests 8:319

Moreau G, Lucarotti CJ (2007) A brief review of the past use of baculoviruses for the management of eruptive forest defoliators and recent developments on a sawfly virus in Canada. For Chron 83:105–112

Mumm R, Hilker M (2006) Direct and indirect chemical defence of pine against folivorous insects. Trends Plant Sci 11:351–358

Murray KD, Elkinton JS (1989) Environmental contamination of egg masses as a major component of transgenerational transmission of gypsy moth nuclear polyhedrosis virus (LdMNPV). J Invertebr Pathol 53:324–334

Netherer S, Schopf A (2010) Potential effects of climate change on insect herbivores in European forests – general aspects and the pine processionary moth as specific example. For Ecol Manag 259:831–838

Niesar M, Zúbrik M, Kunca A (2015) Waldschutz im Klimawandel. Landesbetrieb Wald und Holz Nordrhein-Westfalen, Münster, 202 S

Nola P, Morales M, Motta R, Villalba R (2006) The role of larch budmoth (*Zeiraphera diniana* Gn.) on forest succession in a larch (*Larix decidua* Mill.) and Swiss stone pine (*Pinus cembra* L.) stand in the Susa Valley (Piedmont, Italy). Trees 20:371–382

Oberhuber W (2009) Bäume als Zeugen der Klima- und Umweltgeschichte. In: Hofer R (Hrsg.) Die Alpen. Einblicke in die Natur. Innsbruck University Press, Innsbruck, pp 57–60

Patočka J, Krištin A, Kulfan J, Zach P (1999) Die Eichenschädlinge und ihre Feinde. Institut für Waldökologie der Slowakischen Akademie der Wissenschaften, Zvolen, 396 p

Pernek M, Pilas I, Vrbek B, Benko M, Hrasovec B, Milkovic J (2008) Forecasting the impact of the gypsy moth on lowland hardwood forests by analyzing the cyclical pattern of population and climate data series. For Ecol Manag 255:1740–1748

Quiring D, McKinnon M (1999) Why does early-season herbivory affect subsequent budburst? Ecology 80:1724–1735

Radermacher FJ (2011) Wege zum 2-Grad-Ziel – Wälder als Joker. Polit Ökol 127:136–139

Ray D, Peace A, Moore R, Petr M, Grieve Y, Convery C, Ziesche T (2016) Improved prediction of the climate-driven outbreaks of *Dendrolimus pini* in *Pinus sylvestris* forests. Forestry 89:230–244

Roques A (2015) Processionary moths and climate change: an update. Springer, Dordrecht, 427 p

Ruohomäki K, Tanhuanpää M, Ayres MP, Kaitaniemi P, Tammaru T, Haukioja E (2000) Causes of cyclicity of *Epirrita autumnata* (Lepidoptera, Geometridae): grandiose theory and tedious practice. Popul Ecol 42:211–223

Schafellner C, Schopf A (2014) Massenauftreten der Fichtengebirgsblattwespe in Tieflagen als Folge des Klimawandels? Forstschutz Aktuell 60(61):12–19

Scherzinger W (1996) Naturschutz im Wald: Qualitätsziele einer dynamischen Waldentwicklung. Eugen Ulmer, Stuttgart, 448 p

Schopf R (1986) Laboruntersuchungen zur Disposition unterschiedlich geschädigter Fichten (*Picea abies* Karst.) für den Befall durch *Gilpinia hercyniae* Htg.(Hym., Diprionidae) und *Lymantria monacha* L. (Lep., Lymantriidae). J Appl Entomol 101:389–396

Schröder J, Wenning A, Hentschel R, Möller K (2016) Rückkehr eines Provokateurs: Was steuert die Ausbreitungsdynamik des Eichenprozessionsspinners in Brandenburg? Eberswalder Forstl Schrreihe 62:77–88

Schultz JC, Baldwin IT (1982) Oak leaf quality declines in response to defoliation by gypsy moth larvae. Science 217:149–151

Schulze EE, Rock J, Kroiher F, Egenolf V, Wellbrock N, Irslinger R, Bolte A, Spellmann H (2021) Klimaschutz im Wald. Biol Unserer Zeit 1:46–54

Schwenke W (1978) Die Forstschädlinge Europas. Schmetterlinge. 3. Band. Paul Parey, Hamburg und Berlin, 467 p

Schwerdtfeger F (1981) Die Waldkrankheiten: ein Lehrbuch der Forstpathologie und des Forstschutzes, 4th edn. Paul Parey, Hamburg/Berlin, 486 p

Sierpinska A (1988) Towards an integrated management of *Dendrolimus pini* L. In: McManus ML, Liebhold AM (eds) Population dynamics, impacts, and integrated management of forest defoli- IIIIII)(IIIIIIIII), I I I I III IIIIIIII I II IIIIII)IIIIIIII IIIIIIIIIIII IIIIIIII IIII III I, III I IIII III I

Strasser LJZ, Wolf M, Lobinger G, Petercord R (2013) Die Waldschutzsituation in Bayern 2012. AFZ/Wald 7:12–15

Thomas FM, Blank R, Hartmann G (2002) Abiotic and biotic factors and their interactions as causes of oak decline in Central Europe. For Pathol 32:277–307

Tobin PC, Blackburn LM (2007) Slow the spread: a national program to manage the gypsy moth, General technical reports NRS-6. U.S. Department of Agriculture Forest Service, Northern Research, Newtown Square, 109 p

Vanhanen H, Veteli TO, Päivinen S, Kellomäki S, Niemelä P (2007) Climate change and range shifts in two insect defoliators: gypsy moth and nun moth – a model study. Silva Fenn 41:621

Varley GC (1949) Population changes in German forest pests. J Anim Ecol 18:117–122

Wagenhoff E, Wagenhoff A, Blum R, Veit H, Zapf D, Delb H (2014) Does the prediction of the time of egg hatch of *Thaumetopoea processionea* (Lepidoptera: Notodontidae) using a frost day/temperature sum model provide evidence of an increasing temporal mismatch between the time of egg hatch and that of budburst of *Quercus robur* due to recent global warming? Eur J Entomol 111:207

Weckwerth W (1952) Der Kiefernspinner und seine Feinde. Akademische Verlagsgesellschaft Geest & Portig, Leipzig, 40 p

Wehrli A, Brang P, Maier B, Duc P, Binder F, Lingua E, Ziegner K, Kleemayr K, Dorren L (2007) Schutzwaldmanagement in den Alpen – eine Übersicht/Management of protection forests in the Alps – an overview. Schweiz Z Forstwes 158:142–156

Wenk M (2016) Dokumentation des Schadverlaufs nach Kahlfraß in Kiefernforsten – Einsatz von Satellitenbildern und Bügelschaber. Eberswalder Forstl Schrreihe 62:23–26

Wenk M, Möller K (2013) Prognose Bestandesgefährdung – Bedeutet Kahlfraß das Todesurteil für Kiefernbestände? Eberswalder Forstl Schrreihe 51:9–14

Wermelinger B, Forster B, Nievergelt D (2018) Cycles and importance of the larch budmoth. WSL Fact Sheet 61, 12 p

Wulf A, Graser E (1996) Gypsy moth outbreaks in Germany and neighboring countries. Nachrbl Dtsch Pflanzenschutzd 48:265–269

Wulf A, Kehr R (1996) Eichensterben in Deutschland. Situation, Ursachenforschung und Bewertung. Mitt Biol Bundesanst Land- Forstwirtsch Berl-Dahl 318:157

Zúbrik M, Hajek A, Pilarska D, Špilda I, Georgiev G, Hrašovec B, Hirka A, Goertz D, Hoch G, Barta M, Saniga M, Kunca A, Nikolov C, Vakula J, Galko J, Pilarski P, Csoka G (2016) The potential for *Entomophaga maimaiga* to regulate gypsy moth *Lymantria dispar* (L.) (Lepidoptera: Erebidae) in Europe. J Appl Entomol 140:565–579

Chapter 12
Bark Beetles

Beat Wermelinger ⓘ and Oliver Jakoby ⓘ

Abstract Bark beetles are important components of the natural dynamics of coniferous forests. In Europe, the European spruce bark beetle (*Ips typographus*) in particular has the potential to cause extensive infestations and thus can act as an ecological disturbance. After disturbances such as windthrow or drought, this beetle can increase its population density to such an extent that it is able to successfully colonize vigorous trees. The further development of mass infestations of living trees mainly depends on temperature and the susceptibility of the host trees. From an ecological viewpoint, bark beetles contribute to the natural dynamics of forest ecosystems and create valuable new habitats for many organisms. Socio-economically, however, bark beetle outbreaks often lead to extensive damage.

Keywords Climate change · Damage · Dynamics · Ecological significance · European spruce bark beetle · Host tree · Infestation · Mountain pine beetle · Outbreak · Scolytinae

12.1 Distribution and Ecology

Mass outbreaks of bark beetles (Curculionidae: Scolytinae) are natural events in the long-term dynamics particularly of coniferous forests. European spruce forests and North American pine forests are among the most frequently and most severely affected ecosystems. In European forests, 8% of all forest damage is caused by bark beetles (Schelhaas et al. 2003). Due to their economic significance, such outbreaks are mainly perceived as an economic damage and less as natural disturbances. However, of the more than 250 European bark beetle species, only a few bark

B. Wermelinger (✉) · O. Jakoby
Forest Health and Biotic Interactions Research Unit, Swiss Federal Research Institute WSL, Birmensdorf, Switzerland
e-mail: b.wer@hispeed.ch

© The Author(s), under exclusive license to Springer Nature
Switzerland AG 2022
T. Wohlgemuth et al. (eds.), *Disturbance Ecology*, Landscape Series 32,
https://doi.org/10.1007/978-3-030-98756-5_12

271

breeding specialists are able to mass propagate on vigorous trees – under specific conditions – and to cause extensive disturbances. The most important species of these pest species belong to the genera *Ips* and *Dendroctonus*. In Europe, these are mainly the European spruce bark beetle (*Ips typographus* [L.]) and the pine bark beetle (*I. acuminatus* [Gyll.]), as well as, to a lesser extent, the six-toothed spruce bark beetle (*Pityogenes chalcographus* [L.]) and *Scolytus* and *Tomicus* species (Grégoire and Evans 2004; Wermelinger 2004). In North America, the mountain pine beetle (*Dendroctonus ponderosae* Hopk.) and other species of the genera *Dendroctonus* and *Ips* play a major role (Kleinman et al. 2012; Six and Bracewell 2015). Most of these species attack adult conifers and are monophagous, i.e. restricted to a single tree genus. A comprehensive overview of the general biology and ecology of bark beetles is given by Raffa et al. (2015b).

Although the lifestyles of bark beetles can be very different, the above-mentioned species share some common features. The most important and therefore best studied bark beetle in Europe is the European spruce bark beetle (*I. typographus*). It is used here as an example to discuss the general biology of bark beetles. The so-called pioneer males look for Norway spruce (*Picea abies* [L.] H.Karst.) hosts that are suitable for colonization. They are mainly guided by visual cues and host-tree specific volatiles (Byers 2004; Saint-Germain et al. 2007). After boring into the bark, they release aggregation pheromones consisting of monoterpenoids (Blomquist et al. 2010) that attract other conspecifics. The beetles mate under the bark and the females then deposit their eggs in the maternal galleries. The hatched larvae develop in the nutrient-rich phloem and pupate at the end of their larval galleries. This leads to the characteristic breeding pattern of this species (Fig. 12.1). It takes the European spruce bark beetle 2–3 months to complete its development, including the maturation feeding of the young beetles at the place of their development. Subsequently, the beetles take flight or overwinter underneath the bark. Since the development time of bark beetles is largely controlled by temperature (Wermelinger and Seifert 1998), the number of annual generations also depends on elevation (see Sect. 12.7). In the lowlands of Central Europe, there are usually two generations per year, while at elevations above about 1500 m above sea level, the populations are usually univoltine (one generation per year).

Volatiles play an important role not only in attracting conspecifics but also in regulating the population density. If beetle density in a tree becomes too high, the breeding beetles emit an anti-aggregation pheromone. This pheromone has a repellent effect on conspecifics and thus prevents a too high colonization density leading to intraspecific competition with negative effects (Lindgren and Miller 2002).

Among the most important natural enemies of bark beetles are predatory beetles and flies, parasitic wasps, and woodpeckers (Wegensteiner et al. 2015).

Fig. 12.1 Galleries of the European spruce bark beetle (*Ips typographus*). The parental beetles are visible in the maternal galleries running parallel to the tree stem axis. The larvae develop in the larval galleries branching off perpendicularly to the maternal galleries. High-density larval feeding interrupts the sap flow in the phloem, causing the tree to die. (Photo: B. Wermelinger)

12.2 Colonization Strategies

Conifers can defend against invading insects or fungi by producing resin; this can protect the tree both physically by 'pitching out' the intruders and chemically through the toxic terpenoids in the resin (Franceschi et al. 2005; Krokene 2015). The resistance against the aggressors is composed of the preformed *constitutive* defence and the *induced* defence triggered by the infestation (Phillips and Croteau 1999; Lieutier 2004; Krokene 2015). Depending on the tree species, the constitutive resin is stored in different places in the wood and released when something penetrates the bark. The induced defence consists of a hypersensitive reaction and a delayed resistance: the infested tissue is impregnated with resin and becomes necrotic. This reaction is found both in the phloem and the xylem and precedes the spatial spread of the individual infested zones. The dying tissue and the toxic terpenoids and phenols impair or prevent the development of the bark beetle brood. In the medium term, after a repelled attack, a tree can increase its resistance, e.g. by producing additional 'traumatic resin ducts' around the infested zones. A successful colonization of trees by bark beetles therefore depends not only on the density of attacking beetles but also on the characteristics of these defence mechanisms.

During colonization, the beetles are also assisted by symbiotic fungi, which help to overcome the resin defence and to kill off the infested trees (Paine et al. 1997).

The bark beetle species with the most pronounced potential for disturbance (*I. typographus*, *D. ponderosae*) have a plastic host selection that depends on their population size (Raffa et al. 2015b): in endemic phases with low population densities (latency phase), the beetles live saprophagously on newly dead or severely weakened trees with no defence or only weak defence. Although this 'low-risk' strategy allows the beetles to colonize these trees relatively easily, it has several disadvantages: under normal conditions such trees are relatively rare and randomly distributed in the landscape, the food quality of their degrading phloem is low, and the developing beetle brood has to compete with other saprophagous beetle species (Fig. 12.2). In combination this leads to a low reproductive rate (Raffa et al. 2008).

However, if a beetle population reaches epidemic density (outbreak phase) after extrinsic disturbances such as storms or drought (see Sect. 12.3), it can switch to a 'high-risk' strategy: the many beetles attacking the tree simultaneously can now overcome the resin defence of living, vigorous trees. While their colonization is still risky because of their pronounced resin production, the beetle population – in case of successful establishment – benefits from a higher reproductive rate: potential host trees are abundant, their high phloem quality ensures a favourable food supply, and the developing brood has virtually no competitors, since other beetle species are neither attracted nor can they overcome the intact defence mechanisms of the trees. In the case of the mountain pine beetle, it has been shown that in the epidemic phase, this bark beetle even prefers more vigorous, resin-rich trees over trees producing less resin (Fig. 12.3; Boone et al. 2011). However, this high-risk strategy is only successful at high population levels. The more vigorous a host tree, the higher the minimum number of beetles is required that simultaneously attack the tree (Fig. 12.4; Mulock and Christiansen 1986; Nelson and Lewis 2008). An estimated

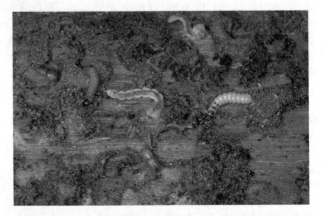

Fig. 12.2 At low population densities in the endemic latency phase, bark beetles (here *Ips sexdentatus* Börner) have to share the resources in weakened host trees with competitors such as jewel beetles (larvae with widened head). The bark beetles are exposed to a high predation pressure (pink and cream-coloured predatory beetle larvae). (Photo: B. Wermelinger)

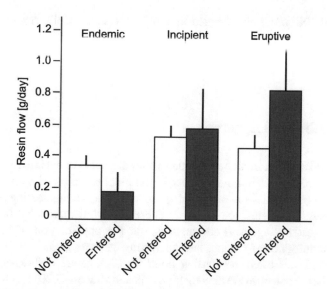

Fig. 12.3 Constitutive resin content (mean ± SE) of lodgepole pine trees (*Pinus contorta* Dougl. ex Loud.) with successful ('Entered') or unsuccessful ('Not entered') colonization by the mountain pine beetle (*Dendroctonus ponderosae*) in different population phases (endemic, incipient, eruptive). At endemic densities, beetles colonize mostly low-resin trees, whereas in the epidemic phase, they prefer trees with high resin content. (Redrawn from Boone et al. 2011 with permission of the authors)

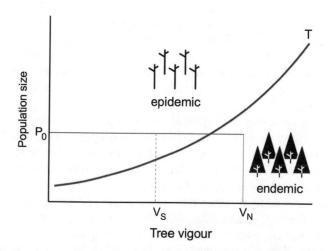

Fig. 12.4 Threshold (T) for the minimum population size that is required to successfully colonize a host tree depending on its vigour. In a 'normal' situation, beetles at endemic population size P_0 are not able to overcome the defence of a host tree with 'normal' vigour V_N. Therefore, the beetle population remains in the endemic phase. If tree vigour – due to stress such as drought – decreases to a lower-level V_S, the same population density P_0 of bark beetles is sufficient for a successful colonization of the weakened trees. As a consequence, bark beetles can reach epidemic densities, which usually leads to the death of many infested trees. (Adapted from Mulock and Christiansen 1986)

minimum of 200 spruce bark beetles are required to successfully colonize a healthy spruce tree (Fahse and Heurich 2011). For the mountain pine beetle, a threshold of around 60–90 beetles per square metre pine bark area has been reported (Berryman et al. 1985).

12.3 Outbreak Triggers

Mass outbreaks of almost all bark beetle species are not cyclical but are mostly triggered by stochastic external disturbances. The most common triggers are climatic disturbances (Six and Bracewell 2015). In addition, foliage-eating insects can weaken trees and predispose them to subsequent infestation by bark beetles. Improper forest management can also lead to the proliferation of bark beetles. Intensive thinning, clear-cutting, and harvesting of trees in already weakened or unstable stands are often the starting point for bark beetle outbreaks, especially when population densities are already high (Six and Bracewell 2015). Outbreaks are often triggered by combinations of abiotic and biotic/anthropogenic disturbances. Irrespective of the trigger type, the conditions for mass infestations are most favourable if large, connected forest areas are concerned, with uniform, mature to overmature stands with a single dominant tree species (often spruce or pine).

Abiotic Triggers

In Europe, *windthrows* were most often the main trigger for outbreaks of the European spruce bark beetle (Table 12.1; Wermelinger 2004; Eriksson et al. 2005). The newly windthrown stems serve the bark beetles as a low-resistance, but still high-quality, substrate in which they can build up their populations to such an extent that mass infestation of more vigorous spruce trees becomes possible (Fig. 12.5). Since a storm usually also weakens the remaining stands (gaps, root damage; Chap. 8), these trees are susceptible to subsequent colonization by bark beetles (Seidl and Blennow 2012).

Another important trigger is *drought*, usually coupled with high temperatures. The trees' increased need for transpiration can no longer be met, which puts them under stress and makes them susceptible to attack. In Europe, for example, the silver fir bark beetles (*Pityokteines* spp.; Meier et al. 2004), but also the spruce bark beetles (Meier et al. 2004; Rouault et al. 2006; Hlásny et al. 2021), react to longer warm and dry periods. In North America this is known from *D. brevicomis* LeConte and *D. rufipennis* Kirby (Six and Bracewell 2015). Severe drought stresses forests abruptly and over large areas, so that outbreaks triggered this way often occur simultaneously at the landscape level (Raffa et al. 2015b, Seidl et al. 2016). Elevated temperatures and drought effects were also the dominant factor in the gigantic mass outbreak of the mountain pine beetle in North America (see Box 12.1).

Occasionally, anthropogenic or natural forest fires (see Chap. 7) can also act as abiotic triggers of bark beetle outbreaks. From North America, mass infestations of *D. pseudotsugae* Hopk. are known after fire (Six and Bracewell 2015). Also snow

Table 12.1 Major mass outbreaks of the European spruce bark beetle (*Ips typographus*) in Central and northern Europe. For comparison, information on the historic outbreak of the mountain pine beetle (*Dendroctonus ponderosae*) in North America is also provided

Region	Period	Trigger	Infested timber (million m³)	Infested area (km²)	Source
Ips typographus L.					
Central Europe	1942–1951	Heat, drought, storms, snow breakage, war effects, clear-cuts	25[a]		Wellenstein (1954)
Southern Norway	1971–1981	Storm, drought	5		Bakke (1989)
Sweden	1990–1996	Storm, drought	4.5		Kärvemo and Schroeder (2010)
NP Bavarian Forest (Germany)	1988–2010	Thunderstorms, drought		58	Lausch et al. (2013)
Switzerland	1990–1996	Storms Vivian/Wiebke	2		Stadelmann et al. (2014b)
Europe	1990–2001	Various storms	31.6		Grégoire and Evans (2004)
Austria	1992–1999	Storms Vivian/Wiebke	10.7[b]		Hoch et al. (2016)
Germany	1999–2005	Storm Lothar, drought	14.5		R. John (unpubl.)
Switzerland	2000–2007	Storm Lothar, drought	8		Stadelmann et al. (2014b)
France	2001–2008	Various storms	3.6		Nageleisen (2009)
Austria	2003–2012	Foehn storm, storm Kyrill	21.6[b]		Hoch et al. (2016)
Sweden	2006–2011	Storm Gudrun	3.2		Kärvemo and Schroeder (2010)
Germany	2009–2015	Storm Kyrill	9		R. John (unpubl.)
Czech Republic	2015–>2019	Heat, drought	>48		Hlásny et al. (2021)
Dendroctonus ponderosae Hopk.					
British Columbia (Canada)	1990–2015	Drought	700	183,000	Corbett et al. (2016)

[a]Total of bark beetles on spruce and fir
[b]Including *Pityogenes chalcographus*

Fig. 12.5 In the overthrown trees in a windthrow, the European spruce bark beetle (*Ips typographus*) can build up sufficiently high population levels to subsequently colonize living trees. (Photo: I. Brauer)

breakage (tree crown fractures) and avalanches can, similar to storm events, provide breeding material as a starting point for outbreaks, but to a more limited spatial extent.

Biotic Triggers

Compared to abiotic factors, biotic stressors are much less often a starting point for bark beetle outbreaks in Europe. They are mostly small scale and therefore tend to trigger local infestations. Most often it is intensive *caterpillar feeding* that can lead to weakened host trees and subsequent infestation by bark beetles. In Finland, for example, infestations by *Tomicus* bark beetles occurred only on Scots pine trees (*Pinus sylvestris* L.) that were virtually completely defoliated by the sawfly *Diprion pini* (L.). This bark beetle attack, however, did not result in large outbreaks (Annila et al. 1999). Larger infestations of bark beetles after defoliation occurred in North America. There, the feeding of moth or sawfly caterpillars on various coniferous tree species has led to subsequent infestations of *Dendroctonus* species (Negrón et al. 2011; Goodsman et al. 2015).

12.4   Infestation Dynamics

The development of an outbreak is characterized by the infestation strategies of the bark beetles. Fresh windthrown logs or trees weakened by drought or other stressors allow the small populations of bark beetles during the endemic latency period to increase above the outbreak threshold (Raffa et al. 2008). In the case of the European spruce bark beetle, this occurs in windthrows of spruce (Fig. 12.5); this phloem is still fresh and rich in nutrients. However, depending on elevation, exposure, temperature, bark thickness, and root contact, the logs desiccate after 1–3 years and are no longer suitable for the development of further beetle generations. Therefore, the masses of beetles emerging from the logs head for the adjacent forest edges. These trees are still weakened by the storm, and their stems, having been suddenly exposed to direct sunlight, are stressed by the intense solar radiation.

At the same time, further infestations may develop around single fallen and weakened trees inside the stand, resulting in the formation of infestation spots (groups of a few dozen infested trees). The local population size then exceeds the threshold necessary for the colonization of vigorous trees (Fig. 12.4) and, pursuing the high-risk strategy, reaches a self-sustaining dynamic with a positive feedback: the larger the population, the more vigorous trees are colonized, and the more successful the reproduction of the beetles is. More than 90% of new infestations occur within 500 m and 66% even within 100 m of the previous year's infestation, independent of the triggering agent and any control measures that have been carried out (Kautz et al. 2011; Stadelmann et al. 2014a; Fig. 12.6). Almost identical distances apply to the first infestations of living trees in stands adjacent to windthrow areas. The radii around uncleared windthrows are significantly smaller (200 m) than those around cleared areas (Havašová et al. 2017). Over time, the infestation spots can merge into one large area, and the infestations may then expand as a front. Usually an outbreak of the European spruce bark beetle lasts several years. In the case of the two largest Central European mass outbreaks of the last decades in the German Bavarian Forest National Park, the infestations – without human intervention – lasted about 10 years. This was favoured by dry and hot summers, recurring minor disturbances such as snow breakage in winter and additional storm events, as well

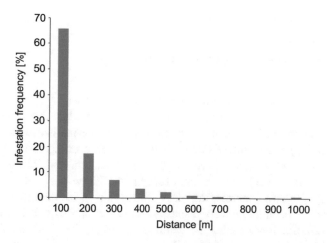

Fig. 12.6 Frequency of new infestation spots of the European spruce bark beetle (*Ips typographus*) at different distances to infestations from previous years. (Data from Kautz et al. 2011)

as by the weakening of the host trees by quickly succeeding years of intensive fruiting ('masts') (Nüsslein et al. 2000; Heurich et al. 2001).

The factors determining the trajectory of an outbreak are primarily related to weather, i.e. temperature and drought (Marini et al. 2017). In an analysis of the European spruce bark beetle outbreaks in Switzerland after the storm Lothar in 1999, the total volume of infested spruce trees depended most on temperature, more so than on windthrown timber volume, precipitation, previous year's infestations, or the proportion of spruce (Stadelmann et al. 2013b). Also for North American *Dendroctonus* and *Ips* species, changes in temperature and precipitation correlated with the beetle population growth rates (Raffa et al. 2008). The decrease in beetle density and thus the decline in infestations can have several reasons: increasing resistance of host trees (mainly due to sufficient precipitation), cooler weather conditions, intraspecific competition, the regulatory effect of natural enemies, decreasing supply of host trees, or a combination of these adverse influences (e.g. Marini et al. 2017). If, as a consequence, beetle populations fall below the critical threshold for the infestation of vigorous trees, the outbreak comes to an end. In managed forests, control measures (cutting and removing infested trees) also play an important role (Stadelmann et al. 2013a).

The type of the trigger also influences the trajectory and extent of a bark beetle outbreak following the disturbance. Forest fire and windthrow are spatially and temporally limited and thus weaken the trees in the adjacent intact forest only once, predominantly in the immediate vicinity of the disturbance. Therefore, the subsequent outbreak is usually limited to a few years, if no further disturbances occur. In contrast, elevated temperatures and drought are often a spatially or temporally extended event, which leads to a greater supply of weakened trees over a certain period (DeRose and Long 2012). This means that bark beetle populations can build up over large areas and remain at epidemic densities for a long time. In general, mass outbreaks decline when the triggering factor is no longer present, and no new disturbances occur.

12.5 Interactions with Other Disturbances

Bark beetles can interact with other disturbances in various ways. As mentioned above, mass infestations are triggered by preceding other disturbance events (Sect. 12.3), and conversely, bark beetle outbreaks can also be the cause for subsequent disturbances. Large-scale infestation often leads to new even-aged stands consisting of only one or a few tree species over larger areas (Coleman et al. 2008). At a certain age, the risk of bark beetle infestation, fire, and windthrow increases again.

Especially in North America, the risk of forest fires in bark beetle-created deadwood stands has been intensively discussed (Hicke et al. 2012; Black et al. 2013; Dhar et al. 2016). Shortly after the death of the trees, the dead needles pose an increased risk for forest fires due to the intense emission of highly inflammable terpenoids. However, most studies have shown that bark beetle mass outbreaks did not significantly increase the risk of crown fires in pine and spruce stands, as this depends primarily on drought and rather less on stand structure (Black et al. 2013). ṭṭṭ ṭṭ ṭṭṭṭṭṭ ṭṭ ṭṭṭ complex interactions between bark beetle infestation and other disturbances in the Yellowstone National Park is shown in Fig. 12.7.

12.6 Significance

Ecological Significance
From the perspective of a forest ecosystem, bark beetles have various ecological functions. They are so-called pioneer insects that are first to colonize weakened living or freshly dead coniferous trees, loosen their bark, and thus start the

Fig. 12.7 Example for complex interactions of disturbances from the North American Yellowstone National Park. The North American whitebark pine (*Pinus albicaulis* Engelm.) is a keystone species of the high elevation ecosystems in this park. In the avalanche path from the Avalanche Peak (photo position, 2009) down to the valley green, whitebark pine regeneration is visible. It originates from seeds from the surrounding pine forests, which were still intact at the time of the avalanche. The seeds were carried by the Clark's nutcracker (*Nucifraga columbiana* Wilson) into the avalanche path. The stand on the slope in the background fell victim to the huge forest fires of 1988, and almost 20 years later, the whitebark pine trees on the slope to the right were killed by the mountain pine beetle (*Dendroctonus ponderosae*). The dark green strip of forest in the photo middle ground consists of spruce (*Picea*) and fir (*Abies*) species which are not attacked by this bark beetle. In the whole catchment, mature whitebark pine trees are missing, which could serve as seed trees until the regeneration in the avalanche path becomes reproductive. If another avalanche were to precede this, the whitebark pine could disappear from this valley for decades, as the Clark's nutcracker, which depends on the whitebark pine seeds, has disappeared from the area. In addition, potential regeneration of whitebark pine is threatened by the introduced blister rust fungus *Cronartium ribicola* (Campbell and Antos 2000). (Photo: B. Wermelinger)

decomposition of wood. They thus initiate an entire succession of many other wood-inhabiting and purely saprophagous insects, fungi, and even vertebrates (Müller et al. 2008; Beudert et al. 2015; Zuo et al. 2016). Bark beetles are also an essential food source for specialized predatory and parasitic insects as well as for woodpeckers (Wegensteiner et al. 2015). Various mites and nematodes use them as vectors for being transported to new habitats (e.g. Moser et al. 2005). The creation of gaps in the forest canopy after the death of bark beetle-colonized trees leads to a new round of forest regeneration. Thus bark beetles also influence the tree species composition and forest structure.

Bark beetles not only act as ecosystem engineers on a small scale but can also alter the composition of and the material and nutrient cycles in entire ecosystems at the landscape level (Figs. 12.7 and 12.8). In the Bavarian Forest National Park, for example, the water runoff into the groundwater and streams of a catchment increased strongly after the extensive attack of spruce by the European spruce bark beetle. The nitrate content of the groundwater was increased for 7 years but returned to the original value after 10–17 years (Huber 2005; Beudert et al. 2015). In the Greater Yellowstone area of North America, a 30-year time series after the infestation by the mountain pine beetle showed that the short pulse of massive needle shedding changed the soil climate and strongly increased the nitrogen mineralization rate, which was also reflected in the longer term as elevated nitrogen content in the new needles of the surviving trees (Griffin et al. 2011). Bark beetle infestations on a regional, or even continental, scale lead to the affected forests binding less carbon and temporarily turning from a carbon sink into a carbon source (Kurz et al. 2008;

Fig. 12.8 A mass outbreak of the European spruce bark beetle (*Ips typographus*) can temporarily markedly alter an ecosystem and lead to losses of both material and non-material nature. (Photo: M. Heurich)

Seidl et al. 2008). The surviving trees and the growing regeneration compensate for this after one to two decades (Hansen 2014).

Economic Importance

Bark beetles such as the European spruce bark beetle or the mountain pine beetle can lead to the large-scale die-off of living trees and thus have a fundamental impact on forest management and forest services for humans. In Europe, the spruce bark beetle has caused extensive damage, especially in the aftermath of the major storms at the turn of the last century (Table 12.1; Grégoire and Evans 2004). These hurricanes produced millions of cubic metres of beetle-infested wood, additionally promoted by the heat wave in 2003 (Fig. 12.9). Although the logs of the dead trees basically can be normally utilized, the sudden oversupply on the market leads to a drop in prices (see Chap. 8), and the sustainable management of a forest may be endangered by mass infestations. An outbreak of mountain pine beetle of an even greater magnitude occurred in the Canadian province of British Columbia. Between 1990 and 2015, this bark beetle infested about 700 million m^3 of pine trees, and the infestation is still ongoing (Corbett et al. 2016). By the middle of this century, a cumulative reduction in the gross domestic product of Can$57 billion is expected for this province. Beetle-infested wood can often only be merchandized at a loss because the market is saturated after major storms. Moreover, large-scale outbreaks strongly affect long-term forest planning. This can also have far-reaching consequences for employment in the timber sector (Abbott et al. 2009; Corbett et al. 2016).

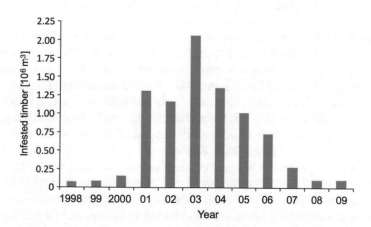

Fig. 12.9 Infestations of living spruce trees by the European spruce bark beetle (*Ips typographus*) after the storm Lothar (December 1999) in Switzerland. Two years after the storm, the infestation reached a first maximum and started to decrease in the following year. The record amount in 2003 was the result of the still exceedingly high populations of bark beetles, which, due to the extremely hot and dry year 2003, had an ample supply of weakened Norway spruce trees available for colonization. (Data from Swiss Forest Protection, WSL)

In mountain forests, the protection against natural hazards can be strongly impaired by bark beetle infestations. Although the dead trees remain standing for some time, they eventually break down, thus reducing the protection of underlying settlements and traffic routes from rockfall, debris flow, and avalanches (Kupferschmid and Bugmann 2005). This presents forest owners and authorities with the difficult question of whether it makes more sense to exploit the dead trees under often non-profitable conditions and to erect temporary artificial constructions, or to leave the dead trees, which still provide protection for some time, and to rely on the natural regeneration to take over the protective function in time (see Chap. 9).

Another consequence of large-scale bark beetle outbreaks is that the affected forest with all trees salvaged or with the bare tree skeletons standing in the forest no longer offers the same recreational values and benefits as an intact forest (Heurich et al. 2001; Fig. 12.8). Although the general public is more and more aware of the ecological significance of deadwood, this can hardly compensate for the absence of an intact shady and relaxing forest. Even if the public understanding of ecological processes on large spatial and temporal scales keeps growing, such disturbances will keep creating conflicts with the continuous human demand for timber, protection, and recreation.

12.7 Bark Beetles and Climate Change

For most of Europe, markedly higher temperatures and increased summer droughts are expected in the second half of the twenty-first century, possibly associated with altered windstorm regimes (Della-Marta and Pinto 2009; Usbeck et al. 2010; Seidl et al. 2017; see Chap. 16). This will generally have a positive effect on the population growth of bark beetles and a negative effect on the resistance of their host trees (Logan et al. 2003; Raffa et al. 2015a; Ramsfield et al. 2016). In addition, such changes can negatively affect the synchronization between bark beetles and their natural enemies (Ayres and Lombardero 2000; Bentz et al. 2010; Raffa et al. 2015a). Climate change will also have numerous direct and indirect effects on bark beetle communities (Økland et al. 2015; Wermelinger et al. 2021). Increased temperature will cause the activity range of bark beetles to spread further north and to higher elevations (Hlásny and Turčáni 2009; Netherer and Schopf 2010). In some of these areas, a complete generation per year will become possible on a regular basis in the future, while at lower elevations the number of generations per year will increase to 2–3 (Jönsson et al. 2011; Jakoby et al. 2019; Fig. 12.10). The infestation period of the European spruce bark beetle will start earlier in spring, and its activity in the lowlands and in warm areas could extend later into autumn, since diapause induction at very warm temperatures (>23°C) is no longer exclusively controlled by day

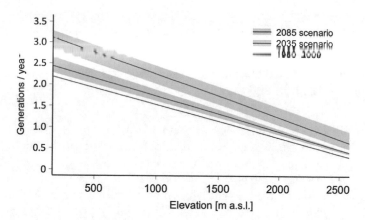

Fig. 12.10 Mean number of annual generations depending on elevation for the time periods 1980–2009, 2020–2049 (2035 scenario), and 2070–2099 (2085 scenario) under the IPCC climate scenario A1B for Switzerland. Mean relationships (lines) and uncertainty bands (coloured areas) based on the climate scenarios are shown

length (Schopf and Kritsch 2010). During the future milder winters, the juvenile stages of initiated generations will more often be able to develop into adult beetles (Štefková et al. 2017), and in many regions the cold-linked mortality of overwintering beetles will become less important. This would increase the size of the start population creating the first infestations in the following spring. Furthermore, increasing drought stress particularly in summer reduces the resistance of spruce trees to bark beetle attack. Netherer et al. (2015) showed that the success of spruce bark beetle colonization increases with increasing drought stress of a tree. At the same time, intense drought stress decreased host acceptance by the beetles.

Since warm and dry conditions generally have a positive effect on the severity of bark beetle outbreaks (Marini et al. 2012), climate change will most likely further increase the infestation risk in European spruce forests as well as in North American forests (Logan et al. 2003; Raffa et al. 2008; Scidl et al. 2014). In Central Europe, for example, increased bark beetle infestation in combination with drought-related spruce mortality could partially or completely eliminate low-elevation secondary spruce stands in the long run (Temperli et al. 2013). Overall, adaptive forest management will play a decisive role in reducing the risk of infestations (Björkman et al. 2015; see Chap. 17).

Box 12.1: Mountain Pine Beetle in North America

Beat Wermelinger ⓘD
Forest Health and Biotic Interactions Research Unit, Swiss Federal Research
Institute WSL, Birmensdorf, Switzerland

The mountain pine beetle (*Dendroctonus ponderosae* Hopk.; MPB) is a bark
beetle native to North America. It has long been known to occasionally
develop mass outbreaks leading to the death of huge amounts of pine trees
(Safranyik et al. 2010; Jarvis and Kulakowski 2015). In most cases, the trigger
has been hot and dry weather, which has impaired the resin defence of the host
trees – primarily lodgepole pine (*Pinus contorta* Dougl. ex Loud.) – and
accelerated and synchronized the dynamics of the bark beetles, thus enabling
large-scale infestations. The outbreaks came to an end when the supply of
suitable host trees was exhausted and larval mortality after cold winters
increased (Safranyik and Linton 1991; Creeden et al. 2014). At high elevations, outbreaks of MPB have not occurred until recently, because a generation of beetles took 2 or even more years to complete because of the low
temperatures in such locations. Moreover, the beetles suffer from high mortality during the cold winter season (Logan and Powell 2001).

Since the end of the twentieth century, a gigantic MPB mass outbreak of
historic dimensions has been going on in western North America, which
seems to have come to an end only in the late 2010s. In the most severely
affected Canadian province of British Columbia alone, pine trees have been
killed across an area of around 183,000 km² (Corbett et al. 2016). In past outbreaks only areas west of the Rocky Mountains have been affected; however,
the infestations have spilled over this climatic barrier into the neighbouring
province of Alberta since around 2003. As a result, the bark beetles have also
attacked jack pine (*Pinus banksiana* Lamb.) for the first time. In the USA, the
infestations stretched south along the Rocky Mountains down to Arizona. In
2010, the total infested area in North America was estimated at over
250,000 km² (Bentz et al. 2010). The infestations expanded not only to the
east but also to the north and to higher elevations compared to the historic
outbreaks. In the Greater Yellowstone Area in the northwestern USA, the
North American five-needle whitebark pines (*Pinus albicaulis* Engelm.) at
elevations of 2000–3000 m were mass attacked for the first time (cf. Fig. 12.1).
For evolutionary reasons, this tree species has little defence against bark beetle attack (Raffa et al. 2013). These infestations were a consequence of
increased temperatures, which allowed the beetles to pass from multi-annual
to annual synchronous generations (Logan and Powell 2001).

(continued)

Box 12.1 (continued)

These continental-scale infestations can be attributed to various causes. The altered weather conditions linked to the ongoing climate change have played a decisive role. The higher temperatures have allowed a faster and synchronous development of the bark beetle brood in summer and have resulted in lower mortality in winter (Logan and Powell 2001; Régnière and Bentz 2007). Also, the more frequent summer droughts reduce the defence capacity of host trees and make them susceptible to infestation (Berg et al. 2006).

Large areas of susceptible pine forest are another important reason for the uninhibited spread of the infestations. The over-mature and uniform forests were the result of many years of fire prevention and control, large-scale clearcutting, or earlier bark beetle infestations (Taylor and Carroll 2004; Youngblood et al. 2009; Jenkins et al. 2014). While natural disturbances such as fire or bark beetle outbreaks used to lead to a mosaic of infestation-prone and unsusceptible stands and thus to a spatial and temporal distribution of MPB infestations, human interventions provided extensive areas of susceptible host trees. After such vast outbreaks, uniform pine forests will usually develop with a tree cohort that will simultaneously become over-mature and thus will be prone again for MPB infestations at the landscape level.

The gigantic infestations also affect the forestry and logging sector. In many places, a sustainable supply of coniferous wood has become impossible for the coming decades. For this reason, the annual allowable cuts have been drastically reduced; this will have a severe impact on the forestry and logging industry (Corbett et al. 2016). The large-scale die-off of trees also affects the global carbon balance, as dead stands turn from a carbon sink to a carbon source. They were estimated to have emitted about 270 megatons of carbon by 2020 (Kurz et al. 2008).

Since 2005, infestations in British Columbia have been slowly decreasing, mainly as a result of the decreasing supply of suitable host trees, but also as a result of low winter temperatures (Sambaraju et al. 2012; Creeden et al. 2014). Nonetheless, it is estimated that by 2023 about 58% of all mature pine trees in British Columbia will have been killed (Walton 2013). With higher winter temperatures in the future, MPB might even have the potential to cross the continental divide and spread across North America's entire boreal pine forest (Safranyik et al. 2010) (Box Fig. 1).

(continued)

Box 12.1 (continued)

Box Fig. 1 Large-scale infestation of white bark pine by the mountain pine beetle in the Yellowstone National Park (USA). Only very young trees or non-*Pinus* species survived. (Photo: B. Wermelinger)

References

Bentz BJ, Régnière J, Fettig CJ, Hansen EM, Hayes JL, Hicke JA, Kelsey RG, Negrón JF, Seybold SJ (2010) Climate change and bark beetles of the western United States and Canada: direct and indirect effects. Bioscience 60:602–613

Berg EE, Henry JD, Fastie CL, De Volder AD, Matsuoka SM (2006) Spruce beetle outbreaks on the Kenai Peninsula, Alaska, and Kluane National Park and Reserve, Yukon Territory: relationship to summer temperatures and regional differences in disturbance regimes. For Ecol Manag 227:219–232

Corbett LJ, Withey P, Lantz VA, Ochuodho TO (2016) The economic impact of the mountain pine beetle infestation in British Columbia: provincial estimates from a CGE analysis. Forestry 89:100–105

Creeden EP, Hicke JA, Buotte PC (2014) Climate, weather, and recent mountain pine beetle outbreaks in the western United States. For Ecol Manag 312:239–251

Jarvis DS, Kulakowski D (2015) Long-term history and synchrony of mountain pine beetle outbreaks in lodgepole pine forests. J Biogeogr 42:1029–1039

Jenkins MJ, Runyon JB, Fettig CJ, Page WG, Bentz BJ (2014) Interactions among the mountain pine beetle, fires, and fuels. For Sci 60:489–501

Kurz WA, Dymond CC, Stinson G, Rampley GJ, Neilson ET, Carroll AL, Ebata T, Safranyik L (2008) Mountain pine beetle and forest carbon feedback to climate change. Nature 452:987–990

(continued)

Box 12.1 (continued)

Logan JA, Powell JA (2001) Ghost forests, global warming, and the mountain pine beetle (Coleoptera: Scolytidae). Am Entomol 47:160–173

Raffa KF, Powell EN, Townsend PA (2013) Temperature-driven range expansion of an irruptive insect heightened by weakly coevolved plant defenses. Proc Natl Acad Sci USA 110:2193–2198

Régnière J, Bentz B (2007) Modeling cold tolerance in the mountain pine beetle, *Dendroctonus ponderosae*. J Insect Physiol 53:559–572

Safranyik L, Carroll AL, Régnière J, Langor DW, Riel WG, Shore TL, Peter BJCB, Cooke BJ, Nealis VG, Taylor SW (2010) Potential for range expansion of mountain pine beetle into the boreal forest of North America. Can Entomol 142:415–442

Safranyik L, Linton D (1991) Unseasonably low fall and winter temperatures affecting mountain pine beetle and pine engraver beetle populations and damage in the British Columbia Chilcotin Region. J Entomol Soc Brit Columb 88:17–21

Sambaraju KR, Carroll AL, Zhu J, Stahl K, Moore RD, Aukema BH (2012) Climate change could alter the distribution of mountain pine beetle outbreaks in western Canada. Ecography 35:211–223

Taylor SW, Carroll AL (2004) Disturbance, forest age, and mountain pine beetle outbreak dynamics in BC: a historical perspective. In: Shore TL, Brooks JE, Stone JE (eds) Mountain pine beetle symposium: challenges and solutions, Information report BC-X-399. Natural Resources Canada, Canadian Forest Service, Pacific Forest Centre, Victoria, pp 41–51

Walton A (2013) Provincial-level projection of the current mountain pine beetle outbreak: update of the infestation projection based on the provincial aerial overview surveys of forest health conducted from 1999 through 2012 and the BCMPB model (year 10). BC Ministry of Forests, Lands and Natural Resource Operations, Victoria, 13 p

Youngblood A, Grace JB, McIver JD (2009) Delayed conifer mortality after fuel reduction treatments: interactive effects of fuel, fire intensity, and bark beetles. Ecol Appl 19:321–337

References

Abbott B, Von Kooten GC, Stennens B (2009) Mountain pine beetle, global markets, and the British Columbia forest economy. Can J For Res 39:1313–1321

Annila E, Långström B, Varama M, Hiukka R, Niemelä P (1999) Susceptibility of defoliated Scots pine to spontaneous and induced attack by *Tomicus piniperda* and *Tomicus minor*. Silva Fenn 33:93–106

Ayres MP, Lombardero MJ (2000) Assessing the consequences of global change for forest disturbance from herbivores and pathogens. Sci Total Environ 262:263–286

Bakke A (1989) The recent *Ips typographus* outbreak in Norway – experiences from a control program. Ecography 12:515–519

Bentz BJ, Régnière J, Fettig CJ, Hansen EM, Hayes JL, Hicke JA, Kelsey RG, Negrón JF, Seybold
    SJ (2010) Climate change and bark beetles of the western United States and Canada: direct and
    indirect effects. Bioscience 60:602–613
Berryman AA, Dennis B, Raffa KF, Stenseth NC (1985) Evolution of optimal group attack, with
    particular reference to bark beetles (Coleoptera: Scolytidae). Ecology 66:898–903
Beudert B, Bässler C, Thorn S, Noss R, Schröder B, Dieffenbach-Fries H, Foullois N, Müller J
    (2015) Bark beetles increase biodiversity while maintaining drinking water quality. Cons Lett
    8:272–281
Björkman C, Bylund H, Nilsson U, Nordlander G, Schroeder M (2015) Effects of new forest man-
    agement on insect damage risk in a changing climate. In: Björkman C, Niemelä P (eds) Climate
    change and insect pests. CAB International, Wallingford, pp 248–266
Black SH, Kulakowski D, Noon BR, Della Sala DA (2013) Do bark beetle outbreaks increase
    wildfire risks in the central US Rocky Mountains? Implications from recent research. Nat
    Areas J 33:59–65
Blomquist GJ, Figueroa-Teran R, Aw M, Song MM, Gorzalski A, Abbott NL, Chang E, Tittiger C
    (2010) Pheromone production in bark beetles. Insect Biochem Mol 40:699–712
Boone CK, Aukema BH, Bohlmann J, Carroll AL, Raffa KF (2011) Efficacy of tree defense physi-
    ology varies with bark beetle population density: a basis for positive feedback in eruptive spe-
    cies. Can J For Res 41:1174–1188
Byers J (2004) Chemical ecology of bark beetles in a complex olfactory landscape. In: Lieutier F,
    Day KR, Battisti A, Grégoire JC, Evans HF (eds) Bark and wood boring insects in living trees
    in Europe: a synthesis. Kluwer Academic Publishers, Dordrecht, pp 89–134
Campbell EM, Antos JA (2000) Distribution and severity of white pine blister rust and mountain
    pine beetle on whitebark pine in British Columbia. Can J For Res 30:1051–1059
Coleman TW, Meeker JR, Clarke SR, Rieske LK (2008) The suppression of *Dendroctonus fronta-
    lis* and subsequent wildfire have an impact on forest stand dynamics. Appl Veg Sci 11:231–242
Corbett LJ, Withey P, Lantz VA, Ochuodho TO (2016) The economic impact of the mountain pine
    beetle infestation in British Columbia: provincial estimates from a CGE analysis. Forestry
    89:100–105
Della-Marta PM, Pinto JG (2009) Statistical uncertainty of changes in winter storms over the
    North Atlantic and Europe in an ensemble of transient climate simulations. Geophys Res Lett
    36:L14703
DeRose RJ, Long JN (2012) Factors influencing the spatial and temporal dynamics of Engelmann
    spruce mortality during a spruce beetle outbreak on the Markagunt Plateau. Utah For
    Sci 58:1–14
Dhar A, Parrott L, Hawkins CDB (2016) Aftermath of mountain pine beetle outbreak in British
    Columbia: stand dynamics, management response and ecosystem resilience. Forests 7:171
Eriksson M, Pouttu A, Roininen H (2005) The influence of windthrow area and timber charac-
    teristics on colonization of wind-felled spruces by *Ips typographus* (L.). For Ecol Manag
    216:105–116
Fahse L, Heurich M (2011) Simulation and analysis of outbreaks of bark beetle infestations and
    their management at the stand level. Ecol Model 222:1833–1846
Franceschi VR, Krokene P, Christiansen E, Krekling T (2005) Anatomical and chemical defenses
    of conifer bark against bark beetles and other pests. New Phytol 167:353–376
Goodsman DW, Goodsman JS, McKenney DW, Lieffers VJ, Erbilgin N (2015) Too much of a good
    thing: landscape-scale facilitation eventually turns into competition between a lepidopteran
    defoliator and a bark beetle. Landsc Ecol 30:301–312
Grégoire J-C, Evans HF (2004) Damage and control of BAWBILT organisms an overview. In:
    Lieutier F, Day KR, Battisti A, Grégoire JC, Evans HF (eds) Bark and wood boring insects in
    living trees in Europe: a synthesis. Kluwer Academic Publishers, Dordrecht, pp 19–37
Griffin JM, Turner MG, Simard M (2011) Nitrogen cycling following mountain pine beetle dis-
    turbance in lodgepole pine forests of Greater Yellowstone. For Ecol Manag 261:1077–1089

Hansen EM (2014) Forest development and carbon dynamics after mountain pine beetle outbreaks. For Sci 60:476–488

Heurich M, Fernández J. Tokis R (?///) Hüfelähnih Hefbüenn Windnäben, Höfz Höhen and forest management in the Tatra national parks. For Ecol Manag 391:349–361

Heurich M, Reinelt A, Fahse L (2001) Waldentwicklung im montanen Fichtenwald nach großflächigem Buchdruckerbefall im Nationalpark Bayerischer Wald. In: Heurich M (ed) Waldentwicklung im Bergwald nach Windwurf und Borkenkäferbefall. Nationalparkverwaltung Bayerischer Wald, Grafenau, pp 99–177

Hicke JA, Johnson MC, Hayes JL, Preisler HK (2012) Effects of bark beetle-caused tree mortality on wildfire. For Ecol Manag 271:81–290

Hlásny T, Turčáni M (2009) Insect pests as climate change driven disturbances in forest ecosystems. In: Střelacová K et al (eds) Bioclimatology and natural hazards. Springer, Dordrecht, pp 165–177

Hlásny T, Zimová S, Merganičová K, Štěpánek P, Modlinger R, Turčáni M (2021) Devastating outbreak of bark beetles in the Czech Republic: drivers, impacts, and management implications. For Ecol Manag 490:119075

Hoch G, Cech TL, Fürst A, Hoyer-Tomiczek U, Krehman HBP, Steyrer G (2016) Waldschutzsituation 2015 in Österreich. AFZ/Wald 7(2016):52–54

Huber C (2005) Long lasting nitrate leaching after bark beetle attack in the highlands of the Bavarian Forest National Park. J Environ Qual 34:1772–1779

Jakoby O, Lischke H, Wermelinger B (2019) Climate change alters elevational phenology patterns of the European spruce bark beetle (Ips typographus). Glob Change Biol 25:4048–4063

Jönsson AM, Harding S, Krokene P, Lange H, Lindelöw Å, Økland B, Ravn HP, Schroeder LM (2011) Modelling the potential impact of global warming on Ips typographus voltinism and reproductive diapause. Clim Chang 109:695–718

Kärvemo S, Schroeder LM (2010) A comparison of outbreak dynamics of the spruce bark beetle in Sweden and the mountain pine beetle in Canada (Curculionidae: Scolytinae). Entomol Tidskrift 131:215–224

Kautz M, Dworschak K, Gruppe A, Schopf R (2011) Quantifying spatio-temporal dispersion of bark beetle infestations in epidemic and non-epidemic conditions. For Ecol Manag 262:598–608

Kleinman SJ, De Gomez TE, Snider GB, Williams KE (2012) Large-scale pinyon ips (Ips confusus) outbreak in southwestern United States tied with elevation and land cover. J Forestry 110:194–200

Krokene P (2015) Conifer defense and resistance to bark beetles. In: Vega FE, Hofstetter RW (eds) Bark beetles: biology and ecology of native and invasive species. Academic, London, pp 177–207

Kupferschmid AD, Bugmann H (2005) Predicting decay and ground vegetation development in Picea abies snag stands. Plant Ecol 179:247–268

Kurz WA, Dymond CC, Stinson G, Rampley GJ, Neilson ET, Carroll AL, Ebata T, Safranyik L (2008) Mountain pine beetle and forest carbon feedback to climate change. Nature 452:987–990

Lausch A, Heurich M, Fahse L (2013) Spatio-temporal infestation patterns of Ips typographus (L.) in the Bavarian Forest National Park. Germany. Ecol Indic 31:73–81

Lieutier F (2004) Host resistance to bark beetles and its variations. In: Lieutier F, Day KR, Battisti A, Grégoire JC, Evans HF (eds) Bark and wood boring insects in living trees in Europe: a synthesis. Kluwer Academic Publishers, Dordrecht, pp 135–180

Lindgren BS, Miller DR (2002) Effect of verbenone on five species of bark beetles (Coleoptera: Scolytidae) in lodgepole pine forests. Environ Entomol 31:759–765

Logan JA, Régnière J, Powell JA (2003) Assessing the impacts of global warming on forest pest dynamics. Front Ecol Environ 1:130–137

Marini L, Ayres MP, Battisti A, Faccoli M (2012) Climate affects severity and altitudinal distribution of outbreaks in an eruptive bark beetle. Clim Chang 115:327–341

Marini L, Økland B, Jönsson AM, Bentz BJ, Carroll AL, Forster B, Grégoire JC, Hurling R, Nageleisen LM, Netherer S (2017) Climate drivers of bark beetle outbreak dynamics in Norway spruce forests. Ecography 40:1426–1435

Meier F, Engesser R, Forster B, Odermatt O, Angst A (2004) Forstschutz-Überblick 2003. Eidgenössische Forschungsanstalt für Wald, Schnee und Landschaft, Birmensdorf, p 22

Moser JC, Konrad H, Kirisits T, Carta LK (2005) Phoretic mites and nematode associates of *Scolytus multistriatus* and *Scolytus pygmaeus* (Coleoptera: Scolytidae) in Austria. Agric For Entomol 7:169–177

Müller J, Bussler H, Gossner M, Rettelbach T, Duelli P (2008) The European spruce bark beetle *Ips typographus* in a national park: from pest to keystone species. Biodivers Conserv 17:2979–3001

Mulock P, Christiansen E (1986) The threshold of successful attack by *Ips typographus* on *Picea abies*: a field experiment. For Ecol Manag 14:125–132

Nageleisen L (2009) L'estimation des dégâts liés aux scolytes après les tempêtes de 1999. In: Birot Y, Landmann G, Bonhême I (eds) La forêt face aux tempêtes. Éd. Quae, Montpellier, pp 69–75

Negrón JF, Schaupp WC Jr, Pederson L (2011) Flight Periodicity of the Douglas—Fir Beetle, *Dendroctonus pseudotsugae* Hopkins (Coleoptera: Curculionidae: Scolytinae) in Colorado, USA. Coleopts Bull 65:182–184

Nelson WA, Lewis MA (2008) Connecting host physiology to host resistance in the conifer-bark beetle system. Theor Ecol 1:163–177

Netherer S, Schopf A (2010) Potential effects of climate change on insect herbivores in European forests—general aspects and the pine processionary moth as specific example. For Ecol Manag 259:831–838

Netherer S, Matthews B, Katzensteiner K, Blackwell E, Henschke P, Hietz P, Pennerstorfer J, Rosner S, Kikuta S, Schume H (2015) Do water – limiting conditions predispose Norway spruce to bark beetle attack? New Phytol 205:1128–1141

Nüsslein S, Faisst G, Weissbacher A, Moritz K, Zimmermann L, Bittersol J, Kennel M, Troycke A, Adler H (2000) Zur Waldentwicklung im Nationalpark Bayerischer Wald 1999. Ber Ber Bay Landesanst Wald Forstwirts 25:47 pp

Økland B, Netherer S, Marini L (2015) The Eurasian spruce bark beetle: the role of climate. In: Björkman C, Niemelä P (eds) Climate change and insect pests. CAB International, Wallingford, pp 202–219

Paine TD, Raffa KF, Harrington TC (1997) Interactions among scolytid bark beetles, their associated fungi, and live host conifers. Annu Rev Entomol 42:179–206

Phillips MA, Croteau RB (1999) Resin-based defenses in conifers. Trends Plant Sci 4:184–190

Raffa KF, Aukema BH, Bentz BJ, Carroll AL, Hicke JA, Turner MG, Romme WH (2008) Cross-scale drivers of natural disturbances prone to anthropogenic amplification: the dynamics of bark beetle eruptions. Bioscience 58:501–517

Raffa KF, Aukema BH, Bentz BJ, Carroll AL, Hicke JA, Kolb TE (2015a) Responses of tree-killing bark beetles to a changing climate. In: Björkman C, Niemelä P (eds) Climate change and insect pests. CAB International, Wallingford, pp 173–201

Raffa KF, Grégoire J-C, Lindgren BS (2015b) Natural history and ecology of bark beetles. In: Vega FE, Hofstetter RW (eds) Bark beetles: biology and ecology of native and invasive species. Academic, London, pp 1–40

Ramsfield TD, Bentz BJ, Faccoli M, Jactel H, Brockerhoff EG (2016) Forest health in a changing world: effects of globalization and climate change on forest insect and pathogen impacts. Forestry 89:245–252

Rouault G, Candau J-N, Lieutier F, Nageleisen L-M, Martin J-C, Warzée N (2006) Effects of drought and heat on forest insect populations in relation to the 2003 drought in Western Europe. Ann For Sci 63:613–624

Saint-Germain M, Buddle CM, Drapeau P (2007) Primary attraction and random landing in host-selection by wood-feeding insects: a matter of scale? Agr For Entomol 9:227–235

Schelhaas MJ, Nabuurs GJ, Schuck A (2003) Natural disturbances in the European forests in the 19th and 20th centuries. Glob Change Biol 9:1620–1633

Oҽłֈʊʊʄ ٌ٬ Ⴌʊ丨ʋ,Ⴌ丨 ᴘ ƮΆ丨ᴑ丨 Ⴇʊ/ɫ丨ᴑ·丨ᴑ丨丨丨 丨ᴑ丨丨丨 ⎾丨丨ᴘ丨ɯ丨丨丨ᴘ丨丨丨丨ᴑ 丨丨ᴘ'丨 ᴋ丨丨丨·丨丨丨丨丨丨丨丨ᴘ'ᴘ丨丨' ᴃ丨丨ᴘ'丨丨丨·'ᴃ丨丨丨丨丨·' 丨丨 丨ᴘ丨丨ᴀ丨丨
ƼU.丨丨·丨ᴑ

Seidl R, Blennow K (2012) Pervasive growth reduction in Norway spruce forests following wind disturbance. PLoS One 7:e33301

Seidl R, Rammer W, Jäger D, Lexer MJ (2008) Impact of bark beetle (*Ips typographus* L.) disturbance on timber production and carbon sequestration in different management strategies under climate change. For Ecol Manag 256:209–220

Seidl R, Rammer W, Blennow K (2014) Simulating wind disturbance impacts on forest landscapes: tree-level heterogeneity matters. Environ Model Softw 51:1–11

Seidl R, Müller J, Hothorn T, Bässler C, Heurich M, Kautz M (2016) Small beetle, large-scale drivers: How regional and landscape factors affect outbreaks of the European spruce bark beetle. J Appl Ecol 53:530–540

Seidl R, Thom D, Kautz M, Martin-Benito D, Peltoniemi M, Vacchiano G, Wild J, Ascoli D, Petr M, Honkaniemi J, Lexer MJ, Trotsiuk V, Mairota P, Svoboda M, Fabrika M, Nagel TA, Reyer CPO (2017) Forest disturbances under climate change. Nat Clim Chang 7:395–402

Six DL, Bracewell R (2015) Dendroctonus. In: Vega FE, Hofstetter RW (eds) Bark beetles: biology and ecology of native and invasive species. Academic, London, pp 305–350

Stadelmann G, Bugmann H, Meier F, Wermelinger B, Bigler C (2013a) Effects of salvage logging and sanitation felling on bark beetle (*Ips typographus* L.) infestations. For Ecol Manag 305:273–281

Stadelmann G, Bugmann H, Wermelinger B, Meier F, Bigler C (2013b) A predictive framework to assess spatio-temporal variability of infestations by the European spruce bark beetle. Ecography 36:1208–1217

Stadelmann G, Bugmann H, Wermelinger B, Bigler C (2014a) Spatial interactions between storm damage and subsequent infestations by the European spruce bark beetle. For Ecol Manag 318:167–174

Stadelmann G, Meier F, Bigler C (2014b) Ursachen und Verlauf von Buchdrucker-Epidemien. Wald Holz 5(14):25–28

Štefková K, Okrouhlík J, Doležal P (2017) Development and survival of the spruce bark beetle, *Ips typographus* (Coleoptera: Curculionidae: Scolytinae) at low temperatures in the laboratory and the field. Eur J Entomol 114:1–6

Temperli C, Bugmann H, Elkin C (2013) Cross-scale interactions among bark beetles, climate change, and wind disturbances: a landscape modeling approach. Ecol Monogr 83:383–402

Usbeck T, Wohlgemuth T, Pfister C, Bürgi A, Dobbertin M (2010) Increasing storm damage to forests in Switzerland from 1858 to 2007. Agric For Meteorol 150:47–55

Wegensteiner R, Wermelinger B, Herrmann M (2015) Natural enemies of bark beetles: predators, parasitoids, pathogens, and nematodes. In: Vega FE, Hofstetter RW (eds) Bark beetles: biology and ecology of native and invasive species. Academic, London, pp 247–304

Wellenstein G (1954) Die grosse Borkenkäferkalamität in Südwestdeutschland 1944–1951. Forstschutzstelle Südwest, Ringingen, 496 p

Wermelinger B (2004) Ecology and management of the spruce bark beetle *Ips typographus*—a review of recent research. For Ecol Manag 202:67–82

Wermelinger B, Seifert M (1998) Analysis of the temperature dependent development of the spruce bark beetle *Ips typographus* (L)(Col., Scolytidae). J Appl Ecol 122:185–191

Wermelinger B, Rigling A, Schneider Mathis D, Kenis M, Gossner MM (2021) Climate change effects on trophic interactions of bark beetles in inner Alpine Scots pine forests. Forests 12:136

Zuo J, Cornelissen JHC, Hefting MM, Sass-Klaassen U, Van Logtestijn RSP, Van Hal J, Goudzwaard L, Liu JC, Berg MP (2016) The (w)hole story: facilitation of dead wood fauna by bark beetles? Soil Biol Biochemist 95:70–77

Chapter 13
Large Herbivores

Josef Senn (iD)

Abstract Large herbivores can influence the population dynamics of their forage plants. Feeding means a disturbance for the affected plants. Although in most cases herbivores consume only a relatively small fraction of the total phytomass, they influence the intra- and interspecific competition between plants. Whether and to what extent herbivores influence the forest–open land distribution has been subject of controversial discussion. Today, species diversity and population density of large herbivores in the temperate zone are lower than they were at the end of the last Ice Age. Therefore, stronger and more diverse influences of these species on the vegetation can be assumed before widespread human activity. This fact must be considered in the modern management of natural landscapes.

Keywords Browsing · Grazing · Forest-grassland distribution · Conservation management · Natural landscapes · Rewilding · Ungulates

13.1 Herbivores as a Disturbance in Plant Communities

Herbivores reduce the biomass of their forage plants by eating them. If small plants are eaten more or less completely or if larger plants are eaten heavily and repeatedly, this may lead to the death of individual plants. If herbivores kill the affected plants, they create space in the vegetation cover that can be colonized by other plants. Therefore, herbivory is a disturbance for the affected plant individuals and populations. Such disturbances may have an important structuring function for plant communities (Grime 1974).

Disturbances influence the competitive relationships between the affected plant individuals and species. When herbivores feed, they reduce the size and density of

J. Senn (✉)
Research Unit Community Ecology, Swiss Federal Research Institute WSL,
Birmensdorf, Switzerland
e-mail: josef.senn@wsl.ch

the food plants. If certain plant species are preferred to others, herbivores affect the relationships between plant species. In this way, herbivores reduce the competitive strength of one plant species to the advantage of another, thus affecting the abundance of plant species. If preferred plant species disappear because of feeding by herbivores, the species composition of the vegetation also changes.

In temperate forests, herbivores consume only a small proportion of the annually produced phytomass. Apart from relatively rare mass outbreaks of herbivorous insects, the quantities of plant material consumed are so small that they can hardly be quantified on a larger spatial scale. Various studies estimate that the amount consumed by insects in forest ecosystems represents on average about 1% of the annual plant production (e.g. Nielsen 1978). In comparison, vertebrate herbivores consume even less, since they cannot reach a large part of the potentially usable phytomass in forests. Nevertheless, vertebrates can locally influence both the forest structure and the composition of tree species (Putman et al. 1989). By eating tree seedlings or saplings, they increase mortality and alter the survival of the affected species. If they eat shoots, and particularly if they terminal shoots of young trees, herbivores reduce the growth of the affected individuals and thus change the competitive relationships between young trees. The selective preference of individual species may affect the tree species composition in the forest, with preferred species becoming locally rare or even disappearing (Putman et al. 1989). Based on a meta-analysis of data from 13 studies from Europe and North America comparing tree species numbers and abundancies in fenced and unfenced areas, Gill and Beardall (2001) showed that certain tree species became rarer in accessible areas with above-average deer densities. However, the high deer densities did not lead to a change in the number of tree species present. Browsing by deer did not lead to the loss of any tree species at any of the sites investigated.

The influence of vertebrate herbivores, particularly ungulates such as red deer (*Cervus elaphus* L.), roe deer (*Capreolus capreolus* L.), and chamois (*Rupicapra rupicapra* L.), on forest development has been the subject of controversial debate (Senn and Suter 2003). Forest owners and forest managers fear that herbivores may have a negative effect on forest development and future stand structure and stability (Gill and Beardall 2001). In a review of browsing by ungulates in Swiss forests, Kupferschmid et al. (2015) found that there was no significant effect of browsing on two thirds of the forest areas monitored, either for individual tree species or for the overall forest regeneration. Negative influences on individual species were found for oak (*Quercus* spp. L.) in the lowlands, for fir (*Abies alba* Mill.) at high elevation sites, and for rowan (*Sorbus aucuparia* L.) and sycamore (*Acer pseudoplatanus* L.) in the subalpine zone. In addition to potential economic losses from not being able to use the timber, loss of tree species through browsing can, for example, impair the protective function of mountain forests against snow avalanches and floods after heavy precipitation (Dorren et al. 2004). In open land, grazing herbivores can also influence the plant species composition. Because, in contrast to forests, on open land the entire vegetation is accessible to herbivores, they may reduce the phytomass much more strongly. By grazing selectively, herbivores may further affect the interspecific competition more strongly than in forests. If grazing pressure is high, whole plants are eaten or torn out of the ground, leading to small openings in the vegetation. Mechanical destruction of the vegetation through trampling by hoofed animals or by other large animals wallowing on the ground sometimes creates larger open spaces in the vegetation cover (Miles and Kinnaird 1979). Seeds dispersed by wind or by animals may germinate in the openings, and thus new plants become established. According to several studies, only a small proportion of the total phytomass is used by wild ungulate populations on open land. Even in relatively unproductive habitats (e.g. arctic tundra), this is in the range of a few percent (Pavlov et al. 1994). In the Swiss National Park, with seasonally high-density populations of red deer, chamois, and Alpine ibex (*Capra ibex* L.), phytomass consumption is significantly higher, ranging from 17% of the annual production in subalpine meadows to 83% in alpine grasslands (Schütz et al. 2006; Risch et al. 2013).

Moderate grazing by both wild and domesticated animals increases both the productivity and the plant species diversity of pastures. In comparison, both ungrazed and overgrazed areas generally contain significantly less species. This phenomenon is explained by the intermediate disturbance hypothesis (IDH) (Grime 1973; Connell 1978; Fox 1979). On ungrazed sites, a few plant species grow faster and spread at the expense of other species. In the case of overgrazing, herbivores prefer certain more palatable plant species, which become rarer as a result. As these species become rarer, the next most palatable species are grazed until finally only unpalatable species remain. These are either mechanically 'defended' (e.g. by spines or thorns) or chemically defended by secondary compounds that make them difficult to digest or even poisonous to herbivores. If overgrazing remains severe and persistent, unpalatable species may spread and finally cover the entire pasture, making it of little value to grazers (DiTomaso 2000). However, in the case of very heavy grazing and mechanical stress on the vegetation by livestock at high densities and depending on soil structure and nutrients, the vegetation may completely disappear over large areas, leading to land degradation over large areas (Foss 1960).

13.2   Comparing Influences of Domesticated Animals and Wildlife on Vegetation

In contrast to wild animals, domesticated animals such as cattle and sheep may negatively affect their pastures more strongly and continuously and may even destroy vegetation completely. In all cases domesticated grazers use a significantly higher proportion of the phytomass than wild animals (Fleischner 1994). The reason for this is that for economic reasons the density of domesticated animals is almost always higher than that of wild animals. More importantly, however, farm animals almost always use a particular area for a longer period than wild animals do, since their grazing areas are usually enclosed.

Even very extensive grazing, e.g. with domesticated reindeer (*Rangifer tarandus* L.) in northern Scandinavia, may lead to large-scale degradation of the vegetation (Moen and Danell 2003). In northern Norway, the density of reindeer was relatively low until the 1950s, with less than two animals per square kilometre. Reindeer used to be herded by semi-nomadic Saami people, who directed the grazing with the aim of not overusing the vegetation. Reindeer density has fluctuated considerably over the years. Severe winters with increased mortality were followed by years of lower density, allowing the vegetation to recover. In later periods, active herding of reindeer by Saami was abandoned to a large extent, and within certain limits, the animals moved independently in the grazing areas for most of the year. More recently, supplementary feeding of the herds with hay has prevented increased mortality as a consequence of adverse winter conditions. This allowed the density of reindeer in northern Scandinavia to increase to more than eight animals per square kilometre by the 1980s (Lempa et al. 2005). In the last few decades, grazing grounds have been increasingly sub-divided by fixed fences. The restriction of grazing reindeer to smaller plots locally led to severe overuse and even to the destruction of the plant cover (Fig. 13.1; Moen and Danell 2003). Once the vegetation is destroyed, it takes decades or even centuries for the vegetation to recover in the subarctic zone.

Historically overgrazing with domesticate animals led to extensive destruction of vegetation in various regions with long-lasting negative effects on the affected human populations (Diamond 2004). Already in the tenth century, grazing sheep had destroyed vegetation over large areas in Iceland (Thorsteinsson et al. 1971). This exposed the fine-grained volcanic soils to precipitation. Consequently, rain washed much of the island's original topsoil into the sea leaving the land barren over large areas. In the Midwest of the USA in the 1930s, overgrazing with cattle and sheep led to the infamous 'Dust Bowl', in which wind blew away fertile soils over hundreds of kilometres (Foss 1960), resulting in the loss of extensive agricultural areas.

Fig. 13.1 In the interior of the Varanger Peninsula (Norway), fences limit migration of reindeer. The vegetation on the left side has been extensively destroyed by strong grazing pressure, while the vegetation on the right side of the fence is largely intact. (Photo: J. Senn)

13.3 Influence of Large Herbivores on the Forest– Grassland Distribution

Whether and to what extent large herbivores (according to Martin and Klein 1984) are capable of opening up closed forests and maintaining open habitats in the long term has been, and still is, the subject of controversial debate (e.g. Owen-Smith 1987; Zoller and Haas 1995; Bunzel-Drüke et al. 1999; Bradshaw et al. 2003; Birks 2005). Several studies from East Africa demonstrated that African bush elephants (*Loxodonta africana* Blumenbach) may clear forest and transform it into grassland (e.g. Buechner and Dawkins 1961). Also, for temperate areas it must be assumed that during interglacials and after the last Ice Age (about 12,000 years before present), the large herbivores, which are extinct today, had a significant influence on the structure and species composition of the vegetation in forests as well as on open land (Gill et al. 2009; Sandom et al. 2014b). This raises the fundamental questions of how densely forested the primeval landscapes of Central Europe were under the natural influence of large herbivorous mammals and how quickly Central Europe reforested after the retreat of the ice sheets. These questions are still relevant today, for example, when defining 'potential natural forest communities' (Tüxen 1956; Ellenberg and Klötzli 1972). Is a 'natural forest community' influenced by herbivores, or is it defined without relevant herbivore influences? If, in modern nature conservation and management, forests and their ecological processes, i.e. primeval, or primary forests with process protection, are to be preserved, the questions arise as to whether herbivore influences are among the processes to conserve and what densities of mammalian herbivores could be reached without human influence. This discussion already began over a hundred years ago with the assumption that the Hungarian 'puzsta' (a vast grassland also known as the Pannonian Steppe) was originally forest-free (Kerner and Marilaun 1863) and with Gradmann (1900), who developed a 'steppe theory' for the Swabian Jura (Gradmann 1933). Already at the beginning of the twentieth century, a controversy arose around the question of the extent of forested habitats in primeval Central Europe (e.g. Tüxen 1931). One side argued that the so-called megaherbivores had created large-scale savannah landscapes in Central Europe and that these herbivores were able to maintain them (Vera 2000). The other side rejected any influence of the megaherbivores on the forest– open land distribution before the Neolithic Revolution about 7500 years ago, when hunter–gatherer societies adopted agriculture and largely sedentary lifestyles (e.g. Zoller and Haas 1995). According to this side of the argument, open habitats in Switzerland were created by anthropogenic fires and thus existed only since then (Gobet et al. 2010).

Until recently, natural densities and population fluctuations of the large herbivores in primeval habitats of temperate latitudes have been unknown. The majority of hoofed species in this zone disappeared during and after the last Ice Age (Owen-Smith 1987), most likely exterminated, or their disappearance at least accelerated, by early human societies (Sandom et al. 2014a). In addition, the tarpan or forest horse (*Equus ferus ferus* Boddaert) and the aurochs (*Bos taurus primigenius*

Bojanus; van Vuure 2005) became extinct as a result of uncontrolled hunting in historic times. In the surviving large herbivore species, almost all populations are now managed to a greater or lesser extent, i.e. population densities and population sizes are controlled, mainly by hunting (Apollonio et al. 2010). In almost all cases, the population densities achieved are well below the carrying capacity of the habitat that may be reached without human intervention (Sinclair 1997). In most cases, wildlife population management aims to prevent large fluctuations in size and density of the animal populations (Kenward and Putman 2011).

On the one hand, sustainable hunting aims not to overuse wildlife populations, i.e. population densities should not fall below a defined density. On the other hand, forest management sets an upper limit for herbivore densities in order to control and limit the influence of wildlife browsing on young trees. However, most wildlife populations today live in habitats influenced by humans to a greater or lesser extent. Herbivores mainly stay in the forest, but they have access to diverse agricultural areas with seasonally rich food supply. Agricultural areas are mainly used for feeding during the night. Landscapes consisting of a small-scale mosaic of interlocking forest and agricultural areas often facilitate easy access to the agricultural fields and thus may sustain locally dense wild ungulate populations.

Nowadays the few remaining larger primeval temperate forests of Eastern Europe and Asia are largely devoid of wild ungulates. While these forests have hardly been used for forestry purposes, the hoofed animals originally found there have been reduced to small remnants through intensive, uncontrolled, or illegal hunting. Thus, there are no longer any 'reference systems' to study the development of forests or natural grassland under the influence of a natural herbivore guild.

Bradshaw et al. (2003) investigated the influence of large herbivore faunas on vegetation, and particularly on the forest–open land distribution, during the last 500,000 years in Central and Northern Europe and discussed the importance of large herbivores for today's forest ecosystems. Earlier interglacial periods were characterized by species-rich large herbivore faunas, which in many studies are referred to as 'megafauna' (Table 13.1). During the last seven interglacial periods (warm periods), the number of large herbivore species remained more or less steady. It was only during and shortly after the last Ice Age that about half of the large herbivore species became extinct (Owen-Smith 1987). The role of humans in the extinction of these species remains controversial (Martin and Klein 1984; Koch and Barnosky 2006; Firestone et al. 2007). However, it is striking that on several continents and large islands, several large-bodied species, or even the entire megafauna, became extinct shortly after the arrival and spread of humans (Martin and Klein 1984; Koch and Barnosky 2006). In contrast, species in regions not used, or little used, by human populations sometimes survived several thousand years longer than in regions populated earlier by humans (Stuart et al. 2004). In Europe and western Asia, the populations of surviving large herbivore species decreased shortly after the beginning of agriculture in the Neolithic era, or they even became locally extinct (e.g. Aaris-Sørensen 1980). In the recent past, the populations of some of the surviving large herbivore species have begun to increase again as a result of efficient legal protection and regulated hunting. Today some of these species occur locally in

Table 13.1 Ungulate faunas from British interglacials. *Alces*, *Palaeoloxodon*, and *Mammuthus* were not recorded at the respective British sites but were abundant at adjacent continental sites at 520,000 years before present and are thus represented by the Y in brackets '(Y)'

Approximate start in thousands of years before present	Interglacials							
	700	620	520	410	330	240	130	11.5
Cervidae								
Megaloceros savini (giant deer)	X	X	X					
M. verticornis (giant deer)	X	X						
M. dawkinsi (giant deer)		X	X					
M. giganteus (Irish giant deer)				X	X	X	X	
Alces latifrons (broad-fronted moose)	X		(X)					
Alces alces (moose)								X
Cervus elaphus (red deer)	X	X	X	X	X	X	X	X
Dama dama (fallow deer)	X	X	X	X	X		X	
Capreolus capreolus (roe deer)			X	X	X	X	X	X
Bovidae								
Bison schoetensacki (woodland bison)	X	X	X	X	X	X	X	
Bos taurus primigenius (aurochs)				X	X	X	X	X
Suina								
Hippopotamus amphibius (hippo)		X					X	
Sus scrofa (wild boar)	X	X		X	X	X	X	X
Rhinocerotidae (rhinoceroses)								
Stephanorhinus hundsheimensis	X	X	X					
S. kirchbergensis				X	X	X		
S. hemitoechus				X	X	X	X	
Equidae								
Equus ferus ferus (wild horse)	X		X	X	X	X		
E. hydruntinus (European wild ass)	X	X		X		X		
Proboscidae (elephant/mammoth)								
Palaeoloxodon antiquus		X	(X)	X	X	X	X	
Mammuthus trogontherii	X	X	(X)			X		
Total number of species	12	12	11	12	11	12	10	5

After Bradshaw et al. (2003)

unprecedented high densities (Bradshaw and Mitchell 1999). Because of reduced species diversity, the influences on vegetation today must be fundamentally different from those of earlier more species-diverse herbivore faunas. Today the influences of the herbivores are mainly determined by density, i.e. with increasing density the influences become stronger and even less specific, while the influences of the more species-rich herbivore faunas of earlier times on the vegetation were structured and more diverse mainly as a result of interactions between the species and species-specific differences in foraging behaviour. These influences ranged from interspecific competition for food to facilitation, i.e. that one herbivore species opened up the food for another species or at least made it more easily accessible (e.g. Fritz et al. 2002). In addition, the different herbivore species making up the fauna of earlier periods differed more in their diet or preferences for woody plants, grasses, and herbs than the few wild ungulate species still found today (Bradshaw and Mitchell 1999).

Diversity and abundance of individual plant species in past vegetation are commonly determined by analysing pollen stratigraphy in sediment cores (Huntley and Birks 1983). To a lesser extent, plant macrofossils are also used for this purpose (Watts 1978). In the Holocene, oak (*Quercus* spp.) and hazel *(Corylus avellana* L.) were very common in Central Europe (Vera 2000). These species indicate open forests. They regenerate only poorly or not at all in closed high forests. While it was previously assumed that small gaps, e.g. of fallen old, dead trees, were sufficient for successful regeneration of these species (Watt 1947), Vera (2000) claimed that larger open areas were needed for these light-demanding species to develop naturally in frequencies indicated by the pollen quantities found in ancient sediments. According to Vera (2000), hoofed animals may clear the forest to such an extent that species such as oak and hazel may establish and survive. However, such open forests would need higher densities of large ungulates over extended periods in order to remain sufficiently open. Pollen analyses as well as findings of macrofossils of herbivores have shown that light-demanding tree and shrub species as well as large herbivores, which could potentially keep the forests open, were present in Europe prior to significant anthropogenic influence (Bradshaw et al. 2003). Vera (2000) derived his hypothesis mainly from observations and experiments in nature reserves where ungulates were used to prevent the development of closed high forest and instead to maintain a structurally rich, relatively open landscape. Such reserves showed a cyclical turnover of the vegetation through four stages (Fig. 13.2). In the first stage, the open landscape contained only a few larger trees. The vegetation consisted mainly of grasses, herbs, and dwarf shrubs. Herbivores grazed most of the

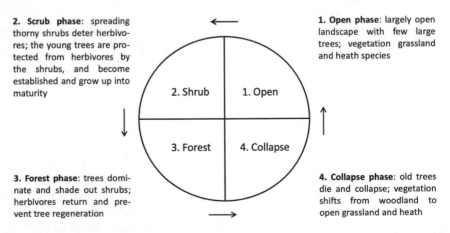

2. Scrub phase: spreading thorny shrubs deter herbivores; the young trees are protected from herbivores by the shrubs, and become established and grow up into maturity

1. Open phase: largely open landscape with few large trees; vegetation grassland and heath species

3. Forest phase: trees dominate and shade out shrubs; herbivores return and prevent tree regeneration

4. Collapse phase: old trees die and collapse; vegetation shifts from woodland to open grassland and heath

Fig. 13.2 Vera's model consists of four development phases ranging from an open landscape with scattered large trees to closed forest, which, through a decay phase, develops back into an open park landscape. (After Kirby 2004)

vegetation but avoided emerging thorny bushes. These shrubs then spread at the expense of the rest of the vegetation which led to the second stage. When the thorny shrubs covered large enough areas, trees could regenerate within the thorny shrub area undisturbed by the herbivores. The number and density of trees increased, and they grew up until their crowns closed to forest. In this third stage, relatively few plants grew in the shade of the canopy, and they were almost completely used as food by the herbivores. Regenerating tree seedlings were also mostly eaten. High forest remained lacking regeneration. In the fourth stage, the overaged forest collapsed, leading again to the first stage, an open landscape with scattered large trees.

Vera (2000) derived his hypothesis from observations in spatially limited, in most cases closed, i.e. fenced, nature reserves. Here high-density herbivore populations exert strong pressure on the vegetation they feed on throughout the whole area – a situation that can be found, for example, in the 6000 ha Oostvaardersplassen (Fig. 13.3) in the Netherlands. However, when applied to a much larger, topographically more structured landscape and more variable grazing pressure, such a scenario seems less realistic.

Kirby (2004) suggests a temporally and spatially variable scenario for larger areas. Accessibility to herbivores varies in different parts of the landscape, and herbivore populations fluctuate strongly over time, depending on, for example, pathogen outbreaks or very snowy winters which limit access to food (Fig. 13.4). The driving force in Kirby's model remains Vera's cycle. However, in addition to Vera's model, Kirby's landscape includes areas that are permanently covered by forest owing to poor accessibility for herbivores. These forests may expand during periods with low-density herbivore populations. On the other hand, there are parts of the

Fig. 13.3 In the 6000 ha large nature reserve Oostvaardersplassen (Netherlands), about 1000 red deer graze together with about 700 Heck cattle and Konik horses largely free from human intervention. (Photo: J. Senn)

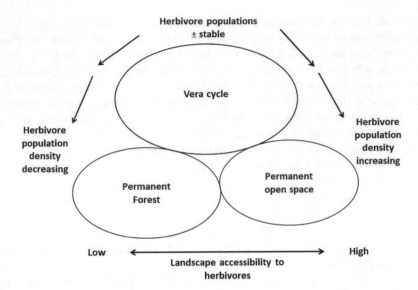

Fig. 13.4 Differences in space (accessibility for herbivores) and time (strongly fluctuating population densities of herbivores) could shift the balance between Vera's cycle, permanently open grassland and permanently closed forest over longer periods and larger areas. (After Kirby 2004)

landscape where herbivores are concentrated for longer periods, preventing growth of trees and thus keeping these sites permanently open. Other parts of the landscape are subject to fluctuating herbivore influences according to the cycle postulated by Vera (2000).

Analyses of Holocene deposits from Ireland revealed large amounts of pollen from light-dependent oaks and hazel (Mitchell 1988). Aurochs, bison, and wild horse or any other large herbivores, however, did not occur in Ireland at that time (Woodman et al. 1997). This shows that, at least in Ireland, these animals could not have created favourable conditions for light-demanding plant species (Mitchell 2005). In deposits from Ireland, England, and continental Europe, large amounts of charcoal particles have been found in addition to pollen (Bradshaw et al. 1997). These indicate that lightning-induced fires may have caused the creation of open areas in forests (Svenning 2002). Recent examples show that high population densities of hoofed animals have been able to keep forests open for extended periods after they were cleared by fire (Scholes and Archer 1997; van Langevelde et al. 2003). The work of Maxwell (2004) has shown that fires are not a new phenomenon and need not be related to human activities. Maxwell studied the vegetation and fire regimes during the last 9300 years in the open monsoon forests of northeast Cambodia. This savannah-like landscape was home to various deer and wild cattle species in high densities until well into the twentieth century (Wharton 1966). The concentration of charcoal particles was highest in the oldest layers (i.e. in the period from 9300 to 8000 years before present), and it was several times higher than today. The frequency of fires was found to be closely correlated with the amount of grass in the forests. The frequency and intensity of fires in northeastern Cambodia have

Fig. 13.5 In temperate forests, only grazing at very high ungulate densities prevents tree regrowth over long periods of time. (Photo: J. Senn)

been lowest in the most recent 2500 years. These data contradict the often-expressed fear by conservation professionals that human-made fires are too frequent today and that these disturbances could threaten the integrity of the monsoon forests and their biodiversity (e.g. Moore et al. 2002). On the contrary, fire is a key factor sustaining these ecosystems. Besides fire, extreme storm events may also open up forests on a large scale (Webb and Walker 1999). If herbivores are present at high densities, they can prevent development of closed forests dominated by a few species for long periods of time (Fig. 13.5; Bond and Keeley 2005). However, large trees uprooted by storms may also provide safe sites protected from browsing herbivores where young trees may regenerate and become established (Moser et al. 2008).

Recent studies from the northeastern USA based on palynological data have demonstrated that open coniferous forests with a low proportion of broadleaved deciduous trees and larger proportions of grasses and herbs prevailed after the end of the last Ice Age (Gill et al. 2009). At the same time, large herbivores at high densities were found in these areas. These megaherbivores were mainly feeding on broadleaved species as well as on grasses and herbs. The density of the herbivores was determined by spores of fungi of the genus *Sporormiella* (Ellis and Everhan) living in the intestines and faeces of ungulates, which were found in the sediments together with the pollen. Analyses of frequencies of *Sporormiella* spores show that with the advancement of the newly immigrated humans in North America, the populations of large mammals decreased rapidly and almost completely disappeared about 10,000 years before present. After the megaherbivores disappeared, the composition of the forests changed dramatically. The abundance of broadleaved species increased strongly, and coniferous species became rarer. After the extinction of the megaherbivores and before the establishment of predominantly broadleaved forests, forest communities existed, whose species composition differed significantly from earlier and later forest communities. In addition, the forests became denser, and most probably because of the lack of herbivores, highly combustible dead and dry plant material accumulated leading to large and frequent forest fires, as indicated by increasing amounts of charcoal particles in the sediments of this period (Gill et al. 2009).

One of the few recent examples that demonstrated that grazing ungulates may open up forests and convert them into grasslands was found in northern Scandinavia. In a meta-analysis, Lempa et al. (2005) investigated the influence of grazing reindeer on vegetation development. Pairwise comparisons of areas with and without grazing from 38 studies were used. The results showed variable influences of reindeer on different vegetation parameters (Fig. 13.6). The strongest influence was found for lichens. If the reindeer grazing density was greater than two per square kilometre, the proportion of soil surface covered by lichen decreased significantly. For downy birch (*Betula pubescens* Ehrh.) and silver birch (*B. pendula* Roth), leaf mass as well as stem number decreased if the number of reindeer was greater than this density. On the other hand, both the coverage of grasses and sedges and their biomass increased with grazing intensity. The density of ericaceous shrubs and even more so the density of birch seedlings varied strongly and seemed not to be related to reindeer feeding. This meta-analysis thus revealed that under the current influence of reindeer, a decrease in forest cover and an increase in open grassland may be expected in the medium to long term.

The proponents of the high forest hypothesis (e.g. Zoller and Haas 1995; Birks 2005) do not assign any significant influence of herbivores on forest cover. According to them, large ungulates have not played any role in forest development, and thus are not relevant for the dynamics in today's natural forests either. The authors mentioned above derived the high forest hypothesis from studies on current semi-natural, i.e. little managed forests (Peterken 1996). They concluded that the forests which developed after the end of the last Ice Age did not differ from the closed, dark primeval/primary forests of the temperate zone, which are still to be found as remnants

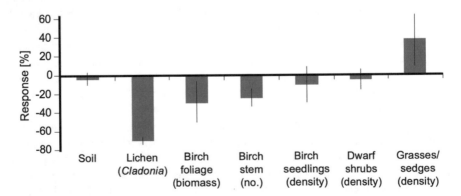

Fig. 13.6 Reactions of various pasture components to reindeer grazing in the downy birch zone of northern Scandinavia. (After Lempa et al. 2005)

in Europe and in the northeastern USA. Open forests should, therefore, be the exception and only occur on sandy, nutrient-poor soils or on seasonally flooded sites. Examples from Denmark show that such light forests are well represented in regional palynological data (Odgaard 1994). In closed high forest, tree regeneration occurred in small openings after single old trees fell to create gaps in the canopy (Yamamoto 2000). In this hypothesis, the sporadic occurrence of pollen of light-demanding plant species found in palynological data from high forests was not considered as evidence for large herbivores affecting forest structure and distribution. However, the high forest hypothesis cannot explain the large-scale and long-lasting occurrence of oaks and hazel in the Holocene and the survival of many other light-dependent plant species in Central Europe (Ellenberg et al. 1991).

13.4 Management of Protected Areas with Wild and Domesticated Ungulates

The extinction of the two largest herbivore species, the wild horse (*Equus ferus ferus*) and the aurochs (*Bos taurus primigenius*), in Central Europe in historic times must have changed both the species composition of the flora and small animal fauna in closed forests as well as in open habitats (examples in Finck et al. 2004).

According to Pykälä (2000), traditional agriculture at least partially compensated for the natural processes that were lost through the reduction of the original ungulate fauna as well as through the extensive suppression of any further disturbance such as fire. The traditionally managed landscape thus preserved many biotic elements of the primeval landscape. Also in forests, after the disappearance of the large herbivores, humans took over the role previously played by the large herbivores by extracting trees and clearing forest (Bocherens 2018). Plants as well as animals depending on sufficient light found adequate habitats in the traditional mowing and grazing systems and in clearings in forests. Most of the species in these human-made open habitats were species originally found in Central Europe. Increasing intensification in modern agriculture has reduced habitat diversity and led to the local extinction of many of these species. Large-scale industrialization in modern agriculture thus depleted species diversity over large areas. Conservation measures at a larger scale were feasible only on alternative land such as nature reserves or on former agricultural land abandoned for economic reasons. Here species and natural processes may be promoted and conserved by using domesticated ungulates to substitute for the functions of their extinct ancestors (Bunzel-Drüke 2001). In the recent past, ancient and robust cattle and horse breeds have been used to manage the vegetation in a targeted manner in an increasing number of nature reserves (WallisDeVries et al. 1998).

Compared to the wild ungulates still found today, domesticated cattle and horses are relatively unselective and adapted to fibre-rich food. They are, therefore, suitable for control of abundant and large plant species. Since the wild ancestors of horses and cattle are extinct and the present wild ungulate species differ in terms of grazing behaviour and food selection, cattle and horses, and particularly the 'back-bred' Heck cattle (Fig. 13.7) and Konik horse (Fig. 13.8), are ideal for use in nature conservation and management (Vera 2000). Woody plant species make up a high proportion of the cattle's diet (Buttenschøn and Buttenschøn 2013), while the horses mainly eat grasses (Cosyns et al. 2001). Thus, the two herbivore species complement each other well (Loucougaray et al. 2004). If these animals are used in large enough areas of pasture, spatially variable grazing pressure will lead to diverse habitats (Pykälä 2000; Gilhaus et al. 2014), which may be used by various plant and animal species leading to a landscape close to the original natural landscape.

While the use of large herbivores so far in most cases has been limited to relatively small, often 'fenced-in' nature reserves, in recent years some large-scale 'rewilding' experiments have been started in North America as well as in Europe (Pereira and Navarro 2015). The aim is to restore large areas in economically marginal areas where agriculture has been abandoned into a semi-natural state with all the original biodiversity being present. The 'reconstruction' of the herbivore fauna including wild cattle and horses, as well as their large predators, such as brown bear (*Ursus arctos* L.) and grey wolf (*Canis lupus* L.), is part of the effort to restore entire landscapes to their original wild state, including the physical and biological disturbance processes that originally took place in them (Navarro et al. 2015).

Fig. 13.7 The Heck cattle back-bred from ancient cattle breeds may closely resemble the aurochs (*Bos taurus primigenius*), which became extinct in the seventeenth century. With a high proportion of woody plants in their diet, these cattle may control regrowth of scrubs and trees in nature reserves. (Photo: J. Senn)

Fig. 13.8 The Konik horse strongly resembles in size and appearance the wild forest horse or tarpan (*Equus ferus ferus*), which became extinct at the beginning of the twentieth century. As a specialized herbivore, which prefers grazing repeatedly on the same patches, it may promote spatial diversity in the vegetation of nature reserves. (Photo: J. Senn)

References

Aaris-Sørensen K (1980) Depauperation of the mammalian fauna of the island of Zealand during the Atlantic period. Vidensk Meddel Dansk Naturhist Foren 142:131–138

Apollonio M, Andersen R, Putman R (2010) European ungulates and their management in the 21st century. Cambridge University Press, Cambridge, 410 p

Birks HJB (2005) Mind the gap: how open were European primeval forests? Trends Ecol Evol 20:154–156

Bocherens H (2018) The rise of the anthroposphere since 50,000 years: an ecological replacement of megaherbivores by humans in terrestrial ecosystems? Front Ecol Evol 6:Article 3

Bond WJ, Keeley JE (2005) Fire as a global 'herbivore': the ecology and evolution of flammable ecosystems. Trends Ecol Evol 20:387–394

Bradshaw RHW, Mitchell FJG (1999) The palaeoecological approach to reconstructing former grazing-vegetation interactions. For Ecol Manag 120:3–12

Bradshaw RHW, Tolonen K, Tolonen M (1997) Holocene records of fire from the boreal and temperate zones of Europe. In: Clark JS, Cachier H, Goldammer JG, Stocks B (eds) Sediment records of biomass burning and global change. Springer, Berlin/Heidelberg, pp 347–365

Bradshaw RHW, Hannon GE, Lister AM (2003) A long-term perspective on ungulate-vegetation interactions. For Ecol Manag 181:267–280

Buechner HK, Dawkins HC (1961) Vegetation change induced by elephants and fire in Murchison Falls National Park, Uganda. Ecology 42:752–766

Bunzel-Drüke M (2001) Ecological substitutes for wild horse (*Equus ferus* Boddaert, 1785 = *E. przewalskii* Poljakov, 1881) and aurochs (*Bos primigenius* Bojanus, 1827). Nat Kulturlandschaft 4:240–252

Bunzel-Drüke M, Drüke J, Hauswirth L, Vierhaus H (1999) Großtiere und Landschaft – Von der Praxis zur Theorie. Nat Kulturlandschaft 3:210–229

Buttenschøn RM, Buttenschøn J (2013) Woodland grazing with cattle – results from 25 years of grazing in acidophilus pedunculate oak (*Quercus robur*) woodland. In: Rotherham ID (ed) Trees, forested landscapes and grazing animals. Routledge, Oxon

Connell JH (1978) Diversity in tropical rain forests and coral reefs: high diversity of trees and corals is maintained only in a non-equilibrium state. Science 199:1302–1310

Cosyns E, Degezelle T, Demeulenaere E, Hoffmann M (2001) Feeding ecology of Konik horses and donkeys in Belgian coastal dunes and its implications for nature management. Belg J Zool 131:111–118

Diamond J (2004) Collapse: how societies choose to fail or succeed. Viking Press, New York, 592 p

DiTomaso JM (2000) Invasive weeds in rangelands: species, impacts, and management. Weed Sci 48:255–265

Dorren LKA, Berger F, Imeson AC, Maier B, Rey F (2004) Integrity, stability and management of protection forests in the European Alps. For Ecol Manag 195:165–176

Ellenberg H, Klötzli F (1972) Waldgesellschaften und Waldstandorte der Schweiz. Mitt Eidg Anst forst Versuchswes 48:589–930

Ellenberg H, Weber HE, Düll R, Wirth V, Werner W, Paulissen D (1991) Zeigerwerte von Pflanzen in Mitteleuropa. Scri Geobot 18:1–248

Finck P, Härdtle W, Redecker B, Riecken U (2004) Weidelandschaften und Wildnisgebiete – Vom Experiment zur Praxis. Schriftreihe Landschpfl Natschutz 78:1–539

Firestone RB, West A, Kennett JP, Becker L, Bunch TE, Revay ZS, Schultz PH, Belgya T, Kennett DJ, Erlandson JM, Dickenson OJ, Goodyear AC, Harris RS, Howard GA, Kloosterman JB, Lechler P, Mayewski PA, Montgomery J, Poreda R, Darrah T, Hee SSQ, Smitha AR, Stich A, Topping W, Wittke JH, Wolbach WS (2007) Evidence for an extraterrestrial impact 12,900 years ago that contributed to the megafaunal extinctions and the Younger Dryas cooling. Proc Natl Acad Sci USA 104:16016–16021

Fleischner TL (1994) Ecological costs of livestock grazing in western North America. Conserv Biol 8:629–644

Foss PO (1960) Politics and grass. Greenwood Press, New York, 236 p

Fox JF (1979) Intermediate-disturbance hypothesis. Science 204:1344–1345

Fritz H, Duncan P, Gordon IJ, Illius AW (2002) Megaherbivores influence trophic guilds structure in African ungulate communities. Oecologia 131:620–625

Gilhaus K, Stelzner F, Holzel N (2014) Cattle foraging habits shape vegetation patterns of alluvial year-round grazing systems. Plant Ecol 215:169–179

Gill RMA, Beardall V (2001) The impact of deer on woodlands: the effects of browsing and seed dispersal on vegetation structure and composition. Forestry 74:209–218

Gill JL, Williams JW, Jackson ST, Lininger KB, Robinson GS (2009) Pleistocene Megafaunal collapse, novel plant communities, and enhanced fire regimes in North America. Science 326:1100–1103

Gobet E, Vescovi E, Tinner W (2010) A paleoecological contribution to assess the natural vegetation of Switzerland. Bot Helv 120:105–115

Gradmann R (1900) Das Pflanzenleben der Schwäbischen Alb, 2 vol, 2nd edn. Schwäbischer Albverein, Tübingen

Gradmann R (1933) Die Steppentheorie. Geogr Z 39:265–278

Grime JP (1973) Competitive exclusion in herbaceous vegetation. Nature 242:344–347

Grime JP (1974) Vegetation classification by reference to strategies. Nature 250:26–31

Huntley B, Birks HJB (1983) An atlas of past and present pollen maps for Europe: 0–13000 years ago. Cambridge University Press, Cambridge, 688 p

Kenward R, Putman RJ (2011) Ungulate management in Europe: towards a sustainable future. In: Apollonio M, Andersen R, Putman RJ (eds) Ungulate management in Europe: problems and practices. Cambridge University Press, Cambridge, pp 376–395

Kerner V, Marilaun A (1863) Das Pflanzenleben der Donauländer. Wagner, Innsbruck, 348 p

Kirby KJ (2004) A model of a natural wooded landscape in Britain as influenced by large herbivore activity. Forestry 77:405–420

Koch PL, Barnosky AD (2006) Late quaternary extinctions: state of the debate. Ann Rev Ecol Evol Syst 37:215–250

Kupferschmid AD, Heiri C, Huber M, Fehr M, Frei M, Gmür D, Imesch N, Rüegg D et al, Heinen P, Olivaz N-C, Odermatt O (2015) Einfluss wildlebender Huftiere auf die Waldverjüngung: ein Überblick für die Schweiz. Schweiz Z Forstwes 166:420–431

Lempa K, Neuvonen S, Tømmervik H (2005) Effects of reindeer grazing on pastures – a necessary basis for sustainable reindeer herding. In: Caldwell MM et al (eds) Plant ecology, herbivory, and human impact in Nordic mountain birch forests. Springer, Berlin/Heidelberg, pp 157–164

Loucougaray G, Bonis A, Bouzille JB (2004) Effects of grazing by horses and/or cattle on the diversity of coastal grasslands in western France. Biol Conserv 116:59–71

Martin PS, Klein RG (1984) Quaternary extinctions. A prehistoric revolution. University of Arizona Press, Tucson, 892 p

Maxwell AL (2004) Fire regimes in north-eastern Cambodian monsoonal forests, with a 9300-year sediment charcoal record. J Biogeogr 31:225–239

Miles J, Kinnaird JW (1979) The establishment and regeneration of birch, juniper and Scots pine in the Scottish Highlands. Scott For 33:102–119

Mitchell FJG (1988) The vegetational history of the Killarney oakwoods, SW Ireland – evidence from fine spatial-resolution pollen analysis. J Ecol 76:415–436

Mitchell FJG (2005) How open were European primeval forests? Hypothesis testing using palaeo-ecological data. J Ecol 93:168–177

Moen J, Danell O (2003) Reindeer in the Swedish mountains: an assessment of grazing impacts. Ambio 32:397–402

Moore P, Ganz D, Lay CT, Enters T, Durst PB (2002) Communities in flames: proceedings of an international conference on community involvement in fire management. FAO Regional Office for Asia and the Pacific, Bangkok, 133 p

Moser B, Schütz M, Hindenlang KE (2008) Resource selection by roe deer: are windthrow gaps attractive feeding places? For Ecol Manag 255:1179–1185

Navarro LM, Proença V, Kaplan JO, Pereira HM (2015) Maintaining disturbance-dependent habitats. In: Pereira HM, Navarro LM (eds) Rewilding European landscapes. Springer, Cham, pp 143–167

Nielsen BO (1978) Above ground food resources and herbivory in a beech forest ecosystem. Oikos 31:273–279

Odgaard BV (1994) The Holocene vegetation history of northern West Jutland. Opera Bot 123:1–171

Owen-Smith N (1987) Pleistocene extinctions: the pivotal role of megaherbivores. Paleobiology 13:351–361

Pavlov BM, Kolpashchikov LA, Zyryanov VA (1994) Population dynamics of the Taimyr reindeer population. Rangifer 9:381–384

Pereira HM, Navarro LM (2015) Rewilding European landscapes. Springer, Cham, 227 p

Peterken GF (1996) Natural woodland. Ecology and conservation in northern temperate regions. Cambridge University Press, Cambridge, 522 p

Putman RJ, Edwards PJ, Mann JCE, How RC, Hill SD (1989) Vegetational and faunal changes in an area of heavily grazed woodland following relief of grazing. Biol Conserv 47:13–32

Pykälä J (2000) Mitigating human effects on European biodiversity through traditional animal husbandry. Conserv Biol 14:705–712

Risch AC, Haynes AG, Busse MD, Filli F, Schütz M (2013) The response of soil CO_2 fluxes to progressively excluding vertebrate and invertebrate herbivores depends on ecosystem type. Ecosystems 16:1192–1202

Sandom C, Faurby S, Sandel B, Svenning JC (2014a) Global late Quaternary megafauna extinctions linked to humans, not climate change. Proc R Soc B Biol Sci 281:1–9

Sandom CJ, Ejrnaes R, Hansen MDD, Svenning JC (2014b) High herbivore density associated with vegetation diversity in interglacial ecosystems. Proc Natl Acad Sci USA 111:4162–4167

Scholes RJ, Archer SR (1997) Tree-grass interactions in savannas. Ann Rev Ecol Syst 28:517–544

Schütz M, Risch AC, Achermann G, Thiel-Egeneter C, Page-Dumroese DS, Jurgensen MF, Edwards PJ (2006) Phosphorus translocation by red deer on a subalpine grassland in the central European Alps. Ecosystems 9:624–633

Senn J, Suter W (2003) Ungulate browsing on silver fir (*Abies alba*) in the Swiss Alps: beliefs in search of supporting data. For Ecol Manag 181:151–164

Sinclair ARE (1997) Carrying capacity and the overabundance of deer. In: McShea WJ, Underwood HB, Rappole JH (eds) The science of overabundance: deer ecology and population management. Smithsonian Institution Press, Washington, DC, pp 380–394

Stuart AJ, Kosintsev PA, Higham TFG, Lister AM (2004) Pleistocene to Holocene extinction dynamics in giant deer and woolly mammoth. Nature 431:684–689

Svenning JC (2002) A review of natural vegetation openness in North-Western Europe. Biol Conserv 104:133–148

Thorsteinsson I, Olafsson G, Van Dyne GM (1971) Range resources of Iceland. J Range Manag 24:86–93

Tüxen R (1931) Die Grundlagen der Urlandschaftsforschung. Ein Beitrag zur Erforschung der Geschichte der anthropogenen Beeinflussung der Vegetation Mitteleuropas. Niedersächs Jahrb Landesgesch 8:59–105

Tüxen R (1956) Die heutige potentielle natürliche Vegetation als Gegenstand der Vegetationskartierung. Angew Pflanezensoz 13:5–42

van Langevelde F, van de Vijver C, Kumar L, van de Koppel J, de Ridder N, van Andel J, Skidmore AK, Hearne JW, Stroosnijder L, Bond WJ, Prins HHT, Rietkerk M (2003) Effects of fire and herbivory on the stability of savanna ecosystems. Ecology 84:337–350

van Vuure C (2005) Retracing the aurochs: history, morphology and ecology of an extinct wild ox. Pensoft Publishers, Sofia/Moscow

Vera FWM (2000) Grazing ecology and forest history. CAB International, Wallingford, 528 p

WallisDeVries MF, Bakker JP, Van Wieren SE (1998) Grazing and conservation management. Kluwer Academic Publishers, Dordrecht

Watt AS (1947) Pattern and process in the plant community. J Ecol 35:1–22

Watts WA (1978) Plant macrofossils and quaternary paleoecology. In: Walker D, Guppy JC (eds) Biology and quaternary environments. Australian Academy of Sciences, Canberra, pp 53–67

Webb SL, Walker LR (1999) Disturbance by wind in temperate-zone forests. In: Walker L (ed) Ecosystems of disturbed ground. Elsevier Science Press, Amsterdam, pp 187–222

Wharton CH (1966) Man, fire and wild cattle in North Cambodia. In: Proceedings of the fifth annual tall timbers fire ecology conference March 24–25, 1966, vol 5. Tallahassee, Florida, pp 23–65

Woodman P, McCarthy M, Monaghan N (1997) The Irish Quaternary fauna project. Quat Sci Rev 16:129–159

Yamamoto S (2000) Forest gap dynamics and tree regeneration. J For Res 5:223–229

Zoller H, Haas JN (1995) War Mitteleuropa ursprünglich von einer halboffenen Weidelandschaft oder von geschlossenen Wäldern bedeckt? Schweiz Z Forstwes 146:321–354

Part V
Anthropogenic Disturbances

Chapter 14
Forest Management

Peter Meyer ⓘ and Christian Ammer ⓘ

Abstract From an ecological perspective, forestry interventions can be defined as disturbances actively implemented at different spatial scales with the aim of obtaining a variety of forest-based ecosystem services. By changing the spatial and temporal distributions of resources, they alter competition between trees at the individual, species, and generational levels. In addition to silvicultural measures in the strict sense, drainage, liming, and pest control also constitute ecological disturbances. The diversity of stand and landscape structures that result from natural and forestry-initiated disturbances has important consequences for biodiversity. Forestry-initiated and natural disturbances have many similarities and also major differences. Ecologically oriented forestry practices are those that integrate the essential elements and attributes of the natural disturbance regime into forest management.

Keywords Emulating natural disturbances · Forest dynamics and management · Forestry-initiated disturbances · Landscape effects · Stand regeneration · Silvicultural systems · Tree removal

P. Meyer (✉)
Department of Forest Nature Conservation, Northwest German Forest Research Institute, Hann. Münden, Germany
e-mail: peter.meyer@nw-fva.de

C. Ammer
Silviculture and Forest Ecology of Temperate Zones, Georg-August-University Göttingen, Göttingen, Germany

T. Wohlgemuth et al. (eds.), *Disturbance Ecology*, Landscape Series 32,
https://doi.org/10.1007/978-3-030-98756-5_14

14.1   Importance of Disturbance Ecology
        for Forest Management

Forestry practices in Central Europe have changed profoundly, especially since the 1980s. Instead of the management of pure, even-aged stands with the primary goal of timber production, forestry has been increasingly understood as the management of complex ecosystems with multiple objectives at different spatial scales (Kohm and Franklin 1997). With the increasing recognition that the complexity and diversity of forests, and thus the ecosystem services they provide, depend largely on the impacts of disturbances, an understanding of the effects of natural and management-related disturbances has become of central importance.

In forestry, the emulation or integration of natural disturbances is often seen as a promising approach for maintaining biodiversity at different levels (from the gene pool level to the species level to the ecosystem level) (Franklin et al. 2002; North and Keeton 2008) and to reduce management efforts according to the principle of "biological rationalization" (Schütz 1996; Puettmann et al. 2009). Close-to-nature forestry practices have a long tradition in Central Europe, especially in Germany (Gayer 1886; Möller 1923), and the corresponding management concepts have still developed further (Pommerening and Murphy 2004; Schmidt 2009). Recently, new concepts have evolved which give priority to natural processes over explicit production targets (Sturm 1993; Otto 1995 with critical discussion, Puettmann et al. 2009).

Studies of European primeval forests (Leibundgut 1993; Korpel 1995) as well as the establishment and monitoring of set-aside forests (Bücking 1997; Meyer 1997) have led to improved understanding and greater appreciation of natural disturbances for the dynamics of forest ecosystems (Meyer et al. 2004; Brang 2005; Svoboda et al. 2012; Trotsiuk et al. 2014; Hobi et al. 2015; Winter 2015). Such insights have formed the basis of close-to-nature forest management, even if, due to climate change, the disturbance regime of the past may be of limited applicability as a reference for the future (Puettmann et al. 2014).

14.2   Historical Changes in Forest Management

Central European forests have a long history of use, dating back thousands of years. Ever since the Neolithic period, forests have been cleared for the establishment of agricultural land and the extraction of firewood and timber and to increase pasture-land and hunting grounds (Grober 2013). The different forms and intensities of forest utilization resulted in changes in the extent, duration, and type of anthropogenic disturbances, and they have changed the tree species composition of forests (Firbas 1949; Ellenberg et al. 2010) (Fig. 14.1). In the Middle Ages, complex management systems, including the establishment of coppice and coppice with standard forests, were already oriented toward sustainable multiple use and ensured simultaneously firewood and timber supply as well as pastureland. With the upswing of

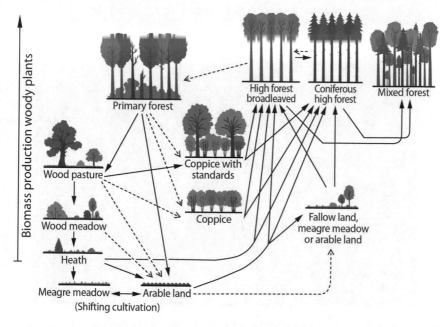

Fig. 14.1 The development of historical and recent forest and agroforestry management systems. The predominant system in use today is high forests. Continuous arrows indicate significant development and dotted arrows less significant development. (Adapted from Ellenberg 1996)

modern forestry during industrialization, the locally and regionally heterogeneous management systems were transformed to high forests on a large scale and managed predominantly for wood production.

14.3 Evaluation of Disturbances in the Forest

Generally, the management of natural resources aims at minimizing the disruptions to production. In particular, the simplification of forest structure in the form of even-aged pure stands and the standardization of interventions aimed at controlled and predictable wood supply (Puettmann et al. 2009). However, it gradually became clear that this management strategy resulted in forests that were more vulnerable to unanticipated natural disturbances (Holling and Meffe 1996). Modern approaches to forest utilization, therefore, aim at diversifying forest area units, interventions, and structure (Wagner 2007). Moreover, the spectrum of tolerated forest disturbances has changed considerably in recent decades. Thus, in the 1970s, area-wide windthrows and bark beetle infestations were perceived exclusively as catastrophes (Kremser 1973), whereas today such disturbances are handled in a more nuanced manner (BUWAL 2000; Brang et al. 2015). After the drought period of 2018–2020, extensive areas of bark beetle infestations characterize the landscape, not only in

Fig. 14.2 Bark beetle infestations at Mount Lusen, Bavarian Forest National Park (Germany). (Photo: NW-FVA)

protected areas (Fig. 14.2) but also in production forests. At the same time, there is less acceptance of larger anthropogenically caused disturbances. The increasingly critical view of activities such as clear-cutting and shelterwood cutting (Fig. 14.3) motivated the development of wood certification systems to ensure that wood production and forest management are carried out in a sustainable manner with respect to economic, environmental, and social criteria and that take into account the interests of various stakeholder groups (FSC Arbeitsgruppe Deutschland 2012).

14.4   Comparison of Natural and Anthropogenic Forest Disturbances

The similarities and differences between natural and anthropogenic forest disturbances can be assessed based on various criteria, especially the type of disturbance, the strength of the disturbance (affected biomass per unit area), the spatial extent of the disturbance, the frequency of recurrence in the same area, and the consequences for the tree population (Tables 14.1 and 14.2).

Both forestry practices and natural disturbances often result in the removal of trees from the living stand and thus reduce stand density. However, while in forestry a large part of the biomass is removed, in natural disturbances all of the biomass typically remains in the ecosystem. Thus, a major difference between natural and

Fig. 14.3 A beech stand after shelterwood cutting favouring regeneration of oak. The image was originally labelled in Kaiser et al. (2012) as representing a clear-cut. (Photo: Andreas Varnhorn/ Greenpeace)

anthropogenic forest disturbances is the amount of deadwood left in the forest. For natural disturbances that do not lead to the death of trees but only to a reduction in their vitality, such as diseases or irregular flooding, there are very few analogies in modern forest management systems. An exception is prescribed burning to reduce combustible biomass or initiate natural regeneration, the effect of which on the soil and tree populations may resemble that of natural fires (Pyne et al. 1996; Kraus and Zeppenfeld 2013). Some forestry measures, such as the "snapping" of trees during thinning (see Sect. 14.6.2), do not directly cause tree death and have a natural equivalent. Conversely, forest interventions like road building, drainage, and soil cultivation have no natural counterparts.

In most cases, natural disturbances occur irregularly and create heterogeneous spatial patterns (Turner 2010). This is only partially the case for forest interventions, which instead tend to create more homogeneous spatial effects (however, see Sect. 14.6). Natural disturbances typically create wide, irregular transition zones (ecotones) between disturbed and undisturbed areas, whereas anthropogenic forest disturbances such as road building have largely linear effects with narrow transition zones. Also, the designation of management units results in linear borders.

In addition, the phases of forest development following natural disturbances differ from those that result from anthropogenic disturbances (Fig. 14.4) (Leibundgut 1993). For many years, trees of advanced age as well as uncleared, deadwood-rich windthrows, and areas affected by bark beetle infestations did not occur in regularly managed high forests. However, silvicultural systems are increasingly being

Table 14.1 Typology of natural disturbances in forests compiled in view of the concepts of Oliver and Larson (1990), Richter (1997) and Roberts (2004)

Type of disturbance	Intensity	Spatial extent of effects				Affected layer			Frequency			Affected by management	Direct effects on the tree stand
		Single tree	Tree group	Stand	Landscape	Canopy	Stem space	Roots	Frequent	Sporadic	Rare		
Storm	◉	◀━ decreasing ━▶				◀━━━▶			◀━ decreasing			●	Windthrow and breakage of whole trees
Hurricane	●		decreasing ▶			◀━━━▶				increasing ▶		◉	Areawise windthrow and breakage
Ground fire	○	◀━ decreasing				◀━▶			?			●	Lower vitality; higher mortality
Crown fire	●		decreasing ▶			◀━━━▶				increasing ▶		●	Lower vitality; much higher mortality
Flood*1	○‑●	decreasing ▶				◀▶					-	-	Lower vitality; higher mortality
Landslide	●	◀━━▶				◀━━▶					increasing ▶	●	Areawise mortality
Rockfall	●	◀━━▶				◀━━▶					increasing ▶	-	Areawise mortality
Avalanche	●		◀━━▶			◀━━▶					increasing ▶	○	Areawise windthrow and breakage
Disease	○‑◉	◀━━━━━▶				◀━━━▶			◀━━━▶			◉	Lower vitality; higher mortality
Pests	○‑●	◀━━━━▶				◀━━▶				increasing ▶		○	Lower vitality; higher mortality

*1 Regular flooding in floodplain forest is not considered a disturbance

Impact/importance: ○ = Low, insignificant, rare, slow   ◉ = Medium   ● = Strong, important, frequent, fast

◀━━━▶ = Range   ◣ = Decreasing   ◢ = Increasing

implemented in which old trees are not harvested but are instead retained as legacy trees (see Gustafsson et al. 2012; Sect. 14.6.3.1). Given the shortened life cycle of a commercial forest, the temporal distribution patterns of forest disturbances will differ depending on the silvicultural system. Harvesting, as the largest disturbance, usually starts in the "optimum" development phase (Fig. 14.4) and may extend over several decades (Fig. 14.5). In natural broadleaved and mixed forests of the temperate zone, disturbances are rare during this phase (Fig. 14.4), and tree mortality is typically low (Holzwarth et al. 2013).

14.5 Silvicultural Systems as Anthropogenic Disturbances

Silvicultural interventions are characterized by the type, strength, and cycles of tree removals or tree enrichments carried out over the course of a stand's life for the purpose of stand maintenance, timber harvesting, and the establishment of tree

Table 14.2 Typology of forest-related disturbances in high forests

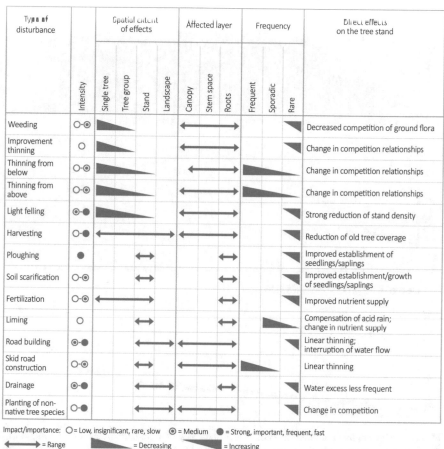

Type of disturbance	Intensity	Spatial extent of effects				Affected layer			Frequency			Direct effects on the tree stand
		Single tree	Tree group	Stand	Landscape	Canopy	Stem space	Roots	Frequent	Sporadic	Rare	
Weeding	○-◉	Decreasing				←→					Increasing	Decreased competition of ground flora
Improvement thinning	○	Decreasing				←→					Increasing	Change in competition relationships
Thinning from below	○-◉	Decreasing				←→			Decreasing			Change in competition relationships
Thinning from above	○-◉	Decreasing				←→			Decreasing			Change in competition relationships
Light felling	◉-●	Decreasing				←→					Increasing	Strong reduction of stand density
Harvesting	○-●	←→				←→					Increasing	Reduction of old tree coverage
Ploughing	●		←→				←→				Increasing	Improved establishment of seedlings/saplings
Soil scarification	○-◉		←→				←→				Increasing	Improved establishment/growth of seedlings/saplings
Fertilization	○-◉	←→					←→				Increasing	Improved nutrient supply
Liming	○		←→				←→		Decreasing			Compensation of acid rain; change in nutrient supply
Road building	◉-●		←→	←→							Increasing	Linear thinning; interruption of water flow
Skid road construction	○-◉		←→			←→			Decreasing			Linear thinning
Drainage	◉-●		←→				←→				Increasing	Water excess less frequent
Planting of non-native tree species	○-●			←→	←→						Increasing	Change in competition

Impact/importance: ○ = Low, insignificant, rare, slow   ◉ = Medium   ● = Strong, important, frequent, fast

←→ = Range   ◣ = Decreasing   ◢ = Increasing

regeneration. A distinction is usually made between coppice, coppice with standards, and high forest silvicultural systems.[1] In all three systems, the average living biomass during the course of the management cycle is generally less than that of an

---

[1] According to Vergani et al. (2017) the systems are defined as follows:

coppice: the cutting of the stems of young trees or shrubs close to the ground, causing them to resprout and to re-establish the canopy, or an area so treated.

coppice with standards: forest or stand consisting of coppice among which a number of trees (standards), that are generally of seedling origin, are retained on a long rotation to provide large material and seeds to regenerate the forests.

high forest: a forest management system which allows the trees to grow to at least two-thirds of their ultimate height, as opposed to earlier cutting or coppicing where a much lower canopy is formed.

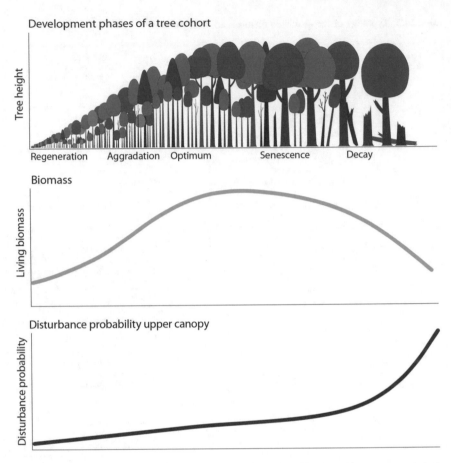

Fig. 14.4 Development of biomass (center) and frequency of natural disturbances (bottom) during the different developmental phases of a tree cohort (top) in broadleaved and mixed broadleaved–coniferous forests of the temperate zones. Derived from conceptual models developed by Watt (1947), Korpel (1982) and Oliver and Larson (1990)

undisturbed forest stand (Fig. 14.4). Actually, "the coppice system involves reproduction by [stump] shoots or suckers. When felled near ground-level, most broadleaved species, up to a certain age, reproduce from shoots sent up from the stump" (Troup 1928). The coppice and coppice with standard systems are typically managed in the form of a spatially coherent system of cutting (felling) areas in which a proportion of the area is cut each year and managed on a rotational basis. Since timber is harvested from a different area each year, the number of fellings corresponds to the length of the rotation period. For coppice forests, rotation periods of 15–20 years are common and for coppice with standards 20–30 years. Coppice forests are mainly used to produce small poles and firewood as the tree stems are typically small.

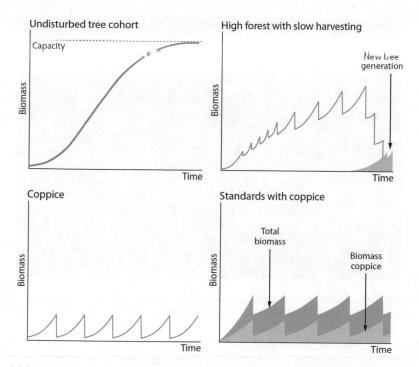

Fig. 14.5 Schematic development of the aboveground woody biomass of an undisturbed tree cohort (**a**) compared to that of a high forest with a temporally extended harvest (**b**), a coppice (**c**), and a coppice with standards (**d**)

Coppice with standards is composed of the vegetatively regenerated coppice layer and individual trees that emerged from seeds. The seedling trees eventually form a loose sub-canopy whose crowns cover 30–50% of the area (Fig. 14.6). Firewood resulting from the coppice layer and sawlogs from canopy trees are the main products of these forests. In the past, interventions were often combined with the establishment of pastures below the canopy trees. In Central Europe, the average aboveground biomass in coppice forests has been estimated at 60 t ha^{-1} (±38 t ha^{1}) and in coppice with standard forests 103 t ha^{-1} (±58 t ha^{-1}) (Albert and Ammer 2012). High forests are distinguished by a significantly higher maximum biomass of >400 t ha^{-1}, depending on the tree species and with large fluctuations over time. Wood growth is significantly higher in high forests than in coppice and coppice with standard forests (Albert and Ammer 2012). The primary goal in high forests is the production of sawtimber.

Because of the mosaic-like divisions resulting from the different felling areas, coppice and coppice with standards offer a wealth of edges and different developmental phases that within a small area provide a high species and structural richness (Schröder 2009; Fartmann et al. 2013). However, the practices that give rise to these areas have become rare in Central Europe. Moreover, this forest management regime cannot be compared to natural disturbances, as neither the regular spatial

Fig. 14.6 Typical structure of a coppice with standards forest 1 year after cutting of the shrub layer, Liebenburg forest of Lower Saxony in the northern Harz region of Germany. (Photo: NW-FVA)

pattern of successive areal disturbances nor the vegetative regeneration within these areas is found in natural forests in Europe. The exceptions are subalpine bushes and mountain pine forests, both of which are exposed to regular rockfall and small avalanches. In addition, consistent coppice and coppice with standard forest management give sprouting tree species a clear competitive advantage (Matula et al. 2012). There is also a certain resemblance between the canopy layer of coppice with standard forests and windthrow areas, where 30–50% of the stand may survive. However, the quality of the remaining trees is very different: In the coppice with standards stand vital trees with large crowns predominate, whereas in natural windthrow areas trees in poor and intermediate condition remain together with partially damaged trees (Fig. 14.7).

A much greater similarity exists between managed high forests and natural broadleaved and coniferous forests. Trees in high forests nearly reach their maximum natural height, and by the time stands are harvested, they can achieve a comparably high biomass. For example, the biomass of a 60-year-old Douglas fir (*Pseudotsuga menziesii* [Mirb.] Franco) stand may be already half that of a 450-year-old natural forest of Douglas fir and western hemlock (*Tsuga heterophylla* [Raf.] Sarg.) (Barnes et al. 1998). Note that a typical rotation for Douglas fir in Central Europe is 80 years. In German high forests of European beech (*Fagus sylvatica* L.), the volume stock at 120 years (yield table of Wiedemann 1931) corresponds to the average stock of a primeval European beech forest (582 m^3 ha^{-1}; Hobi et al. 2015). The average stock of 30- to 120-year-old beech high forests of equal area is about 350 m^3 ha^{-1}. In these comparisons, the respective removals in the course of management and the amounts of deadwood left in the forest must be taken into account. In

Fig. 14.7 An 8-ha windthrow that occurred in 2014 in the primeval beech forest Havešová, eastern Slovakia. (Photo taken in 2015; © Peter Meyer, NW-FVA)

the primary beech forests of the Western Carpathians investigated by Hobi et al. (2015), the amount of deadwood was $162.5 \pm 8.4 \ m^3 \ ha^{-1}$.

The regionally predominant natural disturbance regime will influence the assessment of the closeness to nature of the different types of cutting. While small-scale disturbances dominate in temperate broadleaved forests (Fig. 14.8; Seymour et al. 2002; Hobi et al. 2015), large-scale windthrows, fires, and insect infestation are much more common in boreal forests. In high-elevation montane and subalpine locations, landslides and avalanches also lead to large-scale disturbances (Bebi et al. 2009). However, these can also occur at lower elevations, such as in the European beech forests of southeastern Europe (Nagel et al. 2014; Hobi et al. 2015). Storm-damaged areas in European beech primeval forests (Fig. 14.7) and in natural forest reserves have also been documented (Willig 2002; Schmidt and Meyer 2015). Rare strong interventions over a larger contiguous area also contribute to a greater closenes to nature in nemoral broadleaved forests and are congruent with naturally occurring disturbances (Foster and Boose 1992). In principle, in terms of the amount of disturbed area, no cutting method can be considered unnatural as long as it does not occur significantly more or less often than naturally occurring disturbances (Fig. 14.8). Thus, silvicultural cutting methods do not create light conditions different from those that can result from natural disturbances. However, in all other respects, such as deadwood dynamics, the consequences of almost all interventions in silvicultural systems differ significantly from those of natural disturbances.

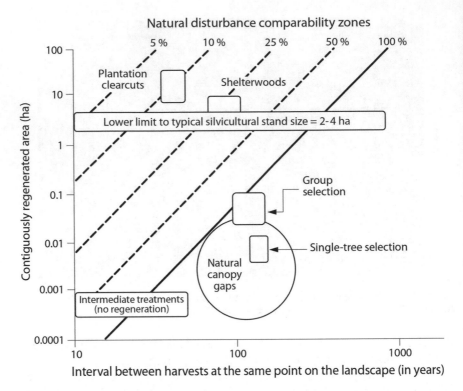

Fig. 14.8 Frequency and extent of the disturbances caused by different cutting regimes compared to natural disturbances in northeast American temperate broadleaved and mixed broadleaved–coniferous forests (Seymour et al. 2002). The comparison zones reflect the recurrence of cutting events compared to the recurrence of natural disturbances in an area of the same dimensions, for example, a recurrence of group shelterwood cutting ("Femelschlag") events every 100 years would be comparable to the natural disturbance regime

14.6   Disturbance Effects of Individual Forest Measures

In the assessment of disturbances, it is primarily their effects on the structures and processes of the concerned ecosystems that are decisive rather than whether they are anthropogenic or natural in their origin (Bazzaz 1983). Disturbances can abruptly change the density and thereby the competitive interactions in forest stands as well as the growth and mortality of trees and regeneration dynamics; consequently, they typically lead to a reorganization of the forest ecosystem (Fig. 14.9). Few forest management measures are so extreme that their impact exceeds that of the natural (undisturbed and disturbed) fluctuation range of the ecosystem. In the long run, many pioneer tree species can survive only if major disturbances occur, as they are unable to successfully regenerate in closed forests.

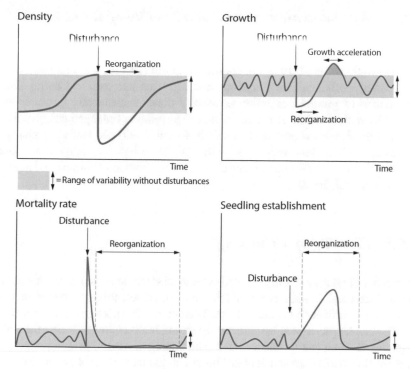

Fig. 14.9 Assumed changes in stand density and population dynamics in forests induced by disturbances: density (**a**), growth (**b**), mortality (**c**), and establishment of seedlings (**d**)

In general, forestry measures reflect two fundamental management decisions: (1) the selection of tree species and (2) the regulation of stand density (Schall and Ammer 2013). From an ecological viewpoint, the selection of tree species may lead to a significant deviation from the species involved in forest development under natural conditions at the same site, while the regulation of stand density influences intra- and interspecific competition and causes a redistribution of resources among the remaining trees (Ammer 2008). The interventions differ depending on a stand's development phase. During the initial phase, they aim at regeneration establishment and securing sapling growth. In this case, the disturbances created by management may also address the competing vegetation. In young stands, trees will be removed by tending measures either because their stem quality is poor or to obtain a desired mixture of tree species. Here, the silvicultural disturbance disrupts natural intra- and inter-species competition because it does not always select trees that would have prevailed in the absence of human interference. In subsequent thinnings, the best quality trees or individuals of certain tree species are promoted by removing competitors. Once the production target is reached, the stand is harvested. The effects of this type of disturbance will vary depending on the progress of the interventions and the amount of wood removed (see Sect. 14.6.3).

14.6.1   Vegetation Management and Tending of Young Stands

Interfering ground vegetation is eliminated in favour of planted, sown, or naturally seeded trees (Davis et al. 1998). In contrast to North America and other parts of the world, where herbicides are frequently used for this purpose, in Central Europe management of ground vegetation is usually done mechanically (Ammer et al. 2011). The redistribution of resources due to vegetation management favours the growth of the desired young trees (Harrington et al. 1999) and leads to a change in biomass allocation. Thus, an increase in the amount of light is usually accompanied by an increase in root biomass in relation to the total biomass (Shipley and Meziane 2002; Schall et al. 2012).

14.6.2   Cleaning and Thinning

The number of plants in a fully stocked even-aged stand decreases as the mean plant biomass increases. This fundamental relationship, the self-thinning line, is independent of human influence (Reineke 1933; Yoda et al. 1963). This density-dependent mortality can be preempted by sufficiently strong interventions. During the thicket phase, that is, from canopy closure to the beginning of natural pruning which usually corresponds to a diameter at breast height of dominant trees of about 15 cm, few interventions are carried out to stop the natural dying-off process, especially in the case of broadleaved trees; instead, clearing consists only of the removal of single, qualitatively unsatisfactory individuals. In the case of coniferous trees, however, for reasons of stability (e.g. to avoid snow-induced breakage), a reduction in the stem number is often carried out already at this stage.

The effects of thinning depend on the frequency, strength, and type of thinning. So-called "thinning from below" essentially intervenes in the suppressed stand layer, so that changes in the competitive environment of dominant trees are largely insignificant. In contrast, interventions in the dominant stratum ("thinning from above" or "crown thinning") have a much larger impact, as they increase the amount of resources available to the remaining trees. Depending on the age of the trees and the growth dynamics of the particular tree species, the remaining trees respond with an increased crown surface. After a certain time lag, this is reflected in a stand productivity that may be higher than at maximum density, assuming that the density has not been excessively reduced (Assmann 1961; Pretzsch 2004; "growth acceleration" in Fig. 14.9b). However, it remains unclear how long this increase in production is maintained and whether the productivity achieved over the entire life span of a stand will be higher than that of an untreated stand (Curtis et al. 1995; Zeide 2001). The stand density at which maximum productivity is achieved depends on the age and mean diameter of the stand (Zeide 2004) and is lower for light-demanding than for shade-tolerant tree species (Pretzsch 2005). Thinning accelerates individual plant development such that a certain target diameter will be reached more quickly. In addition, crown thinning can ensure the survival of suppressed trees as well as the conservation of less competitive species. Examples include the promotion of oaks (*Quercus* spp.) or of other broadleaved tree species such as European ash (*Fraxinus excelsior* L.) and maple species (*Acer pseudoplatanus* L., *A. platanoides* L.) in European beech stands (Nüsslein 1995). This corresponds to natural forests where certain tree species are also favoured, for example, by the emergence of stand gaps during the regeneration phase (Poulson and Platt 1996).

As recent studies have shown (Kohler et al. 2010; Gebhardt et al. 2014), thinning also leads to temporarily reduced drought stress. This is because water loss due to interception is reduced, which allows a larger amount of precipitation to reach the forest floor (Stogsdili et al. 1992; Simonin et al. 2007) and also because stand transpiration decreases significantly. The end result is an increase in the amount of available water (Aussenac and Granier 1988; Gebhardt et al. 2014). Analogous to light extinction, these effects are not proportional to stand density (Bréda et al. 1995). Thus, the effectiveness of an intervention depends on its intensity. However, very strong interventions can also lead to a lush growth of the ground vegetation (Son et al. 2004). The positive effect of thinning on a forest stand continues for several years after a drought event. In the case of more frequent drought events (Lindner et al. 2014), repeated, strong interventions are among the effective measures allowing the adaptation of forest stands to climate change (Ammer 2017).

The effects of thinning on the mechanical stability of the stand and thus on its sensitivity to future disturbances in the forms of wind and snow can be classified in two successive phases. Immediately after thinning, there is a period where the stability of the stand decreases (Richter 2003; Albrecht et al. 2012); however, this is followed by a second, longer period of greater stability, as the morphological adaptations of individuals increase their resistance to mechanical stress. As a result of thinning, the height–diameter ratio decreases, and the crown length increases; both of these characteristics are thought to reduce the risk of windthrow (Mayer and Schindler 2002).

14.6.3   Final Harvest, Stand Regeneration

Harvest of Mature Stands

Depending on the intensity of harvesting and the size of the affected area, harvesting mature trees is accompanied by changes in the abiotic conditions of varying degrees, as the amount of light, the temperature regime, and the availability of belowground resources are affected. The removal of mature trees from a stand is usually the starting point for the establishment and/or development of tree regeneration.

The impact of harvesting operations differs mainly with regard to the affected area, the number of trees removed per intervention, the number of interventions, and

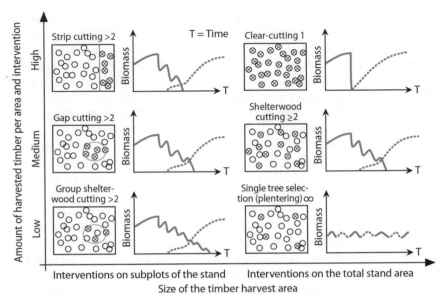

Interventions on subplots of the stand   Interventions on the total stand area
Size of the timber harvest area

Solid line = biomass of the mature trees; dotted line = biomass of the new tree generation;
mixed solid/dotted line = biomass of an uneven-aged stand

Fig. 14.10 Classification of the typical cutting methods used in harvesting wood and regenerating forest stands according to the amount of wood removed per sub-area, intervention, and the area affected by the intervention. The numbers indicate how often the interventions are carried out until the stand is fully harvested. For all cutting methods, the affected area (each circle represents a tree, circles with a cross are the trees removed during an intervention) and the development of the stand's biomass (solid line: biomass of the mature trees (in case of plentering: total stand biomass), dotted line: biomass of the new tree generation) over time are shown (© Abt. Waldbau und Waldökologie der gemäßigten Zonen der Fak. f. Forstwissenschaften und Waldökologie der Georg-August-Univ. Göttingen)

the distribution of the removed trees (Fig. 14.10). The strongest changes are imposed by clear-cutting, in which no mature trees remain on the harvested area. The resulting changes in abiotic conditions are similar to those caused by large-scale natural disturbances (Fig. 14.11) and include increased differences between day and night temperatures, changes in wind speed and light conditions, changed amount of water reaching the forest floor, increased evaporation from the soil surface and transpiration by vegetation on the felled area, decreases in air and soil moisture, and an increase in the litter decomposition rate, which may be accompanied by humus losses. In addition, the rate of nitrogen mineralization increases (Chen et al. 1993; Carlson and Groot 1997), which, depending on the rate at which the ground vegetation is re-established, may be associated with temporary nitrogen losses (Lindo and Visser 2003; Weis et al. 2006; Klinck et al. 2013). In terms of stand regeneration, these conditions favour early successional species and/or those that tolerate high irradiation but also frost, such Norway spruce (*Picea abies* [L.] Karst.) and Scots pine (*Pinus sylvestris* L.) (Fisichelli et al. 2014).

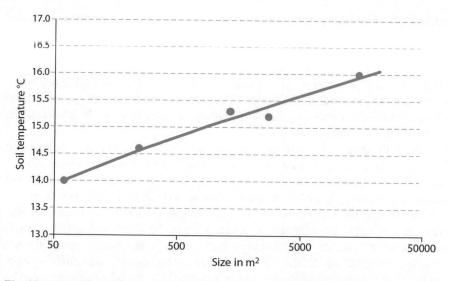

Fig. 14.11 Change in the soil temperature (5 cm depth, average values from June to September) as a function of the size of the area affected by the intervention. Note the exponential scale on the x-axis. The largest area represents a 1.5-ha clear-cut, the smallest a 60-m² gap. (© Abt. Waldbau und Waldökologie der gemäßigten Zonen der Fak. f. Forstwissenschaften und Waldökologie der Georg-August-Univ. Göttingen)

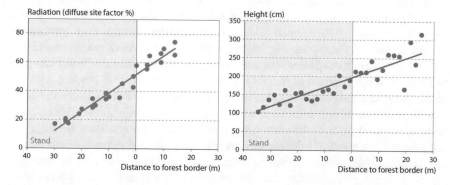

Fig. 14.12 Radiation (DIFFSF = diffuse site factor in %) and the heights of regenerated beech in the inner and outer areas demarcated by a margin. (Figure adapted from Wagner and Spellmann 1993)

The complete removal of trees on a small area is carried out during certain group and strip felling silvicultural systems, such as gap cutting (German: Lochhieb; typical gap diameter of approximately 30 m; see Wagner 1999; Moslonka-Lefebvre et al. 2011) and strip cutting (German: Saumhieb; the width of the strip where all mature trees are removed usually corresponds approximately to the height of the trees; see Röhrig et al. 2006). In these cases, the change in environmental conditions is not as drastic as in clear-cutting. Both gap cutting and strip cutting lead to a gradient of ecological conditions that allows tree species with different light requirements to regenerate simultaneously (Fig. 14.12). In the case of strip cutting, the amount of radiation on the forest floor increases progressively from within the stand to the stand edge to the harvested area. The availability of belowground resources increases accordingly. In the case of gap cutting, significant increases in the amount of light and in belowground resource availability occur toward the center of the gap (Ritter et al. 2005; Herrmann 2014). In the Northern Hemisphere, the highest radiation will be found at the northern edge of the gap, assuming that the ground is level (Wagner 1999). For both gap and strip cutting, it takes years to decades until all mature trees of a stand are removed.

Also in the case of shelterwood (German: Schirmschlag) and group shelterwood (German: Femelschlag) cutting, tree harvest requires several interventions carried out over the same area and usually extending over several decades. As shelterwood cutting is always applied on the entire stand area, it creates relatively homogeneous ecological conditions, whereas in a "Femel" cutting, the initial tree removal is carried out in discrete areas (usually between 500 and 1000 m²) and results in more heterogeneous resource conditions. In both cases, some mature trees are not removed during the initial interventions: in shelterwood cutting, the remaining trees are distributed over the entire area, and in a group shelterwood cutting, they can be found on both the discretely harvested and the untouched remaining areas of the stand. These mature trees reduce the resources available to seedlings (Ammer 2002; Petriţan et al. 2011). The effect of cuttings on canopy closure is mainly used to control competition within tree regeneration and to enhance stem quality of saplings. Von Lüpke and Hauskeller-Bullerjahn (2004) drew on the example of a very unevenly exposed 160-year-old beech stand (light availability at the forest floor ranging from 6% to 67% of open field conditions) to demonstrate the importance of canopy closure on the relationship between the height growth of oak and beech and thus the control of tree species composition in the tree regeneration layer. Five-year-old oak seedlings exposed to radiation above a certain threshold reached a greater height than did young beech of a similar age. However, only 3 years later oak needed a much higher threshold radiation level to outgrow beech, indicating that the amount of light required by oak to remain competitive with beech increases with age. An increase in the amount of light can be guaranteed by harvesting mature trees (Lüpke and Hauskeller-Bullerjahn 2004). Another important function of canopy density is that tree seedlings and sapling trees below the canopy of mature trees form fewer and thinner side branches because of the reduced light availability; this results in more trees with a straight and clear (i.e., with few and thin side branches) stem, meaning an increase in the future value of the timber from the stand (Weidig et al. 2014). Another positive effect of shading mature trees is that the establishment of ground vegetation which competes with tree regeneration is limited (Kuuluvainen and Pukkala 1989; Ammer 1996).

A method in which only a few individual trees are removed and in single interventions is the so-called plentering or single-tree selection (Fig. 14.10; Schütz 1994). By definition, in plentering, the growing stock of a stand should not exceed nor fall below a certain value, as either would result in the loss of the typical multi-layered structure of the stand. The disturbance regime corresponds to a large extent to the small-scale gap dynamics of natural mixed broadleaved forests made up of

shade-tolerant tree species. Interruption of canopy closure presumably ensures the
ꞔontinuous regeneration of the stand,

The light and temperature conditions resulting from application of the plentering
system are more uniform than those produced by the previously mentioned
approaches for harvesting high forests (Burschel and Huss 2003; Ehbrecht et al.
2017). While even-aged stands, for example, resulting from the shelterwood system, give rise to a mosaic of different developmental phases and differ between
stands but much less within stands on the landscape level, Plenter forests are characterized by a high within-stand heterogeneity but a low between-stand heterogeneity. It has been shown that the uneven-aged Plenter forests are economically feasible
and less frequently damaged by storm events than even-aged forests (Knoke 1998).
However, since conditions on the forest floor are relatively dark, the species diversity of various taxa (other than trees) was found to be lower on the landscape level
when compared to the even-aged systems, which provide a greater diversity of abiotic conditions (Schall et al. 2018).

Old and large trees are crucial for biodiversity, as they frequently provide microhabitats (Vuidot et al. 2011; Larrieu and Cabanettes 2012), which are important
habitats for rare species (Hofmeister et al. 2016). This is taken into account in the
so-called retention method (Gustafson et al. 2007; Aubry et al. 2009), in which not
all of the trees of a mature stand are harvested, but rather some are left permanently
in the area, mostly in groups (approximately 10 trees ha^{-1}), and continue to mature.
After their natural death, their positive effects on biodiversity remain in the form of
standing or lying deadwood (Müller and Bütler 2010; Seibold et al. 2015). Thus,
these trees are in part taken up by the next forest generation. This approach is more
close to nature than cutting methods characterized by complete harvesting, and it
can contribute to the preservation of the forest biodiversity that relies on old trees
(Rosenvald and Lõhmus 2008; Fedrowitz et al. 2014).

Regeneration Measures   The harvest of mature timber is usually accompanied by
measures supporting tree regeneration. There are, for example, occasionally additional measures, many of which are a disturbance for the ground vegetation, including tillage and slash removal. While, especially in Scandinavia, soil preparation to
encourage establishment of regeneration over large areas is common (Örlander
et al. 1996), in the temperate zones of Europe, it is used only occasionally, for
example, to support the natural regeneration of pine (Lehnigk and Ammer 2012). In
the past, ploughing, such as conducted in agriculture, was carried out to encourage
tree regeneration, whereas today only the humus layer is "scarified" (i.e., partly
removed to expose the mineral soil). Slash removal is usually carried out after cutting procedures that produce a large quantity of wood per operation if the harvested
area must immediately be made accessible and plantable, or for reasons of pest
control (Lobinger 2006). Whereas the "slash" (i.e., the logging debris left after
removal of stems during harvesting—foliage, smaller branches, etc.) was previously collected into large piles or strips in the stand, today it is occasionally removed
and burned to generate heat and energy. Depending on the nutrient content of the
soils, slash removal can involve a considerable loss of nutrients (Meiwes et al.

2008); consequently, slash removal is often regulated in certification guidelines. As a rule, however, the slash is left in the forest. Planting measures under the canopy of mature trees— to close gaps caused by disturbances through planting fast-growing (possibly non-native) tree species or to convert a pure stand of one species into a mixed-species stand by establishing shade-tolerant tree species—are not disturbances in a strict sense, but they do influence the response of a forest ecosystem to a disturbance.

14.6.4 Indirect Effects of Forest Management

Forest Road/Track Network In actively managed forest landscapes, there is a permanent network of paved, truck-compatible access roads as well as rough tracks within the stands. The creation of a paved road results in several disturbances: An inner edge of the forest is created, the groundcover is removed, road construction material is brought in, and a side ditch is usually dug to provide drainage for the road. In the root zone, the water flow is altered to some extent. In addition, the infrastructure also changes the water flow at the landscape level and has negative effects on running and still waters (Lindenmayer and Franklin 2002). In particular, on forest sites with a high groundwater table, roadside ditches serve as drainage for the adjacent forest stands. The smaller tracks may likewise act as drainage channels during heavy rainfall events and therefore accelerate surface runoff (Witzig et al. 2004). In acidic forest landscapes, roads made of alkaline limestone gravel frequently affect the flora adjacent to the road (Mrotzek et al. 2000). Furthermore, forest roads facilitate the distribution of ruderal species and neophytes, as their seeds (and other propagules) are transported by vehicles (Ehbrecht et al. 2017) and by game stopping along tracks to feed (Heinken et al. 2005).

Unpaved rough tracks have a less pronounced effect because they are narrower and no material is brought in and do not usually have associated ditches. However, the use of heavy machinery (such as harvesters and transport vehicles) may result in soil compaction, which for sensitive soils leads to a restriction of the oxygen supply, waterlogging, and a reduction of the nitrogen supply (Hildebrand 2008; Ehbrecht et al. 2017). Compaction of loamy and clayey soils or of acidic soils with low biological activity is almost irreversible (Ebrecht and Schmidt 2005; von Wilpert and Schäffer 2006). Ebeling et al. (2016) showed recovery of the soil structure of biologically active soils on limestone and on loess over sandstone 20 years after compaction caused by heavy machinery whereas sandy-loamy podzols showed little recovery over the same period.

Drainage Forest stands close to groundwater or on waterlogged sites are often continuously drained to allow their profitable management (see above). In the nineteenth and twentieth centuries, close-meshed drainage systems were frequently installed (Fig. 14.13). Until the 1980s, waterlogged sites in Germany were in many cases ploughed before planting (Dertz 1972; Fig. 14.14), but this practice has since then been discontinued for reasons of soil protection.

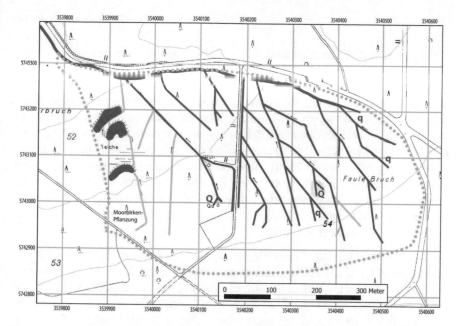

Fig. 14.13 The drainage system (blue lines) in a forest stand in the Solling Mountains, Lower Saxony (Germany). (Figure adapted from Küchler 2011)

Fig. 14.14 Ploughing using a disc harrow to prepare the conversion of a mire in the Reinhardswald (Germany) into a spruce stand. (Photo taken in 1928; photographer unknown; archive Prof. Dr. Gisbert Backhaus)

On peatland sites, drainage results in strong ground subsidence and peat mineralization (Kuntze 1993), and on mineral sites drainage induces large changes both in water and nutrient balances and in the species composition of the ground vegetation. On periodically wet soils, drainage reduces the extremes of water levels. A more consistent water supply can have a positive effect on the vitality of forest trees, by shortening the phases of oxygen deficiency in the topsoil. This may alter the balance of competition since tree species widely differ in their tolerance of waterlogging/flooding (Glenz et al. 2006). However, while drainage increases the growth success of planted saplings, extends the time window when soil-friendly timber harvesting can take place, and thus indirectly reduces soil damage, its effects are often negative with respect to carbon storage, landscape water balance, and nature conservation (Niemelä et al. 2005).

Fertilization and Liming Harvesting of whole trees in combination with nutrient losses caused by acidic nutrient inputs can cause nutrient levels to fall below the amounts needed for a sustainable nutrient supply (Stüber et al. 2008; Waldner et al. 2015). Until the 1980s, application of synthetic fertilizers in forests was not uncommon, and its effectiveness in increasing the vitality and productivity of forests had been demonstrated in numerous studies. From a nature conservation point of view, however, fertilization is a nonnatural measure that can abruptly change the nutrient balance and chemical conditions of the soil and thus cause an unnatural alteration of the species composition.

Compensatory liming can be used to counteract the effects of acid rain (Hüttl 1989). However, the previous practice of widespread liming has been replaced by a more targeted approach (Janssen et al. 2016). Although liming runs counter to the preservation of naturally nutrient-poor ecosystems (Reif et al. 2014), leads to nitrogen release (Kreutzer 1995), and causes changes in the ground vegetation (Schmidt 2002), it nonetheless can compensate for acids in the soil, induce biological activation of the mineral soil, and increase the supply of nutrients to trees (Grüneberg et al. 2017).

Pesticide Use The application of pesticides has been largely reduced in the forests of Central Europe (Kogan 1998; Ammer et al. 2011). However, if insect densities threaten the existence of the forest, insecticides can be applied to control an otherwise self-perpetuating disturbance and thus stabilize the forest development. Nevertheless, the use of pesticides remains controversial, especially because of their effects on nontarget organisms (Petercord and Lobinger 2010).

14.7   Effects of Forest Management at the Landscape Level

14.7.1   Changes in the Spatial Distribution of Forests

Deforestation leads to a fragmentation of forested areas and an increase in the amount of forest edge (Harris 1984). The effects of the latter are strongly dependent on the shape of the edge, the site conditions, and the tree species composition (Murcia 1995). From the interior of the stand toward the edge of the forest, the physical and ecological conditions increasingly take on the characteristics of open space. Investigations based on the microclimate and on water and nutrient balances have shown that this transition zone may be relatively wide (Keenan and Kimmins 1993; Klinck et al. 2013). Open-land species (Schmidt et al. 2011) may invade the previously closed forest stand. For species characteristic of closed forests, both the available habitat areas and the possibilities for dispersal are reduced. Consequently, a formerly large, spatially coherent population may disintegrate into subpopulations that are isolated from each other. In forest borders, the increased amounts of light and heat together with reduced competition among the trees lead to an improvement in the growth conditions of the shrub and herb layers and to increases in abundance and species diversity, especially of arthropods and birds (Reif and Achtziger 2000). The intersections of open-land and forest ecosystems at forest margins can act as dispersal axes for animal and plant species. Given the positive effects of forest margins on biodiversity, their targeted creation and maintenance may serve as important nature conservation measures (Coch 1995).

In addition to increased edge effects, deforestation also results in the increasing isolation of subpopulations and a reduction of habitat area (Schmidt et al. 2011). The latter has a larger impact on populations than does habitat isolation (Bailey 2007; Fahrig 2013). Deforestation also influences water and nutrient budgets and erosion processes within a landscape. For example, forest clearances in Central Europe during the Middle Ages led to an increase in groundwater levels, flooding, and a sharp rise in erosion (Bork et al. 1998; Ellenberg et al. 2010). The widespread loss of fertile soils that followed the St. Mary Magdalene's flood in 1342 is an impressive example of the potentially negative consequences of the agricultural use of what were previously forested landscapes (Bork and Kranz 2008).

Changes in land use can result not only in a decrease in forest area but also an increase in forest area. For example, the raised bogs of northwest Germany became forested after peat cutting and drainage. Forests have also established on post-mining landscapes and on previous military training grounds. Globally, the increasing concentration of human settlements and economic activity along with the intense land use in preferred locations have led to large areas of traditionally cultivated landscapes becoming fallow and subsequently developing into forests (Poyatos et al. 2003).

14.7.2 Landscape Effects of Stand Treatments

Studies in agricultural landscapes indicate that landscape effects are often more important for biodiversity than effects at the stand level (Gámez-Virués et al. 2015). In forests, the different cutting regimes (see Sect. 14.6) lead to the development of different landscape patterns. While the small-scale removal of single stems or groups of stems creates a fine-grained mosaic of different age groups, shelterwood or clear-cutting creates more coarse-grained patterns (Shifley et al. 2008). In a forest landscape with different forest owners and enterprises, the heterogeneity of targets and subsequent management concepts often lead to a high diversity of forest stands in terms of tree species composition, density, and structure (Gustafson et al. 2007; Schaich and Plieninger 2013).

A large-scale survey of older beech forests in a number of German state forest enterprises showed how the landscape pattern is influenced by the cutting regime (Meyer et al. 2016). Under natural conditions, the dominant climax community in many places would be closed old beech forests (Kaiser and Zacharias 2003; Meyer and Schmidt 2008), but today the stands of older beech forests are highly fragmented, and the canopy density is much reduced as a result of harvesting (Fig. 14.15). However, under current beech management practices, legacy trees are usually retained (see Sect. 14.6.3.1) thereby extending marginal zones and increasing the fine-grained character of the stand mosaic while reducing the isolation of habitats of species that depend on older forests.

Different species presumably react differently to the size and spatial distribution of the remaining forest (Fedrowitz et al. 2014). In the absence of reliable and locally

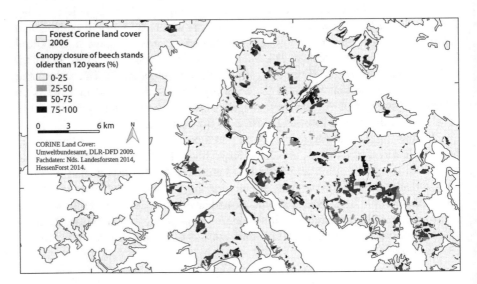

Fig. 14.15 Degree of canopy closure (%) and spatial distribution of >120-year-old beech forests in Solling, Lower Saxony (Germany)

transferable results, a broad range of silvicultural approaches is recommended with the integration of existing biodiversity hotspots as a key element (Meyer et al. 2015).

Overall, the application of a mixture of different silvicultural systems and protection concepts within a landscape is likely to be the best strategy for preserving biodiversity, as it maximizes the heterogeneity of environmental conditions. In their comparison of shelterwood and single-tree (plenter) cutting in beech forests, Schall et al. (2018) showed that the shelterwood systems resulted in a significantly higher diversity for 6 of 15 species groups (vascular plants, beetles, spiders, weavers, birds, and lichens) according to at least one diversity measure. Interestingly, for vascular plants and spiders, this trend was also significant when species restricted to forests were considered. There were no significant differences for bats, mosses, deadwood fungi, lacewings, Hymenoptera, bugs, ectomycorrhiza, and bacterial RNA, while the diversity of bacterial DNA was higher in the Plenter forest (Schall et al. 2018). A possible explanation for these findings is that single-tree harvesting creates relatively homogeneous structures at the landscape level. By contrast, in shelterwood-cut forests, there is little variation in the abiotic conditions within stands, whereas between the stands of different ages, the differences are significant.

14.8 Conclusions

Forest management is characterized by a large variety of disturbances, some of which are similar to natural disturbances. However, managed and unmanaged forest landscapes/stands differ considerably (Lindenmayer and Franklin 2002).

Understanding of the causes and consequences of natural and anthropogenic disturbances is essential for sustainable forest management and nature conservation. In the development of a strategy for the conservation of biological diversity, forestry must be guided by natural disturbances since many species and ecosystems depend on them for their continued survival (Spies and Turner 1999). However, as a direct blueprint, natural disturbances are rarely suitable since the objectives of forestry include economic and other societal interests as well as nature conservation (Lindenmayer and Franklin 2002). Natural disturbances can support but also impede those objectives. Disturbance ecology has enhanced our understanding of the different developmental pathways followed by natural ecosystems and the significantly altered conditions that may arise after random events (DeAngelis and Waterhouse 1987; Perry and Amaranthus 1997). Larger areas of disturbances are of crucial importance for the conservation of biodiversity. Ecologically oriented, close-to-nature forestry is therefore characterized not by uniform and small-scale interventions of varying intensity, but rather by a wide range of interventions of varying intensity and by the integration of essential elements of natural disturbance regimes.

References

Albert K, Ammer C (2012) Biomasseproduktivität ausgewählter europäischer Mittel- und Niederwaldbestände – Ergebnisse einer vergleichenden Metaanalyse. Allg Forst Jagdztg 183:225–237

Albrecht A, Hanewinkel M, Bauhus J, Kohnle U (2012) How does silviculture affect storm damage in forests of south-western Germany? Results from empirical modeling based on long-term observations. Eur J For Res 131:229–247

Ammer C (1996) Konkurrenz um Licht – zur Entwicklung der Naturverjüngung im Bergmischwald. Forstl Forsch Ber München 158:1–189

Ammer C (2002) Response of *Fagus sylvatica* (L.) seedlings to root competition by overstorey *Picea abies* (L.) Karst Scand J For Res 17:408–416

Ammer C (2008) Konkurrenzsteuerung – Anmerkungen zu einer Kernaufgabe des Waldbaus beim Aufbau vielfältiger Wälder. In: Eberswalder Forstl Schr'reihe, vol 36, pp 21–26

Ammer C (2017) Unraveling the importance of inter- and intraspecific competition for the adaptation of forests to climate change. Prog Bot 78:345–367

Ammer C, Balandier P, Bentsen NS, Coll L, Löf M (2011) Forest vegetation management under debate: an introduction. Eur J For Res 130:1–5

Assmann E (1961) Waldertragskunde. Organische Produktion. Struktur. Zuwachs und Ertrag von Waldbeständen. München, Bonn, Wien: BLV Verlagsgesellschaft, 490 p

Aubry KB, Halpern CB, Peterson CE (2009) Variable-retention harvests in the Pacific Northwest: A review of short-term findings from the DEMO study. For Ecol Manag 258:398–408

Aussenac G, Granier A (1988) Effects of thinning on water stress and growth in Douglas-fir. Can J For Res 18:100–105

Bailey S (2007) Increasing connectivity in fragmented landscapes: an investigation of evidence for biodiversity gain in woodlands. For Ecol Manag 238:7–23

Barnes BV, Zak DR, Denton SR, Spurr SH (1998) Forest ecology. Wiley, 774 p

Bazzaz FA (1983) Characteristics of populations in relation to disturbance in natural and man-modified ecosystems. In: Mooney HA, Gordon M (eds) Disturbance and ecosystems. Springer, Berlin/Heidelberg, pp 259–275

Bebi P, Kulakowski D, Rixen C (2009) Snow avalanche disturbances in forest ecosystems: State of research and implications for management. For Ecol Manag 257:1883–1892

Bork H-R, Kranz A (2008) Die Jahrtausendflut des Jahres 1342 prägt Deutschland – Neue Forschungsergebnisse aus dem Einzugsgebiet des Mains. Jahresber Wetterauischen Ges gesamt Naturkunde 158:119–129

Bork HR, Bork H, Dalchow C, Piorr HP, Schatz T, Faust B (1998) Landschaftsentwicklung in Mitteleuropa: Wirkungen des Menschen auf Landschaften. Klett-Perthes, Gotha, 328 p

Brang P (2005) Virgin forests as a knowledge source for central European silviculture: reality or myth? Forest Snow Landsc Res 79:19–32

Brang P, Hilfiker S, Wasem U, Schwyzer A, Wohlgemuth T (2015) Langzeitforschung auf Sturmflächen zeigt Potenzial und Grenzen der Naturverjüngung. Schweiz Z Forstwes 166:147–158

Bréda N, Granier A, Aussenac G (1995) Effects of thinning on soil and tree water relations, transpiration and growth in an oak forest (*Quercus petraea* (Matt.) Liebl.). Tree Physiol 15:295–306

Bücking W (1997) Naturwald, Naturwaldreservate, Wildnis in Deutschland und Europa. Forst Holz 52:515–522

Burschel P, Huss J (2003) Grundriss des Waldbaus. Ein Leitfaden für Studium und Praxis, 3rd edn. Verlag Eugen Ulmer, 487 p

BUWAL (2000) Entscheidungshilfe bei Sturmschäden im Wald. Schweizerisches Bundesamt für Umwelt, Wald und Landschaft. In: BUWAL (ed), Bern, 100 p

Carlson DW, Groot A (1997) Microclimate of clear-cut, forest interior, and small openings in trembling aspen forest. Agric For Meteorol 87:313–329

Chen J, Franklin JF, Spies TA (1993) Contrasting microclimates among clearcut, edge, and interior of old-growth Douglas-fir forest. Agric For Meteorol 63:219 237

Coch T (1995) Waldrandpflege, Grundlagen und Konzepte. Neumann, Radebeul, 340 p

Curtis RO, Marshall DD, Bell JF (1995) Logs: A pioneering example of silvicultural research in coast Douglas-fir. J Forestry 95:19–25

Davis MA, Wrage KJ, Reich PB (1998) Competition between tree seedlings and herbaceous vegetation: support for a theory of resource supply and demand. J Ecol 86:652–661

DeAngelis DL, Waterhouse JC (1987) Equilibrium and non-equilibrium concepts in ecological models. Ecol Monogr 57:1–21

Dertz W (1972) Möglichkeiten und Grenzen forstlicher Bewirtschaftung der Molkenböden des Oberweserberglandes. Allg Forst Jagdztg 143:153–162

Ebeling C, Lang F, Gaertig T (2016) Structural recovery in three selected forest soils after compaction by forest machines in Lower Saxony, Germany. For Ecol Manag 359:74–82

Ebrecht L, Schmidt W (2005) Einfluss von Rückegassen auf die Vegetation. Forstarchiv 76:83–101

Ehbrecht M, Schall P, Ammer C, Seidel D (2017) Quantifying stand structural complexity and its relationship with forest management, tree species diversity and microclimate. Agric For Meteorol 242:1–9

Ellenberg H (1996) Vegetation Mitteleuropas mit den Alpen in ökologischer, dynamischer und historischer Sicht, 5th edn. Ulmer, Stuttgart, 1095 p

Ellenberg H, Leuschner C, Dierschke H (2010) Vegetation Mitteleuropas mit den Alpen in ökologischer, dynamischer und historischer Sicht, 6th edn. Eugen Ulmer, Stuttgart, 1334 p

Fahrig L (2013) Rethinking patch size and isolation effects: the habitat amount hypothesis. J Biogeogr 40:1649–1663

Fartmann T, Müller C, Poniatowski D (2013) Effects of coppicing on butterfly communities of woodlands. Biol Conserv 159:396–404

Fedrowitz K, Koricheva J, Baker SC, Lindenmayer DB, Palik B, Rosenvald R, Beese W, Franklin JF, Kouki J, Macdonald E, Messier C, Sverdrup-Thygeson A, Gustafsson L (2014) Can retention forestry help conserve biodiversity? A meta-analysis. J Appl Ecol 51:1669–1679

Firbas R (1949) Spät- und nacheiszeitliche Waldgeschichte Mitteleuropas. Gustav Fischer, Jena, 480 p

Fisichelli N, Vor T, Ammer C (2014) Broadleaf seedling responses to warmer temperatures "chilled" by late frost that favors conifers. Eur J For Res 133:587–596

Foster DR, Boose ER (1992) Patterns of forest damage resulting from catastrophic wind in central New England, USA. J Ecol 80:79–98

Franklin JF, Spies TA, Van Pelt R, Carey AB, Thornburgh DA, Berg DR, Lindenmayer DB, Harmon ME, Keeton WS, Shaw DC, Bible K, Chen J (2002) Disturbances and structural development of natural forest ecosystems with silvicultural implications, using Douglas-fir forests as an example. For Ecol Manag 155:399–423

FSC Arbeitsgruppe Deutschland (2012) Deutscher FSC-Standard – Deutsche übersetzte Fassung. Forest Stewardship Council, 51 p

Gámez-Virués S, Perović DJ, Gossner MM, Börschig C, Blüthgen N, de Jong H, Simons NK, Klein AM, Krauss J, Maier G, Scherber C, Steckel J, Rothenwöhrer C, Steffan-Dewenter I, Weiner CN, Weisser W, Werner M, Tscharntke T, Westphal C (2015) Landscape simplification filters species traits and drives biotic homogenization. Nature Commun 6:article 8568

Gayer K (1886) Der gemischte Wald. Parey, Berlin, 168 p

Gebhardt T, Häberle K-H, Matyssek R, Schulz C, Ammer C (2014) The more, the better? Water relations of Norway spruce stands after progressive thinning. Agric For Meteorol 197:235–243

Glenz C, Schlaepfer R, Iorgulescu I, Kienast F (2006) Flooding tolerance of Central European tree and shrub species. For Ecol Manag 235:1–13

Grober U (2013) Die Entdeckung der Nachhaltigkeit – Kulturgeschichte eines Begriffs. Kunstmann, München, 303 p

Grüneberg E, von Wilpert K, Meesenburg H, Evers J, Ziche D, Andrae H, Wellbrock N (2017) Was nützt die Waldkalkung? AFZ/Wald 72:15–17

Gustafson EJ, Lytle DE, Swaty R, Loehle C (2007) Simulating the cumulative effects of multiple forest management strategies on landscape measures of forest sustainability. Landsc Ecol 22:141–156

Gustafsson L, Baker SC, Bauhus J, Beese WJ, Brodie A, Kouki J, Lindenmayer DB, Lõhmus A, Pastur GM, Messier C, Neyland M, Palik B, Sverdrup-Thygeson A, Volney WJA, Wayne A, Franklin JF (2012) Retention forestry to maintain multifunctional forests: a world perspective. Bioscience 62:633–645

Harrington TB, Edwards MB, Boyd M (1999) Understory vegetation, resource availability, and litterfall responses to pine thinning and woody vegetation control in longleaf pine plantations. Can J For Res 29:1055–1064

Harris LD (1984) The fragmented forest. The University of Chicago Press, Chicago, London, p 211

Heinken T, von Oheimb G, Schmidt M, Kriebitzsch W-U, Ellenberg H (2005) Schalenwild breitet Gefäßpflanzen in der mitteleuropäischen Kulturlandschaft aus – ein erster Überblick. Natur Landsch 80:141–147

Herrmann I (2014) Räumliche Optimierung der Bestandesstruktur unter Berücksichtigung von Einzelbaumeffekten. Universität Dresden; Fakultät Umweltwissenschaften, Dresden, 233 p

Hildebrand EE (2008) Lässt sich das „Großraumexperiment Waldbodenverformung" stoppen? AFZ/Wald 63:291–293

Hobi ML, Commarmot B, Bugmann H (2015) Pattern and process in the largest primeval beech forest of Europe (Ukrainian Carpathians). J Veg Sci 26:323–336

Hofmeister J, Hosek J, Malicek J, Palice Z, Syrovatkova L, Steinova J, Cernajova I (2016) Large beech (*Fagus sylvatica*) trees as 'lifeboats' for lichen diversity in central European forests. Biodivers Conserv 25:1073–1090

Holling CS, Meffe GK (1996) Command and control and the pathology of natural resource management. Conserv Biol 10:328–337

Holzwarth F, Kahl A, Bauhus J, Wirth C (2013) Many ways to die – partitioning tree mortality dynamics in a near-natural mixed deciduous forest. J Ecol 101:220–230

Hüttl RF (1989) Liming and fertilization as mitigation tools in declining forest ecosystems. Water Air Soil Pollut 44:93–118

Janssen A, Schäffer J, von Wilpert K, Reif A (2016) Flächenbedeutung der Waldkalkung in Baden-Württemberg – Extent of forest liming in Baden-Württemberg. Waldökol Landschaftsforsch Natursch 15:5–33

Kaiser T, Zacharias D (2003) PNV-Karten für Niedersachsen auf Basis der BÜK50. Niedersächsisches Landesamt für Ökologie, Hannover

Kaiser M, Sadik WJ, Kunkel M (2012) Potenzial und Gefährdung der Urwälder von morgen. Der Bayerische Spessart. Abschlussbericht der Kartierung im BaySF Forstbetrieb Rothenbuch Winter 2011/12. Greenpeace, 51 p

Keenan RJ, Kimmins JP (1993) The ecological effects of clear-cutting. Environ Rev 1:121–144

Klinck U, Fröhlich D, Meiwes KJ, Beese F (2013) Entwicklung der Stoffein- und -austräge nach einem Fichten-Kleinkahlschlag. Forstarchiv 84:93–101

Knoke T (1998) Analyse und Optimierung der Holzproduktion in einem Plenterwald zur Forstbetriebsplanung in ungleichaltrigen Wäldern. Forstliche Forschungsberichte München 170

Kogan M (1998) Integrated pest management: Historical perspectives and contemporary developments. Annu Rev Entomol 43:243–270

Kohler M, Sohn J, Nägele G, Bauhus J (2010) Can drought tolerance of Norway spruce (*Picea abies* (L.) Karst.) be increased through thinning? Eur J For Res 129:1109–1118

Kohm KA, Franklin JF (1997) Creating a forestry for the 21st century: The science of ecosystem management. Island Press, Washington, DC, 475 p

Korpel S (1982) Degree of equilibrium and dynamical changes of the forest on example of natural forest of Slovakia, Act Fac For 24:9–31

Korpel S (1995) Die Urwälder der Westkarpaten. Gustav Fischer, Stuttgart, 310 p

Kraus D, Zeppenfeld T (2013) Feuer als Störfaktor in Wäldern AFZ/Wald 68:8–9

Kremser W (1973) Lacerati turbine ventorum – vom Sturm zerfetzt! Ein Orkan verheerte Niedersachsens Wälder, Neues Arch Niedersachsen 22,219–241

Kuntze K (1996) Effects of forest liming on soil processes. Plant Soil 168,117–170

Küchler P (2011) Vorstudie als Grundlage zur zukünftigen Behandlung der Moore und Feuchtwälder im Solling. Gutachten im Auftrag des Forstamtes Neuhaus, Holzminden-Neuhaus, Niedersachsen, 172 p

Kuntze H (1993) Niedermoore als Senke und Quelle für Kohlenstoff und Stickstoff. Wasser Boden 9:699–702

Kuuluvainen T, Pukkala T (1989) Effect of Scots pine seed trees on the density of ground vegetation and tree seedlings. Silva Fenn 23:159–167

Larrieu L, Cabanettes A (2012) Species, live status, and diameter are important tree features for diversity and abundance of tree microhabitats in subnatural montane beech fir forests. Can J For Res 42:1433–1445

Lehnigk M, Ammer C (2012) Erfolgreiche Naturverjüngung in Kiefernbeständen: Plätzeweise Bodenbearbeitung mit dem MOHEDA-Kultivator. AFZ/Wald 67:22–24

Leibundgut H (1993) Europäische Urwälder: Wegweiser zur naturnahen Waldwirtschaft. Verlag Paul Haupt, Bern und Stuttgart, 260 p

Lindenmayer DB, Franklin JF (2002) Conserving forest biodiversity: a comprehensive multiscaled approach. Island Press, Washington, DC, 351 p

Lindner M, Fitzgerald JB, Zimmermann NE, Reyer C, Delzon S, Van Der Maaten E, Schelhaas M-J, Lasch P, Eggers J, van der Maaten-Theunissen M, Suckow F, Psomas A, Poulter B, Hanewinkel M (2014) Climate change and European forests: What do we know, what are the uncertainties, and what are the implications for forest management? J Environ Manag 146:69–83

Lindo Z, Visser S (2003) Microbial biomass, nitrogen and phosphorus mineralization, and meso-fauna in boreal conifer and deciduous forest floors following partial and clear-cut harvesting. Can J For Res 33:1610–1620

Lobinger G (2006) Entwicklung neuer Strategien im Borkenkäfermanagement. Forstschutz Aktuell 37:11–13

Matula R, Svatek M, Kurova J, Uradnicek L, Kadavy J, Kneifl M (2012) The sprouting ability of the main tree species in Central European coppices: implications for coppice restoration. Eur J For Res 131:1501–1511

Mayer H, Schindler D (2002) Forstmeteorologische Grundlagen zur Auslösung von Sturmschäden im Wald in Zusammenhang mit dem Orkan „Lothar". Allg Forst Jagdztg 173:200–208

Meiwes KJ, Asche N, Block J, Kallweit R, Kölling C, Raben G, von Wilpert K (2008) Potenziale und Restriktionen der Biomassenutzung im Wald. AFZ/Wald 63:598–603

Meyer P (1997) Probleme und Perspektiven der Naturwaldforschung am Beispiel Niedersachsens. Forstarchiv 68:87–98

Meyer P, Schmidt M (2008) Aspekte der Biodiversität von Buchenwäldern – Konsequenzen für eine naturnahe Bewirtschaftung. Beitr Nordwestdeutsch Forstl Versuchsanst 3:159–192

Meyer P, Bücking W, Schmidt S, Schulte U, Willig J (2004) Stand und Perspektiven der Untersuchung von Naturwald-Vergleichsflächen. Forstarchiv 75:167–179

Meyer P, Lorenz K, Engel F, Spellmann H, Boele-Keimer C (2015) Wälder mit natürlicher Entwicklung und Hotspots der Biodiversität – Elemente einer systematischen Schutzgebietsplanung am Beispiel Niedersachsen. Nat Schutz Landsch Plan 47:275–282

Meyer P, Blaschke M, Schmidt M, Sundermann M, Schulte U (2016) Wie entwickeln sich Buchen- und Eichen-FFH-Lebensraumtypen in Naturwaldreservaten? – Eine Bewertung anhand von Zeitreihen. Nat Schutz Landsch Plan 48:5–14

Möller A (1923) Der Dauerwaldgedanke. Erich Degreif Verlag, Oberteuringen, 127 p

Moslonka-Lefebvre M, Finley A, Dorigatti I, Dehnen-Schmutz K, Harwood T, Jeger MJ, Xu X, Holdenrieder O, Pautasso M (2011) Networks in plant epidemiology: from genes to landscapes, countries, and continents. Phytopathology 101:392–403

Mrotzek R, Pfirrmann H, Barge U (2000) Einfluss von Wegebaumaterial und Licht auf die Vegetation an Waldwegen und im angrenzenden Bestand – dargestellt an Wegen im niedersächsischen Forstamt Bramwald. Forstarchiv 71:234–244

Müller J, Bütler R (2010) A review of habitat thresholds for dead wood: a baseline for management recommendations in European forests. Eur J For Res 129:981–992

Murcia C (1995) Edge effects in fragmented forests: implications for conservation. Trends Ecol Evol 10:58–62

Nagel TA, Svoboda M, Kobal M (2014) Disturbance, life history traits, and dynamics in an old-growth forest landscape of southeastern Europe. Ecol Appl 24:663–679

Niemelä J, Young J, Alard D, Askasibar M, Henle K, Johnson R, Kurttila M, Larsson T-B, Matouch S, Nowicki P, Paiva R, Portoghesi L, Smulders R, Stevenson A, Tartes U, Watt A (2005) Identifying, managing and monitoring conflicts between forest biodiversity conservation and other human interests in Europe. Forest Policy Econ 7:877–890

North MP, Keeton WS (2008) Emulating natural disturbance regimes: an emerging approach for sustainable forest management. In: Lafortezza R, Chen J, Sanesi G, Crow TR (eds) Patterns and processes in forest landscapes: multiple use and sustainable management. Springer, Dordrecht, pp 341–372

Nüsslein S (1995) Struktur und Wachstumsdynamik jüngerer Buchen-Edellaubholz-Mischbestände in Nordbayern. Forstl Forschber München 151:1–295

Oliver CD, Larson BC (1990) Forest stand dynamics. McGraw-Hill, New York, 467 p

Örlander G, Egnell G, Albrektson A (1996) Long-term effects of site preparation on growth in Scots pine. For Ecol Manag 86:27–37

Otto HJ (1995) Zielorientierter Waldbau und Schutz sukzessionaler Prozesse. Forst Holz 50:203–209

Perry DA, Amaranthus MP (1997) Disturbance, recovery, and stability. In: Kohm KA, Franklin JF (eds) Creating a forestry for the 21st century: the science of ecosystem management. Island Press, Washington, DC, pp 31–56

Petercord R, Lobinger G (2010) Diflubenzuron: ein notwendiger Wirkstoff für einen integrierten Pflanzenschutz. Forstschutz Aktuell 50:23–26

Petriţan IC, Lüpke VB, Petriţan AM (2011) Effects of root trenching of overstorey Norway spruce *(Picea abies)* on growth and biomass of underplanted beech *(Fagus sylvatica)* and Douglas fir *(Pseudotsuga menziesii)* saplings. Eur J For Res 130:813–828

Pommerening A, Murphy ST (2004) A review of the history, definitions and methods of continuous cover forestry with special attention to afforestation and restocking. Forestry 77:27–44

Poulson TL, Platt WJ (1996) Replacement patterns of beech and sugar maple in Warren Woods, Michigan. Ecology 77:1234–1253

Poyatos R, Latron J, Llorens P (2003) Land use and land cover change after agricultural abandonment: the case of a Mediterranean mountain area (Catalan Pre-Pyrenees). Mt Res Dev 23:362–368

Pretzsch H (2004) Gesetzmäßigkeit zwischen Bestandesdichte und Zuwachs. Lösungsansatz am Beispiel von Reinbeständen aus Fichte *(Picea abies* (L.) Karst.) und Buche *(Fagus sylvatica* L.). Allg Forst Jagdztg 175:225–234

Pretzsch H (2005) Stand density and growth of Norway spruce *(Picea abies* (L.) Karst.) and European beech *(Fagus sylvatica* L.): evidence from long-term experimental plots. Eur J For Res 124:193–205

Puettmann KJ, Coates KD, Messier C (2009) A critique of silviculture: managing for complexity. Island Press, 189 p

Puettmann K, Messier C, Coates KD (2014) Managing forests as complex adaptive systems: introductory concepts and applications. In: Messier C, Puettmann K, Coates KD (eds) Managing forests as complex adaptive systems. Routledge Taylor & Francis Group, London/New York, pp 3–14

Pyne SJ, Andrews PL, Laven RD (1996) Introduction to wildland fire, 2nd edn. Wiley, New York, 808 p

Reif A, Achtziger R (2000) Gebüsche, Hecken, Waldmäntel, Feldgehölze (Strauchformationen): XI-2.2. Handbuch Naturschutz und Landschaftspflege 3. Erg Lfg 11:1–46

Ⅱⅰ ⅠⅠ Ⅱ, Ⅱ ⅠⅠⅠⅠⅠ ⅠⅠⅡ, ⅠⅠⅠ,ⅠⅠⅠ Ⅰ, Ⅱ,ⅠⅠⅠ. Ⅱ ⅠⅠⅠ Ⅱ) ⅠⅠ ⅠⅠⅠⅠⅠ ⅠⅠ ⅠⅠ Ⅱ ⅠⅠ ⅠⅠ ⅠⅠⅠⅠⅠⅠ ⅠⅠ ⅠⅠ Ⅱ ⅠⅠ ⅠⅠⅠⅠ Ⅱ ⅠⅠ ⅠⅠⅠ Ⅱ Ⅰ Waldökol. Landschaftsforsch Natursch:5–29

Reineke LH (1933) Perfecting a stand density index for even aged forests. J Agric Res 46:627–638

Richter M (1997) Allgemeine Pflanzengeographie. Teubner Verlag, Stuttgart, 256 p

Richter J (2003) Wurf- und Bruchschäden in Fichtenbeständen. Forstarchiv 74:166–170

Ritter E, Dalsgaard L, Einhorn KS (2005) Light, temperature and soil moisture regimes following gap formation in a semi-natural beech-dominated forest in Denmark. For Ecol Manag 206:15–33

Roberts MR (2004) Response of the herbaceous layer to natural disturbance in North American forests. Can J Bot 82:1273–1283

Röhrig E, Bartsch N, Lüpke VB (2006) Waldbau auf ökologischer Grundlage (7. Aufl.). Verlag Eugen Ulmer, Stuttgart, 479 p

Rosenvald R, Lõhmus A (2008) For what, when, and where is green-tree retention better than clear-cutting? A review of the biodiversity aspects. For Ecol Manag 255:1–15

Schaich H, Plieninger T (2013) Land ownership drives stand structure and carbon storage of deciduous temperate forests. For Ecol Manag 305:146–157

Schall P, Ammer C (2013) How to quantify forest management intensity in Central European forests. Eur J For Res 132:379–396

Schall P, Lödige C, Beck M, Ammer C (2012) Biomass allocation to roots and shoots is more sensitive to shade and drought in European beech than in Norway spruce seedlings. For Ecol Manag 266:246–253

Schall P, Gossner MM, Heinrichs S, Fischer M, Boch S, Prati D, Jung K, Baumgartner V, Blaser S, Böhm S, Buscot F, Daniel R, Goldmann K, Kaiser K, Kahl T, Lange M, Müller J, Overmann J, Renner SC, Schulze E-D, Sikorski J, Tschapka M, Türke M, Weisser WW, Wemheuer B, Wubet T, Ammer C (2018) The impact of even-aged and uneven-aged forest management on regional biodiversity of multiple taxa in European beech forests. J Appl Ecol 55:267–278

Schmidt W (2002) Einfluss der Bodenschutzkalkungen auf die Waldvegetation. Forstarchiv 73:43–54

Schmidt UE (2009) Wie erfolgreich war das Dauerwaldkonzept bislang: eine historische Analyse I Continuous cover forests – a success? A historical analysis. Schweiz Z Forstwes 160:144–151

Schmidt M, Meyer P (2015) Hessische Naturwaldreservate im Portrait: Weiherskopf. Nordwestdeutsche Forstliche Versuchsanstalt, Göttingen, 43 p

Schmidt M, Kriebitzsch W-U, Ewald J (2011) Waldartenliste der Farn- und Blütenpflanzen, Moose und Flechten Deutschlands. BfN-Skripten 299:116

Schröder K (2009) Der Mittelwald als waldbauliche Option in Deutschland. Handbuch Naturschutz und Landschaftspflege 22:1–14

Schütz JP (1994) Geschichtlicher Hergang und aktuelle Bedeutung der Plenterung in Europa. Allg Forst Jagdztg 165:106–114

Schütz JP (1996) Bedeutung und Möglichkeiten der biologischen Rationalisierung im Forstbetrieb. Schweiz Z Forstwes 147:315–344

Seibold S, Bässler C, Brandl R, Gossner MM, Thorn S, Ulyshen MD, Müller J (2015) Experimental studies of dead-wood biodiversity – a review identifying global gaps in knowledge. Biol Conserv 191:139–149

Seymour RS, White AS, Demaynadier PG (2002) Natural disturbance regimes in northeastern North America–evaluating silvicultural systems using natural scales and frequencies. For Ecol Manag 155:357–367

Shifley SR, Thompson FR, Dijak WD, Fan Z (2008) Forecasting landscape-scale, cumulative effects of forest management on vegetation and wildlife habitat: a case study of issues, limitations, and opportunities. For Ecol Manag 254:474–483

Shipley B, Meziane D (2002) The balanced-growth hypothesis and the allometry of leaf and root biomass allocation. Funct Ecol 16:326–331

Simonin K, Kolb TE, Montes-Helu M, Koch GW (2007) The influence of thinning on components of stand water balance in a ponderosa pine forest stand during and after extreme drought. Agric For Meteorol 143:266–276

Son Y, Lee YY, Jun YC, Kim Z-S (2004) Light availability and understory vegetation four years after thinning in a *Larix leptolepis* plantation of central Korea. J For Res 9:133–139

Spies TA, Turner MG (1999) Dynamic forest mosaics. In: Hunter ML (ed) Maintaining biodiversity in forest ecosystems. Cambridge University Press, Cambridge, pp 95–160

Stogsdili WR, Wittwer RF, Hennessey TC, Dougherty PM (1992) Water use in thinned loblolly pine plantations. For Ecol Manag 50:233–245

Stüber V, Meiwes KJ, Mindrup M (2008) Nachhaltigkeit und Vollbaumnutzung: Bewertung aus Sicht der forstlichen Standortskartierung am Beispiel Niedersachsen. Forst Holz 63:28–33

Sturm K (1993) Prozeßschutz – ein Konzept für naturschutzgerechte Waldwirtschaft. Z Ökol Nat Schutz 2:181–192

Svoboda M, Janda P, Nagel TA, Fraver S, Rejzek J, Bače R (2012) Disturbance history of an old-growth sub-alpine *Picea abies* stand in the Bohemian Forest, Czech Republic. J Veg Sci 23:86–97

Trotsiuk V, Svoboda M, Janda P, Mikolas M, Bače R, Rejzek J, Samonil P, Chaskovskyy O, Korol M, Myklush S (2014) A mixed severity disturbance regime in the primary *Picea abies* (L.) Karst. forest of the Ukrainian Carpathians. For Ecol Manag 334:144–153

Troup RS (1928) Silvicultural systems. Clarendon Press, Oxford, 199 p

Turner MG (2010) Disturbance and landscape dynamics in a changing world. Ecology 91:2833–2849

Vergani C, Giadrossich F, Buckley P, Conedera M, Pividori M, Salbitano F, Rauch H, Lovreglio R, Schwarz M (2017) Root reinforcement dynamics of European coppice woodlands and their effect on shallow landslides: a review. Earth-Sci Rev 167:88–102

von Lüpke B, Hauskeller-Bullerjahn K (2004) Beitrag zur Modellierung der Jungwuchsentwicklung am Beispiel von Traubeneichen-Buchen-Mischverjüngungen. Allg Forst Jagdztg 175:61–69

von Wilpert K, Schäffer J (2006) Ecological effects of soil compaction and initial recovery dynamics: a preliminary study. Eur J For Res 125:129–138

Vuidot A, Paillet Y, Archaux F, Gosselin F (2011) Influence of tree characteristics and forest management on tree microhabitats. Biol Conserv 144:441–250

Wagner S (1999) Ökologische Untersuchungen zur Initialphase der Naturverjüngung in Eschen-Buchen-Mischbeständen. Schr. Forstl. Fak. Univ. Gött. Niedersächs. forstl. Vers Anst 129:1–262

Wagner S (2007) Rationaler Waldumbau: Fragen und Anregungen. Forst Holz 62:12–17

Wagner S, Spellmann H (1993) Entscheidungshilfen für die Verjüngungsplanung in Fichtenbeständen zum Voranbau der Buche im Harz. Forst Holz 48:483–490

Waldner P, Thimonier A, Graf Pannatier E, Etzold S, Schmitt M, Marchetto A, Rautio P, Derome K, Nieminen TM, Nevalainen S, Lindroos A-J, Merilä P, Kindermann G, Neumann M, Cools N, de Vos B, Roskams P, Verstraeten A, Hansen K, Pihl Karlsson G, Dietrich H-P, Raspe S, Fischer R, Lorenz M, Iost S, Granke O, Sanders TGM, Michel A, Nagel H-D, Scheuschner T, Simončič P, von Wilpert K, Meesenburg H, Fleck S, Benham S, Vanguelova E, Clarke N, Ingerslev M, Vesterdal L, Gundersen P, Stupak I, Jonard M, Potočić N, Minaya M (2015) Exceedance of critical loads and of critical limits impacts tree nutrition across Europe. Ann For Sci 72:929–939

Watt AS (1947) Pattern and process in the plant community. J Ecol 35:1–22

Weidig J, Wagner S, Huth F (2014) Qualitätsentwicklung von Buchenvoranbauten (*Fagus sylvatica* L.) im Thüringer Wald nach unplanmäßigem sturmbedingtem Verlust des Fichtenschirms. Forstarchiv 85:122–133

Weis W, Rotter V, Göttlein A (2006) Water and element fluxes during the regeneration of Norway spruce with European beech: effects of shelterwood-cut and clear-cut. For Ecol Manag 224:304–317

Wiedemann E (1931) Die Rotbuche. Mitt. Forstwirtsch. Forstwiss 2:1–90

Willig J (2002) Natürliche Entwicklung von Wäldern nach Sturmwurf: 10 Jahre Forschung im Naturwaldreservat Weiherskopf. Naturwaldreservate Hessen 8:1–167

Winter MB (2015) Regeneration dynamics and resilience of unmanaged mountain forests in the northern limestone alps following bark beetle induced spruce dieback. Eur J For Res 134:949–968

Witzig J, Badoux A, Hegg C, Lüscher P (2004) Waldwirkung und Hochwasserschutz: eine standörtlich differenzierte Betrachtung. Forst Holz 59:476–479

Yoda K, Kira T, Ogawa H, Hozumi K (1963) Self-thinning in overcrowded pure stands under cultivated and natural conditions. J Biol 14:107–129

Zeide B (2001) Thinning and growth: a full turnaround. J Forestry 99:20–25

Zeide B (2004) Optimal stand density: a solution. Can J For Res 34:846–854

Chapter 15
Grasslands

Michael Bahn ⓘ, Johannes Ingrisch ⓘ, and Anke Jentsch ⓘ

Abstract Anthropogenic disturbances in temperate grasslands include grazing and mowing, as well as associated management measures such as fertilization, irrigation, and fire. In addition to the mechanical disturbances associated with the removal of biomass, grassland management changes the availability of resources and thus biogeochemical cycles, species composition, and biodiversity. In addition to their importance for livestock farming, grassland systems are also important as a cultural landscape. Grasslands regenerate faster than forests after drought, heat waves, and fire. On the other hand, overuse including overgrazing can promote erosion and lead to grassland degradation.

Keywords Biodiversity · Canopy structure · Carbon and nutrient cycles · Ecosystem services · Grassland management · Microclimate · Grazing · Mowing · Productivity · Species adaptation

15.1 Grasslands and Land Use

For centuries, grasslands have been subject to anthropogenic management such as grazing by cattle, horses, and sheep, as well as mowing for fodder production. Grasslands are widespread throughout the world, especially in temperate, Mediterranean, and subtropical regions where livestock farming is practised and where climatic or orographic conditions and soil quality are unfavourable for arable farming. From a global perspective, pastoralism is the most important form of land utilization. In many regions, including Central Europe, mowing plays an important

M. Bahn (✉) · J. Ingrisch
Department of Ecology, University of Innsbruck, Innsbruck, Austria
e-mail: michael.bahn@uibk.ac.at

A. Jentsch
Bayreuth Center of Ecology and Environmental Research (BayCEER), University of Bayreuth, Bayreuth, Germany

T. Wohlgemuth et al. (eds.), *Disturbance Ecology*, Landscape Series 32,
https://doi.org/10.1007/978-3-030-98756-5_15

role in addition to grazing (Fig. 15.1). The production of hay and silage ensures and supports the supply of livestock outside the grazing season. In the Mediterranean and subtropical grassland areas, "transhumance" (Poschlod and WallisDeVries 2002; Suttie et al. 2005) is a common practice of livestock management that involves the movement of animals over long distances between different locations for summer and winter grazing. Fire management is sometimes used to keep grasslands open and to ensure a faster turnover of nutrients (Fernandes et al. 2013; Valese et al. 2014; Reinhart et al. 2016).

Grazing and mowing are the most important anthropogenic disturbance types in grasslands. Their effects on ecosystem dynamics are similar in that they both lead to a loss and subsequent recovery of aboveground biomass, but they differ in their impact on biodiversity. Grazing increases heterogeneity through selective removal

Fig. 15.1 Anthropogenic disturbance regimes in grasslands: (**a**) grazing by cattle or sheep, (**b**) mowing of meadows for hay and silage production for stable and winter feeding, (**c**) fertilization of meadows with organic residues from livestock farming, (**d**) irrigation of meadows in dry areas, and (**e**) fire management to keep the pastures open and to fertilize them. Mowing and grazing are usually a requirement for the conservation of grasslands in Europe. After abandonment of managed grasslands, secondary succession leads to shrub encroachment. (Photos: A. Jentsch **a, c, d, e**; L. Hörtnagl **b**)

or promotion of plant species and the creation of small-scale differences in nutrient availability, soil compaction, canopy structure, and community structure (Milchunas und Lauenroth 1993, Hastienheim et al 1999; Adler et al. 2001). Furthermore, grazing often takes place over a period of several days or weeks (e.g. in alpine pastures), while mowing is a pulsed disturbance of limited duration. Grazing and mowing, as well as related management measures such as fertilization, irrigation, and controlled burning (Fig. 15.1), have direct effects on canopy structure, microclimate, and biogeochemical cycles. However, they can also have long-term consequences by selecting for adaptations of species and by changing community composition and soil properties (Fig. 15.2).

European grasslands have developed under the influence of centuries of extensive human use and harbour a large part of the biodiversity of the Eurasian continent. Many rare and endangered plant species are dependent on anthropogenic disturbance regimes, because both abandonment and intensification of management can lead to changes or loss of the habitats those species depend on. More recently, European grasslands have been strongly affected by land-use change, climate change, and, to a lesser extent, biotic invasions. In many places, the abandonment of grassland management leads to shrub encroachment and a slow recovery of the autochthonous vegetation. In other regions, the intensification of grassland management—through strong nutrient inputs, through the use of species-poor sowing mixtures, and through increased mowing frequency—has led to a massive decline in local biodiversity (Blüthgen et al. 2012).

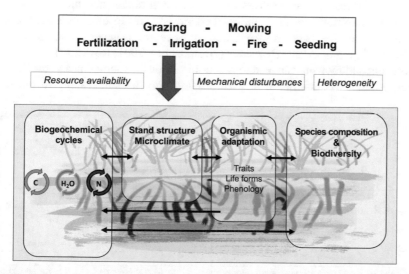

Fig. 15.2 Grazing and mowing as well as associated management measures, such as fertilization, irrigation, fire and seeding influence organisms, the stand structure and microclimate, as well as nutrient cycles via mechanical disturbances and sometimes rapid changes in resource availability and their spatial heterogeneity. Changes in species composition and plant functional traits lead to the long-term formation of grassland systems that are well adapted to the prevailing disturbance regime

15.2 Effects of Disturbance on the Structure and Function of Grasslands

15.2.1 Canopy Structure, Microclimate, and Water Balance

Grazing and mowing influence plant growth and canopy structure, reducing the aboveground biomass and thereby also the leaf area index (LAI) and the total photosynthetically active area, including the green stems (green area index GAI; Figs. 15.3 and 15.5). This affects both ecosystem CO_2 fluxes and the microclimate (Wan et al. 2002; Klein et al. 2005). Thus, a reduction of the LAI leads to a reduction of the radiation absorbed by the stand and, depending on soil colour and soil moisture, the albedo (reflectivity; Fig. 15.3). In consequence, grazing and mowing alter important components of the energy balance, including soil heat flux, sensible heat flux, and evapotranspiration.

Mowing and grazing influence the water balance and its components by affecting transpiration and evaporation, precipitation infiltration, surface runoff, and soil water content. The reduction of the LAI by grazing or mowing leads to a reduction of transpiration and increases the available energy for evaporation due to the increased irradiation at the soil surface (Fig. 15.4), leading to reduction in soil moisture. In addition to LAI, aerodynamic effects and leaf physiology can also alter water loss from grasslands (Fig. 15.4). For example, the transpiration rate per unit leaf area tends to decrease with increasing leaf age (Larcher 2001). Grazed or mown stands composed of young leaves in the regrowth phase can, therefore, show a higher transpiration rate than undisturbed grasslands. In consequence, the total amount of water lost from grasslands across the season is generally not much influenced by grazing or cutting frequency (Bremer et al. 2001; Rose 2004).

Grazing and mowing can lead to soil compaction and sometimes to turf tearing, which reduces the infiltration of precipitation into the soil leading to increased

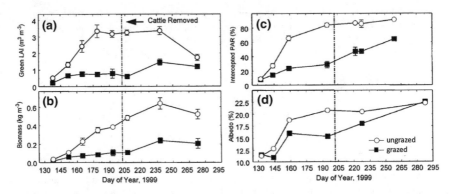

Fig. 15.3 Seasonal developments on grazed (black dots) and ungrazed (white dots) grasslands: (**a**) leaf area index (LAI), (**b**) aboveground biomass, (**c**) photosynthetically active radiation absorbed by the stand, and (**d**) albedo. The dotted vertical line shows the point in time when grazing stops. (After Bremer et al. 2001)

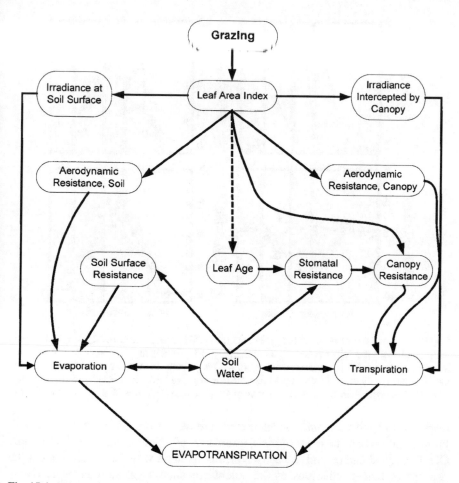

Fig. 15.4 Effects of grazing on evapotranspiration. (From Bremer et al. 2001)

surface runoff and ultimately increased soil erosion. When grazing pressure is high and vegetation cover is low, the risk of erosion under heavy precipitation events increases and can lead to widespread ecosystem degradation, especially on mountain slopes (Asner et al. 2004).

15.2.2   *Carbon Balance*

The carbon balance in ecosystems is determined by the amount of photosynthetic CO_2 fixation and the amount of carbon released by respiration processes. Under anaerobic conditions, for example, in wetlands or during fermentation in the rumen of ruminants, carbon is primarily released in the form of methane. In grasslands, respiration processes take place mainly belowground and are determined by the

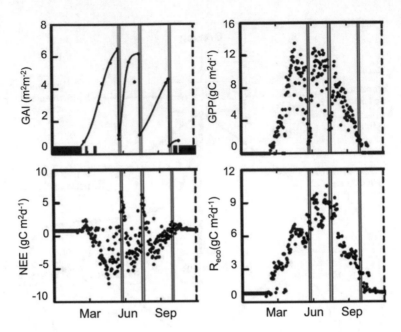

Fig. 15.5 Impacts of mowing on the green area index (GAI, photosynthetically active plant area per ground area), the daily net ecosystem exchange of CO_2 (NEE), and its components (gross primary productivity, GPP; ecosystem respiration, R_{eco}) in a managed meadow in the Alps. The three mowing dates are indicated by vertical lines. NEE is negative from an atmospheric point of view when GPP is larger than R_{eco} and a net uptake of CO_2 occurs. (Data from Wohlfahrt et al. 2008a)

respiration of the roots and the microbial turnover of organic matter in the soil. From a carbon balance point of view, grasslands often represent a weak CO_2 sink. CO_2 fluxes and carbon sink capacity typically increase with fertilization but at the expense of higher emissions of the potent greenhouse gas nitrous oxide (N_2O) (Velthof and Oenema 1995; Allard et al. 2007; Soussana et al. 2007; Schmitt et al. 2010; Chang et al. 2015; Hörtnagl et al. 2018; Harris et al. 2021). As a consequence of the abandonment of grasslands, larger amounts of litter may accumulate on the soil surface, which may also result in a temporary, but overall very minor increase of the carbon stored in the soil (Meyer et al. 2012; Poeplau and Don 2013). Increased grazing intensity, on the other hand, leads to a reduction of the carbon content in European grassland soils (Klumpp et al. 2011; Paz-Ferreiro et al. 2012; McSherry and Ritchie 2013).

Grazing and mowing remove a major part of the photosynthetically active biomass and thereby lead to a temporarily negative carbon balance (Fig. 15.5; Wohlfahrt et al. 2008b). While soil respiration can be stimulated by an increase in soil temperature due to the lack of a shading canopy, the activity of the rhizosphere is frequently restricted by a reduced supply of fresh assimilates from photosynthesis (Bahn et al. 2006). Depending on the vigour of the stand and the weather conditions, regrowth leads again to a net uptake of CO_2 within one to three weeks (Fig. 15.5).

15.2.3 Nutrient Balance

Grazing and mowing, as well as management-related measures (such as fertilization, liming, or sowing of legumes), influence the nutrient balance through nutrient removal and nutrient supply. In addition, they change the amount and quality of the litter (e.g. Semmartin et al. 2004; Bai et al. 2012) and the environmental conditions (pH, microclimate, stoichiometry) for microorganisms (Robson et al. 2007; Le Roux et al. 2008; Legay et al. 2016), which can affect nutrient turnover. Grazing leads to a strong spatial heterogeneity of nutrient availability in a largely closed nutrient cycle, whereas mowing and fertilization of meadows results in a more homogeneous distribution of nutrients in a mostly open nutrient cycle.

Fertilization increases productivity as well as the nutrient content in plant tissue and the quality of forage (Mosier et al. 2004). It promotes fast-growing species whose litter decomposes more quickly, which in turn accelerates the nutrient cycle. Heavy or unbalanced fertilization leads to an accumulation of nutrients in the soil (eutrophication) and to a shift in nutrient ratios, which changes biodiversity (see Sect. 15.3.2) and its response to disturbances (Bakker et al. 2006). Strongly increased nitrogen inputs through fertilization can, for example, lead to nitrogen leaching into the groundwater and to increased N_2O emissions (Di and Cameron 2002; Soussana et al. 2007; Harris et al. 2021). Therefore, the amount and timing of fertilization in relation to the actual nutrient requirements and, thus, also to the growth dynamics of the plant stand, are essential features of environmentally friendly grassland management.

Liming is a management measure to increase the pH value of acid soils (Goulding 2016; Heyburn et al. 2017). It leads to an increase in microbial turnover of nitrogen and nutrient availability in soils. Since limed soils require less fertilization, addition of lime provides an opportunity to reduce N_2O emissions in grasslands. However, the increased CO_2 emissions from lime can worsen the overall greenhouse gas balance of grasslands.

Nutrient availability in ecosystems is strongly determined by the complex interactions in the food web, in which not only aboveground biomass loss but also root herbivory and plant–microorganism interactions play an important role (Fig. 15.6; Bardgett and Wardle 2003). For example, plant roots can exude energy-rich carbon compounds in case of increased nutrient demand, thereby promoting the microbial conversion of nutrients fixed in soil organic matter (Kuzyakov 2010). On the other hand, microbial growth rapidly recaptures (immobilizes) a large part of the released nitrogen. Especially in the regeneration phase of plants immediately after grazing and mowing and also after disturbances, microbes play an important role determining the availability of nutrients (Hautier et al. 2014; Kübert et al. 2019; Seabloom et al. 2020).

The nutrient balance of grasslands is strongly influenced by legumes (family Fabaceae; e.g. clover, *Trifolium* spp.) (Lüscher et al. 2014). Because of their symbiosis with nodule bacteria, legumes improve not only their own nitrogen supply but also that of neighbouring plants (Fig. 15.7). Thus, they contribute to an increase in

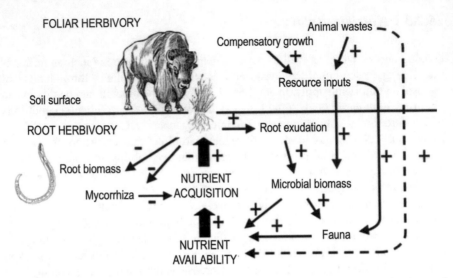

Fig. 15.6 Interactions between grazing and the nutrient cycle across trophic levels. (Bardgett and Wardle 2010)

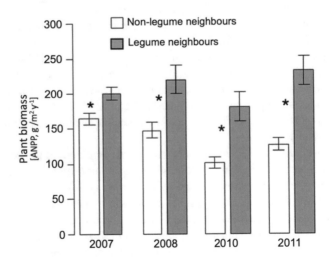

Fig. 15.7 Growth-stimulating legume effect on biomass production in grasslands (Arfin Khan et al. 2014). Species growing together with nitrogen-fixing legumes (grey bar) show a higher aboveground biomass production than species growing without legumes

primary production and fodder value. In order to increase the legume share in grass-lands, reseeding is often considered as a useful management measure.

15.2.4 Effects of Specific Management Forms and Accompanying Measures

Besides fertilization and liming (see Sect. 15.3), there are a number of regionally important accompanying measures associated with grassland management. New sowing, reseeding, and oversowing play an important role for soil cover and can also introduce new species and varieties into the stand (Opitz von Boberfeld 1994). In Bavaria, for example, the State Institute for Agriculture recommends climate-specific and regionally differentiated overseeding mixtures and makes them avail-able to farmers. These consist of a mixture of grass and legume species of different origins, which also differ in terms of growth, nutrient requirements, and flowering period in a complementary way to ensure continuity of yield under different condi-tions. In Switzerland, recommendations for sowing in meadows with suitability assessments for different elevations are regularly published (Suter et al. 2017).

As accompanying measures of grassland management, also fire (see Chap. 7) and irrigation play an important role in some regions. Controlled burning is still a widespread method to prevent scrub encroachment on landscapes and to accelerate the return of nutrients into the soil, although nutrient losses may occur during heavy fire events as a result of volatilization and increased leaching (Wang et al. 2001; Alcañiz et al. 2018). In Mediterranean and semi-arid regions, wildfires can also occur. In subtropical savannahs fire regimes are a central component of vegetation dynamics. The risk of fire occurrence is often increased by climate change (espe-cially drought events and heatwaves) and by invasive species (e.g. *Bromus tectorum* L. in the United States, Ziska et al. 2005). Up until the twentieth century, irrigation was an important measure for increasing grassland productivity and also for pest control in many parts of Central Europe and in the Mediterranean regions. Regular irrigation leads to an increase in the proportion of rosette plants and legumes and to an improvement in the structural richness of the vegetation (Müller et al. 2016).

15.3 Adaptation of Species and Plant Communities to Grassland Management

15.3.1 Adjustments of Species Characteristics

Plant species are usually well adapted to the disturbance regime in grasslands. This concerns, among other traits, their growth form and height, their phenology, and also their ability to regenerate after disturbances. Globally, grazing favours plants that regenerate from runners and plants with a rosette growth form, as well as small or prostrate species (Díaz et al. 2007), and thus species with meristems near the ground. As a result of the fertilization often associated with mowing, more fast- and tall-growing plant species with higher leaf nitrogen concentrations are found in mown meadows, and this is associated with soil microbial communities dominated by bacteria. However, on extensively managed and abandoned meadows and pastures, there are more slow-growing plant species and soil microbial communities that are more strongly dominated by fungi; these are associated with reduced productivity and ecosystem turnover rates of carbon and nitrogen (Fig. 15.8; De Deyn et al. 2008; Grigulis et al. 2013; Smith et al. 2014).

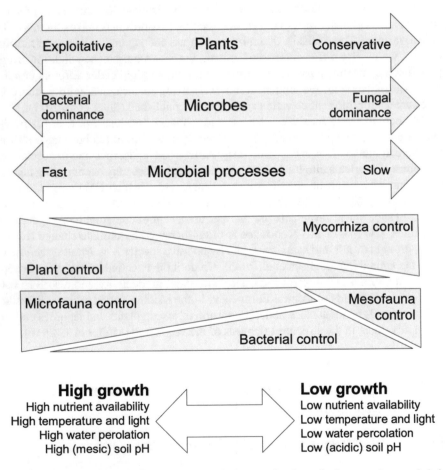

Fig. 15.8 Trends in the dominant strategies of plants, animals, and microorganisms and their influences on ecosystem processes between intensively (left) and lightly (right) managed grasslands (Schade et al. 2005; Grigulis et al. 2013; Mulder et al. 2013)

Evolutionary adaptations of phenology and morphology to grazing and mowing can be clearly seen in the example of the ecotypes of the annual species *Rhinanthus alectorolophus* (Scop.) Pollich, which is typical of Central European grasslands (Fig. 15.9); a later cutting time combined with fertilization leads to a later flowering and seed ripening, and because of the increased light competition, the internodes and shoots are longer. Grazing also results in a lower plant height and a pronounced branching of the shoots (Fig. 15.9). A strong branching of the shoots reduces the probability that all aboveground biomass is lost at moderate grazing intensity. In meadows, plant height and phenology are sometimes closely linked; plants with a rosette growth form develop and usually flower rapidly at the beginning of the growing season, whereas tall-growing herbaceous species and grasses invest more biomass in the shoot and thus dominate the upper stand layers during the ripening phase (McIntyre et al. 1995; Klimešová et al. 2008).

For rapid regeneration after grazing and mowing, perennial grassland species can often rely on considerable carbohydrate stores in rhizomes and tubers (Larcher 2001). Through compensatory growth, lost plant parts can often be, at least partially, replaced; in rare cases, defoliated plants can even develop a larger aboveground biomass than undisturbed plants (Ferraro and Oesterheld 2002). The regeneration of plant species and their biomass depends strongly on the intensity and frequency of the disturbance, the availability of nutrients, and the life history stage of the plant (Strauss and Agrawal 1999; Del-Val and Crawley 2005; Wise and Abrahamson 2005).

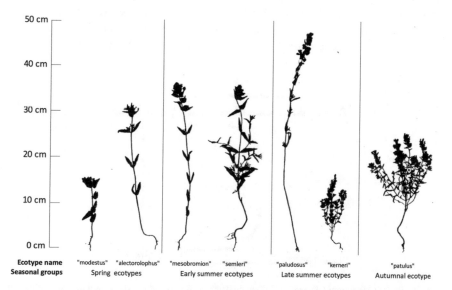

Fig. 15.9 Eco-morphotypes of the annual grassland species European yellow-rattle (*Rhinanthus alectorolophus* (Scop.) Pollich) in response to mowing and fertilization (ecotypes "alectorolophus" and "mesobromion") and grazing ("kerneri" and "patulus"). (Zopfi 1993)

15.3.2 Grassland Management and Biodiversity

For many animal and plant species characteristic of open habitats, the once widespread extensively managed meadow and pasture landscapes in Europe represent important semi-natural habitats prone to secondary succession. However, over the course of the land-use changes that have occurred over recent decades, extensively used areas have often been converted into intensively used grasslands. Thereby, anthropogenic land use as human disturbance regimes have had a significant impact on the biodiversity and productivity of grasslands. For example, high soil fertility allows for high aboveground biomass production, but high productivity increases the competitive pressure on slow-growing and small plants, resulting in a significant reduction in species richness (Fig. 15.10; Gross 2016; Harpole et al. 2016). Evidence from a global study suggests that the highest species numbers are found in grassland systems of moderate productivity (Fig. 15.11, "humped-back model"; Adler et al. 2011; Fraser et al. 2015).

A high number of species can stabilize the biomass production of grasslands in the long term (Hector et al. 2010; Gross et al. 2014). In doing so, species richness increases functional resilience (see Chap. 5) and the resistance of grasslands to climatic changes, including extreme weather events (Isbell et al. 2015). The beneficial effect of high biodiversity on grassland productivity and stability is attributed to various mechanisms, including (a) the degree of asynchronous behaviour of species of a plant community after disturbance, (b) insurance effects through complementary plant strategies from fast-growing to stress-tolerant species, (c) the overcompensation of individual species in the event of disturbances by reducing competitive pressure, and (d) the probability that a particularly productive species and a greater degree of overall species complementarity will occur when the number of species is

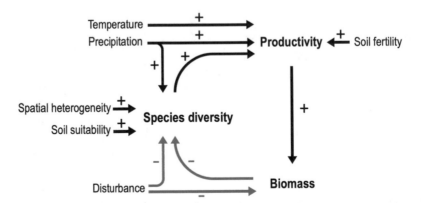

Fig. 15.10 Anthropogenic and natural disturbance regimes significantly influence the biodiversity and productivity of grasslands. Disturbances can reduce species diversity or indirectly increase it by reducing the standing biomass. A higher spatial heterogeneity, for example, promoted by moderate grazing intensity, also favours higher biodiversity. (After Gross 2016, redrawn)

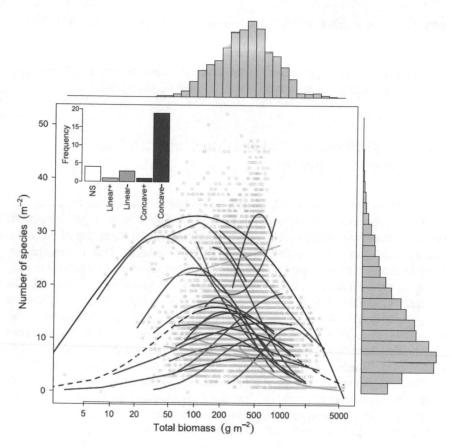

Fig. 15.11 Biomass production in grasslands in relation to the number of species for 28 locations on five continents. The black line shows the general pattern, while the coloured lines represent the regression curves for the individual sites and are predominantly concave in shape (red). (From Fraser et al. 2015)

high (Yachi and Loreau 1999; Lehman and Tilman 2000; Loreau and de Mazancourt 2008; de Mazancourt et al. 2013; Hautier et al. 2014). Measures to maintain biodiversity in grasslands, such as seed supplementation, fertilizer reduction, or changes in the frequency of mowing thus serve the continuity of biomass production in times of changing environmental conditions (see also Moog et al. 2002; Socher et al. 2013).

In addition to the number of species, the functional species composition of grassland vegetation contributes significantly to the continuity of performance under changing disturbance regimes resulting from, for example, land use and climate changes. A distinction is made between species that have developed fast, acquisitive strategies with respect to resource uptake and conversion—often accompanied by high specific leaf area, high leaf nitrogen concentrations, and low leaf mass—and species that have developed slow, conservative strategies in this respect (Grime

1979; Wright et al. 2004; Reich 2014; Díaz et al. 2016; see also Fig. 15.8). Thus, a higher diversity of functional traits in grassland plant communities also increases the complementarity of temporal strategies for the uptake of limiting resources such as water, nutrients, and light, or for regeneration after disturbances, thereby enabling the continuity of biomass production over time and between years.

15.4 Interactions with Other Disturbance Regimes (Weather Extremes)

Climate change is expected to lead to increased variability in the frequency and amount of precipitation in the coming decades, leading to both longer dry periods and also more heavy rainfall events. Both the extent and the frequency of precipitation extremes can influence the productivity of grasslands. This is increasingly being researched in the context of nationally and internationally coordinated experiments on the effects of climate change on ecosystem functions (Fig. 15.12). The resilience of biomass production to drought events is strongly influenced by species composition and biodiversity (Grime et al. 2000; Kahmen et al. 2005; Jentsch et al. 2011; Isbell et al. 2015; Craven et al. 2018). In addition, precipitation anomalies can alter biotic interactions between plant species and across different trophic levels, leading to shifts in the competitive balance (Gilgen et al. 2010; Grant et al. 2014).

Water scarcity and the composition of plant communities have been suggested to change the availability of plant nutrients and the accumulation of secondary metabolites supporting defence against herbivorous insects (Walter et al. 2012a; Van Sundert et al. 2020), though the effects of drought stress are still largely unexplored.

Fig. 15.12 Experimental disturbance ecology: Research on the effects of extreme weather events, land-use changes, and biodiversity on grassland ecosystem functions (Photo: C. Schaller)

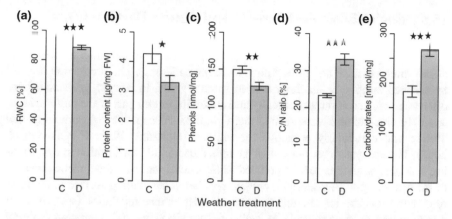

Fig. 15.13 Influence of drought stress (D; compared to control C) on (**a**) relative leaf water content, (**b**) leaf protein content, (**c**) soluble phenol content, (**d**) C/N ratio, and (**e**) carbon content in the leaves of the meadow soft grass (*Holcus lanatus* L.). (From Walter et al. 2012b)

Drought reduces the relative water content, soluble protein content, and nitrogen and phenol content, and increases the proportion of carbohydrates in the grass (Fig. 15.13). As a consequence, larvae that feed on drought-affected leaves have a longer larval phase, increased pupal weight, and higher pupation rates (Walter et al. 2012b).

Drought and heavy rainfall can also reduce litter decomposition rates by affecting litter quality and soil biological activity (Joos et al. 2010; Walter et al. 2013). Different microbial groups react differently to drought events. For example, fungal communities, including mycorrhizae, are more drought resistant than bacterial communities (Fuchslueger et al. 2014; Walter et al. 2016). Since grassland management promotes bacterial versus fungal communities, managed meadows are more affected by drought than abandoned grasslands (Karlowsky et al. 2018). In the course of soil rewetting after drought, significant amounts of microbially bound nitrogen are often released. This can promote the regrowth of vegetation and, thus, productivity after drought events (Ingrisch et al. 2018), especially under the scenario of future increased atmospheric CO_2 concentrations (Roy et al. 2016).

In regions with frequent drought events, irrigation plays an important role as a compensatory management measure. Increased mowing frequency can also buffer the adverse effects of increased precipitation variability on productivity, but the effects are weak and usually only temporary.

15.5   Social-Ecological Aspects and Future Developments

Worldwide, grasslands are of great importance for livestock farming, for which the added value via meat and milk production is considerable. The production of wool from sheep farming is also important in some regions. In Europe, grasslands have often been created anthropogenically and are an important part of the cultural landscape. In mountainous areas such as the Alps, the landscape is richly structured and diverse and particularly attractive for tourism (Hunziker 1995; Hunziker et al. 2008). The ecosystem services of grasslands thus include, in addition to provisioning services, numerous services related to the conservation of the genetic diversity of the biota (supporting service), climate regulation (regulating service in the form of carbon sequestration and energy balance), disturbance regulation (such as water retention and erosion control), as well as aesthetic value and recreation (cultural service), the latter mainly through extensively used grasslands.

Depending on the type and intensity of management, the different ecosystem services are affected to different degrees (Fig. 15.14) and are further affected by global change (Lamarque et al. 2014; Egarter Vigl et al. 2016; Thonicke et al. 2020). Thus, socio-economic changes in the past decades have often led to an intensification of land use but also to the abandonment of less accessible grassland areas. On the one hand, intensification increases productivity, but on the other hand, it increases greenhouse gas emissions and nitrogen pollution of groundwater (see Sect. 15.2), and it reduces biodiversity. The abandonment of grasslands in turn leads to shrub encroachment and reforestation, increasing biodiversity in the initial phases of secondary succession but then reducing species numbers as secondary succession progresses and the increased dominance of shrubs and trees increases. Because of the relatively high resilience of grassland systems with respect to climate change, their importance for livestock farming, their biodiversity, and their cultural services, grasslands will continue to play an important social role in the future, although its development in Europe will continue to be strongly influenced by the socio-economic framework conditions (see Box 15.1 "Too much of a good thing—the decline of species-rich hay meadows in Central Europe").

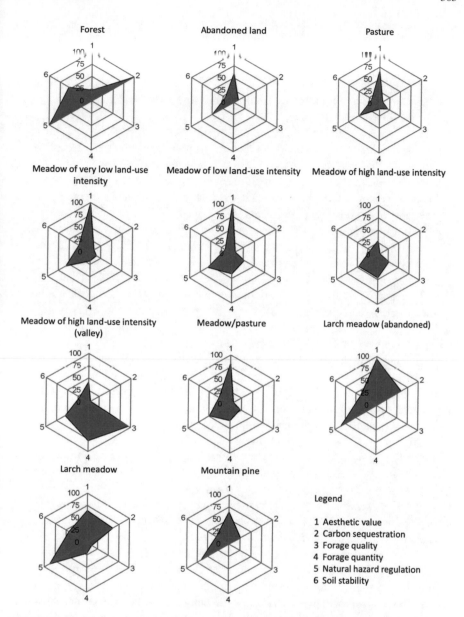

Fig. 15.14 Ecosystem services of differently managed and abandoned grassland systems in a cultural landscape in the Alps. (Schirpke et al. 2013)

Box 15.1: Too Much of a Good Thing: The Decline of Species-Rich Hay Meadows in Central Europe

Andreas Bosshard

Ö+L Ökologie und Landschaft GmbH, Oberwil-Lieli, Switzerland

e-mail: ab@agraroekologie.ch

The development of grasslands in Central Europe is a good example of the important role that disturbances play in biodiversity. Below the timberline, natural grasslands cover only small areas. In contrast, anthropogenic grasslands represent one of the region's most important habitats—with regard to both area and biodiversity. Grassland types have evolved and diversified as a function of infrequent to frequent human disturbances. These disturbances consist mainly of mowing or targeted grazing. Accordingly, the anthropogenic grasslands can be subdivided into three types: meadows, pastures, and mowed pastures.

These more or less extensively managed grassland types are clearly more species-rich than undisturbed forests in terms of plant biodiversity (e.g. Wilson et al. 2012). In mountain regions, where many meadows and pastures have been abandoned by farmers, forests have spread rapidly in recent decades, and as a consequence, biodiversity has declined sharply in some areas (Stöcklin et al. 2007). For centuries, anthropogenic meadows and pastures of a variety of habitat types have accounted for the largest proportion of biodiversity of Central Europe. It is assumed that grassland biodiversity increased steadily until the end of the nineteenth century, especially because the meadow uses became more diverse. For example, by the eighteenth century, the improved three-field system of crop rotation[1] generated an intensification in agriculture, which resulted in new types of meadows and thus new habitats—false oat-grass (*Arrhenatherum elatius* (L.) P.Beauv. ex J.Presl & C.Presl,) or lowland hay meadows (association *Arrhenatheretum elatioris*) and golden oat-grass (*Trisetum flavescens* (L.) P.Beauv.) or mountain hay meadows (association *Trisetetum flavescentis*).

Until the 1950s, the false oat-grass meadow was the most abundant meadow type on productive soils at lower elevations in Central Europe. These meadows were usually cut twice a year for traditional hay production. The meadows were regularly fertilized with dung and, when available, slurry. Still, in the first half of the twentieth century, this meadow type covered large

---

[1] In the three-field rotation, the available land was divided into three sections. Each year, one section was planted with cereals, one section was planted with legumes (e.g. peas, beans, clover), and one section was left fallow. The sections were rotated on a 3-year rotation.

(continued)

Box 15.1 (continued)

parts of the cultural landscape of Central Europe and formed the basis for fodder production in dairy farming.

False oat-grass as well as the golden oat-grass meadows were not only productive but also rich in species—more diverse than most of the meadows that are subsidized by, for example, the Swiss government for biodiversity reasons today. Within a few decades after the end of World War II, false oat-grass meadows had almost disappeared as a consequence of agricultural intensification (Bosshard et al. 2011). No other habitat type in Switzerland and many other parts of Central Europe has declined so rapidly in such a short time.

Today, more or less typical false oat-grass meadows in Switzerland cover on average only 1–2% of the agricultural area (Box Fig. 1; Bosshard 2016). They have been widely replaced by more productive but also conspicuously species-poor, intensive meadows. Instead of two to three annual cuts, they are now mowed or grazed intensively four to six times a year. Many formerly characteristic and widespread plant species, including a large number of attractive meadow flowers such as oxeye daisy/marguerite (*Leucanthemum vulgare* Lam.), sage (*Salvia pratensis* L.), bell flowers (species of the genus *Campanula*) or knapweeds (species of the genus *Centaurea*), have disappeared completely from intensively used grasslands.

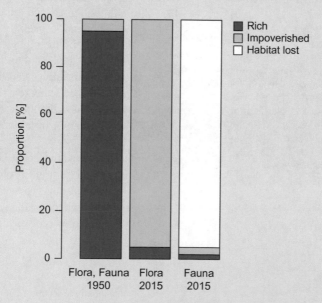

Box Fig. 1 Biodiversity in hay meadows of the lowlands in Switzerland in 1950 and 2015

(continued)

Box 15.1 (continued)

Even more than the flora, the fauna has been affected by the intensified use of meadowland. Species from practically all animal groups that used the false oat-grass meadows as the main habitat until the 1950s, and which accounted for a large part of the species diversity in the cultural landscape, are not able to survive in today's intensively managed meadows. These include grasshoppers and crickets, butterflies, vegetation-inhabiting spiders, and ground-breeding bird species. The populations of these groups in the meadows at lower elevations have collapsed over the last 100 years to an average of about 1% of the ancestral populations. Many once typical meadowland species such as the tree pipit (*Anthus trivialis* L.) or the whinchat (*Saxicola rubetra* L.) have completely disappeared over large areas.

The history of grasslands in Central Europe demonstrates the crucial role of disturbances for biodiversity. While abandoned or underutilized grasslands develop into shrub and forests where few species prevail, also intensified use of grasslands (with four to six annual cuts) reduces biodiversity as only a few plant species can tolerate the intensive disturbance regime. On the other hand, considerably more species can thrive at the moderate disturbance intensity between these two extremes—in the extensively managed meadows that are, on rich soils, represented by false oat-grass communities. Both too much and too little disturbance can lead to a reduction in species diversity and, in particular, if such disturbance regimes are applied over large areas.

References

Bosshard A (2016) Das Naturwiesland der Schweiz und Mitteleuropas. Mit besonderer Berücksichtigung der Fromentalwiesen und des standortgemässen Futterbaus (Bristol-Schriftenreihe, Bd. 50). Haupt Verlag, Bern, 265 p

Bosshard A, Schläpfer F, Jenny M (2011) Weissbuch Landwirtschaft Schweiz: Analysen und Vorschläge zur Reform der Agrarpolitik (2nd edn). Haupt Verlag, Bern, 272 p

Stöcklin J, Bosshard A, Klaus G (2007) Landnutzung und biologische Vielfalt in den Alpen: Fakten, Perspektiven, Empfehlungen. vdf Hochschulverlag, Zürich, 192 p

Wilson JB, Peet RK, Dengler J, Pärtel M (2012) Plant species richness: the world records. J Veg Sci 23:796–802

References

Adler PB, Raff D, Lauenroth WK (2001) The effect of grazing on the spatial heterogeneity of vegetation. Oecologia 128:465–479

Adler PB, Seabloom EW, Borer ET, Hillebrand H, Hautier Y, Hector A, Harpole WS, O'Halloran LR, Grace JB, Anderson TM, Bakker JD, Biederman LA, Brown CS, Buckley YM, Calabrese LB, Chu CJ, Cleland EE, Collins SL, Cottingham KL, Crawley MJ, Damschen EI, Davies KF, DeCrappeo NM, Fay PA, Firn J, Frater P, Gasarch EI, Gruner DS, Hagenah N, Lambers JHR, Humphries H, Jin VL, Kay AD, Kirkman KP, Klein JA, Knops JMH, La Pierre KJ, Lambrinos JG, Li W, MacDougall AS, McCulley RL, Melbourne BA, Mitchell CE, Moore JL, Morgan JW, Mortensen B, Orrock JL, Prober SM, Pyke DA, Risch AC, Schütz M, Smith MD, Stevens CJ, Sullivan LL, Wang G, Wragg PD, Wright JP, Yang LH (2011) Productivity is a poor predictor of plant species richness. Science 333:1750–1753

Alcañiz M, Outeiro L, Francos M, Úbeda X (2018) Effects of prescribed fires on soil properties. Sci Total Environ 613–614:944–957

Allard V, Soussana J-F, Falcimagne R, Berbigier P, Bonnefond JM, Ceschia E, D'hour P, Hénault C, Laville P, Martin C, Pinarès-Patino C (2007) The role of grazing management for the net biome productivity and greenhouse gas budget (CO_2, N_2O and CH_4) of semi-natural grassland. Agric Ecosyst Environ 121:47–58

Arfin Khan MAS, Grant K, Beierkuhnlein C, Kreyling J, Jentsch A (2014) Climatic extremes lead to species-specific legume facilitation in an experimental temperate grassland. Plant Soil 379:161–175

Asner GP, Elmore AJ, Olander LP, Martin RE, Harris AT (2004) Grazing systems, ecosystem responses, and global change. Annu Rev Environ Resour 29:261–299

Bahn M, Knapp M, Garajova Z, Pfahringer N, Cernusca A (2006) Root respiration in temperate mountain grasslands differing in land use. Glob Chang Biol 12:995–1006

Bai Y, Wu J, Clark CM, Pan Q, Zhang L, Chen S, Wang Q, Han X, Wisley B (2012) Grazing alters ecosystem functioning and C. J Appl Ecol 49:1204–1215

Bakker ES, Ritchie ME, Olff H, Milchunas DG, Knops JMH (2006) Herbivore impact on grassland plant diversity depends on habitat productivity and herbivore size. Ecol Lett 9:780–788

Bardgett RD, Wardle DA (2003) Herbivore-mediated linkages between aboveground and belowground communities. Ecology 84:2258–2268

Bardgett RD, Wardle DA (2010) Aboveground-belowground linkages: biotic interactions, ecosystem processes, and global change. Oxford University Press, Oxford, 301 p

Blüthgen N, Dormann CF, Prati D, Klaus VH, Kleinebecker T, Hölzel N, Alt F, Boch S, Gockel S, Hemp A, Müller J, Nieschulze J, Renner SC, Schöning I, Schumacher U, Socher SA, Wells K, Birkhofer K, Buscot F, Oelmann Y, Rothenwöhrer C, Scherber C, Tscharntke T, Weiner CN, Fischer M, Kalko EKV, Linsenmair KE, Schulze E-D, Weisser WW (2012) A quantitative index of land-use intensity in grasslands. Basic Appl Ecol 13:207–220

Bremer DJ, Auen LM, Ham JM, Owensby CE (2001) Evapotranspiration in a Prairie ecosystem. Agron J 93:338–348

Chang J, Ciais P, Viovy N, Vuichard N, Sultan B, Soussana J-F (2015) The greenhouse gas balance of European grasslands. Glob Chang Biol 21:3748–3761

Craven D, Eisenhauer N, Pearse WD, Hautier Y, Isbell F, Roscher C et al (2018) Multiple facets of biodiversity drive the diversity–stability relationship. Nat Ecol Evo 2:1579–1587

De Deyn GB, Cornelissen JHC, Bardgett RD (2008) Plant functional traits and soil carbon sequestration in contrasting biomes. Ecol Lett 11:516–531

de Mazancourt C, Isbell F, Larocque A, Berendse F, Luca E, Grace JB, Haegeman B, Wayne Polley H, Roscher C, Schmid B, Tilman D, Van Ruijven J, Weigelt A, Wilsey B, Loreau M (2013) Predicting ecosystem stability from community composition and biodiversity. Ecol Lett 16:617–625

Del-Val EK, Crawley MJ (2005) Are grazing increaser species better tolerators than decreasers? J Ecol 93:1005–1016

Di HJ, Cameron KC (2002) Nitrate leaching in temperate agroecosystems: sources, factors and mitigating strategies. Nutr Cycl Agroecosyst 64:237–256

Díaz S, Lavorel S, McIntyre S, Falczuk V, Casanoves F, Milchunas DG, Skarpe C, Rusch G, Sternberg M, Noy-Meir I, Landsberg J, Zhang W, Clark H, Campbell BD (2007) Plant trait responses to grazing—a global synthesis. Glob Chang Biol 13:313–341

Díaz S, Kattge J, Cornelissen JHC, Wright IJ, Lavorel S, Dray S, Reu B, Kleyer M, Wirth C, Colin Prentice I, Garnier E, Bönisch G, Westoby M, Poorter H, Reich PB, Moles AT, Dickie J, Gillison AN, Zanne AE, Chave J, Joseph Wright S, Sheremet'ev SN, Jactel H, Baraloto C, Cerabolini B, Pierce S, Shipley B, Kirkup D, Casanoves F, Joswig JS, Günther A, Falczuk V, Rüger N, Mahecha MD, Gorné LD (2016) The global spectrum of plant form and function. Nature 529:167–171

Egarter Vigl L, Schirpke U, Tasser E, Tappeiner U (2016) Linking long-term landscape dynamics to the multiple interactions among ecosystem services in the European Alps. Landsc Ecol 31:1903–1918

Fernandes PM, Davies GM, Ascoli D, Fernandez C, Moreira F, Rigolot E, Stoof CR, Vega JA, Molina D (2013) Prescribed burning in southern Europe: developing fire management in a dynamic landscape. Front Ecol Environ 11:E4–E14

Ferraro DO, Oesterheld M (2002) Effect of defoliation on grass growth. A quantitative review. Oikos 98:125–133

Fraser LH, Pither J, Jentsch A, Sternberg M, Zobel M, Askarizadeh D et al (2015) Worldwide evidence of a unimodal relationship between productivity and plant species richness. Science 349:302–305

Fuchslueger L, Bahn M, Fritz K, Hasibeder R, Richter A (2014) Experimental drought reduces the transfer of recently fixed plant carbon to soil microbes and alters the bacterial community composition in a mountain meadow. New Phytol 201:916–927

Gilgen AK, Signarbieux C, Feller U, Buchmann N (2010) Competitive advantage of *Rumex obtusifolius* L. might increase in intensively managed temperate grasslands under drier climate. Agric Ecosyst Environ 135:15–23

Goulding KWT (2016) Soil acidification and the importance of liming agricultural soils with particular reference to the United Kingdom. Soil Use Manag 32:390–399

Grant K, Kreyling J, Heilmeier H, Beierkuhnlein C, Jentsch A (2014) Extreme weather events and plant–plant interactions. Ecol Res 29:991–1001

Grigulis K, Lavorel S, Krainer U, Legay N, Baxendale C, Dumont M, Kastl E-M, Arnoldi C, Bardgett RD, Poly F, Pommier T, Schloter M, Tappeiner U, Bahn M, Clément J-C, Hutchings M (2013) Relative contributions of plant traits and soil microbial properties to mountain grassland ecosystem services. J Ecol 101:47–57

Grime JP (1979) Plant strategies and vegetation processes. Wiley, Chichester, 222 p

Grime JP, Brown VK, Thompson K, Masters GJ, Hillier SH, Clarke IP, Askew AP, Corker D, Kielty JP (2000) The response of two contrasting limestone grasslands to simulated climate change. Science 289:762–765

Gross K (2016) Biodiversity and productivity entwined. Nature 529:293–294

Gross K, Cardinale BJ, Fox JW, Gonzalez A, Loreau M, Polley HW, Reich PB, Van Ruijven J (2014) Species richness and the temporal stability of biomass production: a new analysis of recent biodiversity experiments. Am Nat 183:1–12

Harpole WS, Sullivan LL, Lind EM, Firn J, Adler PB, Borer ET et al (2016) Addition of multiple limiting resources reduces grassland diversity. Nature 537:93–96

Harris E, Diaz-Pines E, Stoll E, Schloter M, Schulz S, Duffner C et al (2021) Denitrifying pathways dominate nitrous oxide emissions from managed grassland during drought and rewetting. Sci Adv 7(6). https://doi.org/10.1126/sciadv.abb7118

Hautier Y, Seabloom EW, Borer ET, Adler PB, Harpole WS, Hillebrand H, Lind EM, MacDougall AS, Stevens CJ, Bakker JD, Buckley YM, Chu C-J, Collins SL, Daleo P, Damschen EI, Davies KF, Fay PA, Firn J, Gruner DS, Jin VL, Klein JA, Knops JMH, La Pierre KJ, Li W, McCulley RL, Melbourne BA, Moore JL, O'Halloran LR, Prober SM, Risch AC, Sankaran M, Schuetz

M, Hector A (2014) Eutrophication weakens stabilizing effects of diversity in natural grasslands. Nature 508:521–525

Hector A, Hautier Y, Saner P, Wacker L, Bagchi R, Joshi J, Scherer-Lorenzen M, Spehn EM, Bazeley-White E, Weilenmann M, Caldeira MC, Dimitrakopoulos PG, Finn JA, Huss-Danell K, Jumpponen A, Mulder CPH, Palmborg C, Pereira JS, Siamantziouras ASD, Terry AC, Troumbis AY, Schmid B, Loreau M (2010) General stabilizing effects of plant diversity on grassland productivity through population asynchrony and overyielding. Ecology 91:2213–2220

Heyburn J, McKenzie P, Crawley MJ, Fornara DA (2017) Long-term belowground effects of grassland management. Ecol Appl 27:2001–2012

Hörtnagl L, Barthel M, Buchmann N, Eugster W, Butterbach-Bahl K, Díaz-Pinés E, Zeeman M, Klumpp K, Kiese R, Bahn M, Hammerle A, Lu H, Ladreiter-Knauss T, Burri S, Merbold L (2018) Greenhouse gas fluxes over managed grasslands in Central Europe. Glob Chang Biol 24:1843–1872

Hunziker M (1995) The spontaneous reafforestation in abandoned agricultural lands. Landsc Urban Plan 31:399–410

Hunziker M, Felber P, Gehring K, Buchecker M, Bauer N, Kienast F (2008) Evaluation of landscape change by different social groups. Mt Res Dev 28:140–147

Ingrisch J, Karlowsky S, Anadon-Rosell A, Hasibeder R, König A, Augusti A, Gleixner G, Bahn M (2018) Land use alters the drought responses of productivity and CO_2 fluxes in mountain grassland. Ecosystems 21:689–703

Isbell F, Craven D, Connolly J, Loreau M, Schmid B, Beierkuhnlein C, Bezemer TM, Bonin C, Bruelheide H, De Luca E, Ebeling A, Griffin JN, Guo Q, Hautier Y, Hector A, Jentsch A, Kreyling J, Lanta V, Manning P, Meyer ST, Mori AS, Naeem S, Niklaus PA, Polley HW, Reich PB, Roscher C, Seabloom EW, Smith MD, Thakur MP, Tilman D, Tracy BF, van der Putten WH, van Ruijven J, Weigelt A, Weisser WW, Wilsey B, Eisenhauer N (2015) Biodiversity increases the resistance of ecosystem productivity to climate extremes. Nature 526:574–577

Jentsch A, Kreyling J, Elmer M, Gellesch E, Glaser B, Grant K, Hein R, Lara M, Mirzae H, Nadler SE, Nagy L, Otieno D, Pritsch K, Rascher U, Schädler M, Schloter M, Singh BK, Stadler J, Walter J, Wellstein C, Wöllecke J, Beierkuhnlein C (2011) Climate extremes initiate ecosystem-regulating functions while maintaining productivity. J Ecol 99:689–702

Joos O, Hagedorn F, Heim A, Gilgen AK, Schmidt MWI, Siegwolf RTW, Buchmann N (2010) Summer drought reduces total and litter-derived soil CO_2 effluxes in temperate grassland—clues from a 13C litter addition experiment. Biogeosciences 7:1031–1041

Kahmen A, Perner J, Buchmann N (2005) Diversity-dependent productivity in semi-natural grasslands following climate perturbations. Funct Ecol 19:594–601

Karlowsky S, Augusti A, Ingrisch J, Hasibeder R, Lange M, Lavorel S, Bahn M, Gleixner G (2018) Land use in mountain grasslands alters drought response and recovery of carbon allocation and plant-microbial interactions. J Ecol 106:1230–1243

Klein JA, Harte J, Zhao X-Q (2005) Dynamic and complex microclimate responses to warming and grazing manipulations. Glob Chang Biol 11:1440–1451

Klimešová J, Latzel V, De Bello F, van Groenendael JM (2008) Plant functional traits in studies of vegetation changes in response to grazing and mowing: towards a use of more specific traits. Preslia:245–253

Klumpp K, Tallec T, Guix N, Soussana J-F (2011) Long-term impacts of agricultural practices and climatic variability on carbon storage in a permanent pasture. Glob Chang Biol 17:3534–3545

Kübert A, Götz M, Kuester E, Piayda A, Werner C, Rothfuss Y, Dubbert M (2019) Nitrogen loading enhances stress impact of drought on a semi-natural temperate grassland. Front Plant Sci 10:1051. https://doi.org/10.3389/fpls.2019.01051

Kuzyakov Y (2010) Priming effects: interactions between living and dead organic matter. Soil Biol Biochem 42:1363–1371

Lamarque P, Lavorel S, Mouchet M, Quetier F (2014) Plant trait-based models identify direct and indirect effects of climate change on bundles of grassland ecosystem services. Proc Natl Acad Sci USA 111:13751–13756

Larcher W (2001) Ökophysiologie der Pflanzen: Leben, Leistung und Stressbewältigung der Pflanzen in ihrer Umwelt (6th edition). UTB; Ulmer, Stuttgart, 408 p

Le Roux X, Poly F, Currey P, Commeaux C, Hai B, Nicol GW, Prosser JI, Schloter M, Attard E, Klumpp K (2008) Effects of aboveground grazing on coupling among nitrifier activity, abundance and community structure. ISME J 2:221–232

Legay N, Lavorel S, Baxendale C, Krainer U, Bahn M, Binet M-N, Cantarel AAM, Colace M-P, Foulquier A, Kastl E-M, Grigulis K, Mouhamadou B, Poly F, Pommier T, Schloter M, Clément J-C, Bardgett RD (2016) Influence of plant traits, soil microbial properties, and abiotic parameters on nitrogen turnover of grassland ecosystems. Ecosphere 7:e01448

Lehman CL, Tilman D (2000) Biodiversity, stability, and productivity in competitive communities. Am Nat 156:534–552

Loreau M, de Mazancourt C (2008) Species synchrony and its drivers: neutral and nonneutral community dynamics in fluctuating environments. Am Nat 172:E48–E66

Lüscher A, Mueller-Harvey I, Soussana J-F, Rees RM, Peyraud JL (2014) Potential of legume-based grassland-livestock systems in Europe. Grass Forage Sci 69:206–228

McIntyre S, Lavorel S, Tremont RM (1995) Plant life-history attributes. J Ecol 83:31–44

McSherry ME, Ritchie ME (2013) Effects of grazing on grassland soil carbon: a global review. Glob Chang Biol 19:1347–1357

Meyer ST, Leifeld J, Bahn M, Fuhrer J (2012) Free and protected soil organic carbon dynamics respond differently to abandonment of mountain grassland. Biogeosciences 9:853–865

Milchunas DG, Lauenroth WK (1993) Quantitative effects of grazing on vegetation and soils over a global range of environments. Ecol Monogr 63:327–366

Moog D, Poschlod P, Kahmen S, Schreiber KF (2002) Comparison of species composition between different grassland management treatments after 25 years. Appl Veg Sci 5:99–106

Mosier A, Syers JK, Freney JR (2004) Agriculture and the nitrogen cycle. Island Press, Washington, DC, 296 p

Mulder C, Ahrestani FS, Bahn M, Bohan DA, Bonkowski M, Griffiths BS, Guicharnaud RA, Kattge J, Krogh PH, Lavorel S, Lewis OT, Mancinelli G, Naeem S, Peñuelas J, Poorter H, Reich PB, Rossi L, Rusch GM, Sardans J, Wright IJ (2013) Connecting the green and brown worlds. Allometric and stoichiometric predictability of above- and below-ground networks. Adv Ecol Res 49:69–175

Müller MM, Hamberg L, Hantula J (2016) The susceptibility of European tree species to invasive Asian pathogens: a literature based analysis. Biol Invasions 18:2841–2851

Oesterheld M, Loreti J, Semmartin M, Paruelo JM (1999) Grazing, fire and climate effects on primary productivity of grasslands and savannas. In: Walker LR (ed) Ecosystems of disturbed ground. Elsevier, Amsterdam/New York, pp 287–306

Opitz Von Boberfeld W (1994) Grünlandlehre. E. Ulmer, Stuttgart, 336 p

Paz-Ferreiro J, Medina-Roldán E, Ostle NJ, McNamara NP, Bardgett RD (2012) Grazing increases the temperature sensitivity of soil organic matter decomposition in a temperate grassland. Environ Res Lett 7:014027

Poeplau C, Don A (2013) Sensitivity of soil organic carbon stocks and fractions to different land-use changes across Europe. Geoderma 192:189–201

Poschlod P, WallisDeVries MF (2002) The historical and socioeconomic perspective of calcareous grasslands—lessons from the distant and recent past. Biol Conserv 104:361–376

Reich PB (2014) The world-wide 'fast-slow' plant economics spectrum: a traits manifesto. J Ecol 102:275–301

Reinhart KO, Dangi SR, Vermeire LT (2016) The effect of fire intensity, nutrients, soil microbes, and spatial distance on grassland productivity. Plant Soil 409:203–216

Robson TM, Lavorel S, Clement J-C, Le Roux X (2007) Neglect of mowing and manuring leads to slower nitrogen cycling in subalpine grasslands. Soil Biol Biochem 39:930–941

Rose A (2004) Economic principles, issues, and research priorities in hazard loss estimation. In: Okuyama Y, Chang SE (eds) Modeling spatial and economic impacts of disasters. Springer, Berlin, pp 13–36

Roy J, Picon-Cochard C, Augusti A, Benot M-L, Thiery L, Darsonville O, Landais D, Piel C, Defossez M, Devidal S, Escape C, Ravel O, Fromin N, Volaire F, Milcu A, Bahn M, Soussana

J-F (2016) Elevated CO_2 maintains grassland net carbon uptake under a future heat and drought extreme. Proc Natl Acad Sci USA 113:6224–6229

Schade JD, Espeleta JF, Klausmeier CA, McGroddy ME, Thomas SA, Zhang L (2005) A conceptual framework for ecosystem stoichiometry. Oikos 109:40–51

Schirpke U, Leitinger G, Tasser E, Schermer M, Steinbacher M, Tappeiner U (2013) Multiple ecosystem services of a changing Alpine landscape. Int J Biodiv Sci Ecosyst Serv Manag 9:123–135

Schmitt M, Bahn M, Wohlfahrt G, Tappeiner U, Cernusca A (2010) Land use affects the net ecosystem CO_2 exchange and its components in mountain grasslands. Biogeosciences 7:2297–2309

Seabloom EW, Borer ET, Tilman D (2020) Grassland ecosystem recovery after soil disturbance depends on nutrient supply rate. Ecol Lett 23:1756–1765

Semmartin M, Aguiar MR, Distel RA, Moretto AS, Ghersa CM (2004) Litter quality and nutrient cycling affected by grazing-induced species replacements along a precipitation gradient. Oikos 107:148–160

Smith SW, Woodin SJ, Pakeman RJ, Johnson D, van der Wal R (2014) Root traits predict decomposition across a landscape-scale grazing experiment. New Phytol 203:851–862

Socher SA, Prati D, Boch S, Müller J, Baumbach H, Gockel S, Hemp A, Schöning I, Wells K, Buscot F, Kalko EKV, Linsenmair KE, Schulze E-D, Weisser WW, Fischer M (2013) Interacting effects of fertilization, mowing and grazing on plant species diversity of 1500 grasslands in Germany differ between regions. Basic Appl Ecol 14:126–136

Soussana J-F, Allard V, Pilegaard K, Ambus P, Amman C, Campbell C, Ceschia E, Clifton-Brown J, Czobel S, Domingues R, Flechard C, Fuhrer J, Hensen A, Horvath L, Jones M, Kasper G, Martin C, Nagy Z, Neftel A, Raschi A, Baronti S, Rees RM, Skiba UM, Stefani P, Manca G, Sutton M, Tuba Z, Valentini R (2007) Full accounting of the greenhouse gas (CO_2, N_2O, CH_4) budget of nine European grassland sites. Agric Ecosyst Environ 121:121–134

Strauss SY, Agrawal AA (1999) The ecology and evolution of plant tolerance to herbivory. Trends Ecol Evol 14:179–185

Suter D, Frick R, Hirschi H-U, Bertossa M (2017) Liste der empfohlenen Sorten von Futterpflanzen 2017–2018. Agrarforsch Schweiz 8(1):1–16

Suttie JM, Reynolds SG, Batello C (2005) Grasslands of the world. Food and Agriculture Organization of the United Nations (FAO), Rome, 514 p

Thonicke K, Bahn M, Lavorel S, Bardgett RD, Erb K-H, Giamberini M, Reichstein M, Vollan B, Rammig A (2020) Advancing the understanding of adaptive capacity of social-ecological systems to absorb climate extremes. Earth's Futures 8:e2019EF001221

Valese E, Conedera M, Held AC, Ascoli D (2014) Fire, humans and landscape in the European Alpine region during the Holocene. Anthropocene 6:63–74

Van Sundert K, Brune V, Bahn M, Deutschmann M, Hasibeder R, Nijs I, Vicca S (2020) Post-drought rewetting triggers substantial K release and shifts in leaf stoichiometry in managed and abandoned mountain grasslands. Plant Soil 448:353–368

Velthof GL, Oenema O (1995) Nitrous oxide fluxes from grassland in the Netherlands. Eur J Soil Sci 46:541–549

Walter J, Grant K, Beierkuhnlein C, Kreyling J, Weber M, Jentsch A (2012a) Increased rainfall variability reduces biomass and forage quality of temperate grassland largely independent of mowing frequency. Agric Ecosyst Environ 148:1–10

Walter J, Hein R, Auge H, Beierkuhnlein C, Löffler S, Reifenrath K, Schädler M, Weber M, Jentsch A (2012b) How do extreme drought and plant community composition affect host plant metabolites and herbivore performance? Arthropod-Plant Interact 6:15–25

Walter J, Hein R, Beierkuhnlein C, Hammerl V, Jentsch A, Schädler M, Schuerings J, Kreyling J (2013) Combined effects of multifactor climate change and land-use on decomposition in temperate grassland. Soil Biol Biochem 60:10–18

Walter J, Kreyling J, Singh BK, Jentsch A (2016) Effects of extreme weather events and legume presence on mycorrhization of *Plantago lanceolata* and *Holcus lanatus* in the field. Plant Biol 18:262–270

Wan S, Luo Y, Wacker L (2002) Changes in microclimate induced by experimental warming and clipping in tallgrass prairie. Glob Chang Biol 8:754–768

Wang C, Gower ST, Wang Y, Zhao H, Yan P, Bond-Lamberty BP (2001) The influence of fire on carbon distribution and net primary production of boreal *Larix gmelinii* forests in north-eastern China. Glob Chang Biol 7:719–730

Wise MJ, Abrahamson WG (2005) Beyond the compensatory continuum. Oikos 109:417–428

Wohlfahrt G, Anderson-Dunn M, Bahn M, Balzarolo M, Berninger F, Campbell C, Carrara A, Cescatti A, Christensen T, Dore S, Eugster W, Friborg T, Furger M, Gianelle D, Gimeno C, Hargreaves K, Hari P, Haslwanter A, Johansson T, Marcolla B, Milford C, Nagy Z, Nemitz E, Rogiers N, Sanz MJ, Siegwolf R, Susiluoto S, Sutton M, Tuba Z, Ugolini F, Valentini R, Zorer R, Cernusca A (2008a) Biotic, abiotic, and management controls on the net ecosystem CO_2 exchange of European mountain grassland ecosystems. Ecosystems 11:1338–1351

Wohlfahrt G, Hammerle A, Haslwanter A, Bahn M, Tappeiner U, Cernusca A (2008b) Seasonal and inter-annual variability of the net ecosystem CO_2 exchange of a temperate mountain grassland: effects of climate and management. J Geophys Res Atmos 113:D08110. https://doi.org/10.1029/2007JD009286

Wright IJ, Reich PB, Westoby M, Ackerly DD, Baruch Z, Bongers F, Cavender-Bares J, Chapin T, Cornelissen H, Diemer M, Flexas J, Garnier E, Groom PK, Gulias J, Hikosaka K, Lamont BB, Lee T, Lee W, Lusk C, Midgley JJ, Navas M-L, Niinemets U, Oleksyn J, Osada N, Poorter H, Poot P, Prior L, Pyankov VI, Roumet C, Thomas SC, Tjoelker MG, Veneklaas EJ, Villar R (2004) The worldwide leaf economics spectrum. Nature 428:821–827

Yachi S, Loreau M (1999) Biodiversity and ecosystem productivity in a fluctuating environment: the insurance hypothesis. Proc Natl Acad Sci USA 96:1463–1468

Ziska LH, Reeves JB, Blank B (2005) The impact of recent increases in atmospheric CO2 on biomass production and vegetative retention of Cheatgrass *(Bromus tectorum)*: implications for fire disturbance. Glob Chang Biol 11:1325–1332

Zopfi H-J (1993) Ecotypic variation in *Rhinanthus alectorolophus* (Scopoli) Pollich (Scrophulariaceae) in relation to grassland management. Flora 188:15–39

Part VI
Disturbances and Global Change

Chapter 16
Impacts of Climate Change on Disturbances

Rupert Seidl (ID) **and Markus Kautz** (ID)

Abstract Disturbances are climate-sensitive processes. If climatic conditions continue to change in the future as predicted, this will also cause changes in disturbance regimes. Direct climatic effects such as accelerated life cycles of bark beetles caused by warmer temperatures may result in larger and/or more frequent disturbances. Interactions between disturbances can further intensify the effects of climate change. Indirectly, however, climate change can also alter the structure and composition of vegetation, which could dampen future disturbances.

Keywords Amplifying and dampening feedbacks · Climate warming · Direct and indirect climate effects · Disturbance change · Disturbance interactions

16.1 Introduction

Both abiotic and biotic disturbances are strongly linked to the prevailing climate conditions (Chaps. 7, 8, 9, 10, 11, 12, and 13). Changes in mean and extreme values as well as in climatic variability, as currently observed and increasingly expected for the coming decades (IPCC 2013), thus have a significant impact on natural disturbance regimes in ecosystems (Dale et al. 2001; Seidl et al. 2017). While ecosystems are generally well adapted to gradually changing climate conditions such as increasing annual average temperatures or precipitation, extreme climate events often cause disturbances (Jentsch et al. 2007). The effects of climate change on

R. Seidl (✉)
Ecosystem Dynamics and Forest Management Group, School of Life Sciences, Technical University of Munich, Freising, Germany

Berchtesgaden National Park, Berchtesgaden, Germany
e-mail: rupert.seidl@tum.de

M. Kautz
Department of Forest Health Protection, Forest Research Institute Baden-Württemberg, Freiburg, Germany

© The Author(s), under exclusive license to Springer Nature
Switzerland AG 2022
T. Wohlgemuth et al. (eds.), *Disturbance Ecology*, Landscape Series 32,
https://doi.org/10.1007/978-3-030-98756-5_16

disturbances are often nonlinear since disturbances frequently occur only when certain climatic thresholds are exceeded. In this chapter, we first identify important climatic drivers of disturbance and then describe their various effects on disturbance regimes. The analysis of the complex interactions between climate, vegetation, and disturbances can ultimately provide insights into expected future trends in disturbance dynamics.

16.2 A Changing Climate

Temperature, precipitation, and wind are among the most important climatic drivers that influence disturbance regimes worldwide (Seidl et al. 2017). These climatic drivers and their expected changes will therefore be discussed in more detail below.

16.2.1 Temperature

Temperature is an important factor influencing a large number of ecological processes. The effects of human-induced climate change include increases in global temperature, combined with the more frequent occurrence of regional weather extremes such as periods of heat and drought. Globally, the annual mean temperature already increased by +0.9 °C between 1901 and 2012. Depending on the future development of human greenhouse gas emissions (Representative Concentration Pathways; RCP scenario), a further increase of between +1.0 °C and +3.7 °C is expected for the end of the twenty-first century (2081–2100, relative to 1986–2005; IPCC 2013). It is therefore very likely that the next decades will be warmer than the previous ones. Even under the assumption of an immediate and complete cessation of anthropogenic greenhouse gas emissions, the climate would continue to warm because of the already increased atmospheric CO_2 concentration and the inertia of the climate system—the effects of human-induced climate change are thus largely irreversible, at least for the next millennium (Solomon et al. 2009).

In the context of climate impacts on ecosystems we note that warming is not evenly distributed across space and time. Land areas and higher latitudes, for instance, warm more strongly than oceans and equatorial regions. In recent decades, the European Alps have warmed about twice as much as the average for the Northern Hemisphere (Auer et al. 2007). For Europe, a temperature increase of +1 °C to +3 °C is predicted for the end of the twenty-first century (2071–2100 relative to 1971–2000, under the moderate emission scenario RCP4.5; Jacob et al. 2014). Strong temperature changes (> + 3 °C) will occur in Southern Europe mainly in summer, while in Northern and Eastern Europe the winter months could become significantly milder in the future (Fig. 16.1). This warming would almost double under a more extreme emissions scenario (RCP8.5) (Jacob et al. 2014). Parallel to a

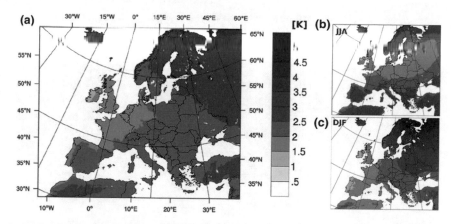

Fig. 16.1 Expected temperature increase in Europe. The maps show (**a**) the change of mean annual temperature and (**b**) mean summer (June–July-August) and (**c**) winter temperature (December–January-February) under the moderate emission scenario RCP4.5 (2071–2100 relative to 1971–2000). (Source: Jacob et al. 2014)

predicted increase in mean temperatures, the probability of heat waves increases, especially in Southern Europe. In addition, the number of annual frost days decreases (Jacob et al. 2014) and the variability of summer temperatures increases, both within a year and between years (Schär et al. 2004; Fischer et al. 2012). Extreme heat, occurring on average only once every 20 years in the past (1961–1990), will become the new normal in Southern Europe (i.e., occurring every 1–2 years in 2071–2100) and will also become significantly more frequent in Northern Europe (occurring every 5 years in 2071–2100). At the same time, past cold extremes will almost completely disappear by the end of the twenty-first century (Nikulin et al. 2011).

Rising temperatures have a predominantly positive effect on a variety of biotic and abiotic disturbance processes, that is, they promote the disturbance activity of insects, pathogens, and fire in particular. Disturbance-inhibiting effects of a warmer climate are less frequent, such as a possible decrease of snow-related disturbances (Gobiet et al. 2014; Seidl et al. 2017). In general, the influence of climate on disturbances is consistent between temperate and boreal ecosystems (Seidl et al. 2020). However, as the northern latitudes will warm faster than the global average (Fig. 16.1; IPCC 2013), an increase of disturbances is expected especially in boreal forests.

16.2.2   Precipitation

Changes in precipitation are much less consistent globally than changes in temperature, and show a slight increase in the global annual precipitation sum of +1.0 to +2.8 mm per decade between 1901 and 2008 (IPCC 2013). For the future, three different patterns of precipitation change are expected in Europe (Fig. 16.2): While

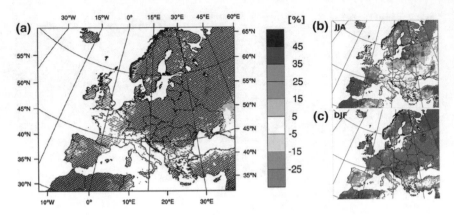

Fig. 16.2 Expected precipitation change in Europe. The maps show (**a**) the change in mean annual precipitation sum and (**b**) mean summer (June–July-August) and (**c**) winter precipitation sum (December–January-February) under the moderate emission scenario RCP4.5 (2071–2100 relative to 1971–2000). The hatched area indicates robust/significant changes. (Source: Jacob et al. 2014)

increasing mean annual precipitation is expected in Eastern, Northern, and parts of Central Europe in the future (up to +25%), it is expected to decrease in Southwestern Europe (up to −15%) and remain largely stable in Western Europe and parts of Central Europe (2071–2100 relative to 1971–2000, under RCP4.5; Jacob et al. 2014). These differences are mainly due to changes in spring, summer, and autumn precipitation. In the winter months, increased precipitation is expected for all of Europe with the exception of Andalusia, Sicily, and southern Greece (Fig. 16.2). These regional differences in the patterns of expected precipitation change are similar also under a more extreme emission scenario (RCP8.5), but the changes are somewhat more pronounced (Jacob et al. 2014). In addition, more frequent heavy precipitation events are expected for large parts of Europe in the future, and a significant increase in the length of summer dry periods is predicted particularly for Southern Europe (Fig. 16.3; Jacob et al. 2014).

Changes in the quantity and distribution of precipitation can strongly influence disturbance activity, but the relationship is less clear than that with temperature. While decreasing precipitation and longer dry periods increase the probability of some disturbances (e.g. bark beetle outbreaks, fire), other disturbances are negatively influenced by these changes (e.g. pathogens, floods, avalanches, debris flows). In general, the effect of reduced precipitation is amplified by warmer temperatures (Seidl et al. 2020). The role of precipitation in the disturbance regime generally decreases with increasing latitude (Seidl et al. 2017). Climate-related changes in the water balance will therefore become disturbance-relevant, especially in areas affected by increasing drought, that is, Mediterranean and (sub-)tropical ecosystems.

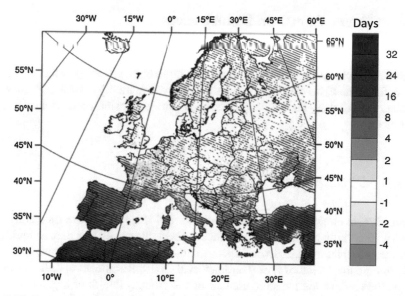

Fig. 16.3 Expected change in the duration of drought periods in Europe. The map shows the change in the 95th percentile of the duration of drought periods (in days) under the moderate emission scenario RCP4.5 (2071–2100 relative to 1971–2000). The hatched area indicates robust/significant changes. (Source: Jacob et al. 2014)

16.2.3   Wind

The changes in wind climate are considerably more uncertain compared to temperature and precipitation. Climate change could cause a slight decrease in both average wind speed and the number of strong wind events globally and in Europe (Ulbrich et al. 2009; Kjellström et al. 2011). At the same time, the future intensity of strong wind events, and thus the probability of wind disturbances, may increase significantly in certain regions. In the tropics, for example, increases in maximum wind speeds of 2–11% are expected by the end of the twenty-first century (Knutson et al. 2010). For Europe, slight increases in wind speed are expected for Northern Europe, whereas slight decreases are expected in the south (Nikulin et al. 2011; Pryor et al. 2012). In general, the projections of changes in wind climate and especially of strong wind events relevant for disturbance regimes are still subject to large uncertainty (Nikulin et al. 2011; Shaw et al. 2016).

Higher maximum wind speeds have positive effects on disturbance activity as they increase the probability of windthrow events. A simulation study for a mountain forest ecosystem in Europe, for example, showed that even small increases in peak wind speed (10%) can lead to a disproportionate increase in disturbances (384%) (Seidl and Rammer 2017). Climate scenarios for the end of the twenty-first century (+2.5 °C compared to current temperature levels) generally predict increasing storm damage both in the tropics (28–63%) and in Europe (23%) (Ranson et al. 2014).

16.3   Climate Effects on Disturbances

Three different types of climate effects on disturbance activity can be distinguished:
(1) direct effects, (2) indirect effects, and (3) interaction effects (Seidl et al. 2017).
These different effects can occur in parallel, they can mutually reinforce, or they can
attenuate each other. Therefore, they should not be considered in isolation from
each other.

16.3.1   Direct Effects

Direct climate effects describe the immediate influence of climate on disturbance
activity. Especially for abiotic disturbances, direct climate effects play a significant
role, for example, increased wind speeds can lead to larger windthrows in forests
and promote the spread of fires (Gardiner et al. 2010; Billmire et al. 2014). Longer
dry periods reduce the moisture content of the combustible dead organic material
and thus increase fire susceptibility (Williams and Abatzoglou 2016; Fig. 16.4).
Another example of direct climatic effects is disturbances by ice storms, caused by

Fig. 16.4 Longer periods of drought can lead to an increase in forest fires, even in areas which
previously had low fire risk, such as Central Europe. The picture shows a burned area in a pine
forest in eastern Austria in the dry summer of 2013. (Photo: R. Seidl)

freezing rain at temperatures <0 °C near the ground and >0 °C in higher layers of the atmosphere (Nagel et al. 2016).

Also biotic disturbances are directly influenced by climate, mainly through the effect temperature has on the metabolism of disturbance-causing organisms. Warmer temperatures can shorten the reproductive period of bark beetles and thus promote the development of multiple beetle generations per year (Jönsson and Bärring 2011; see Chap. 12) and also reduce the winter mortality of individuals (Koštál et al. 2011). In addition to an increase in insect populations, this will allow for a range expansion of insects to higher elevations and latitudes that were previously too cold for these insects (Battisti et al. 2005; Fig. 16.5a). Similar relationships exist for a number of disturbance-relevant insects and pathogens (Aguayo et al. 2014).

The disturbance response to direct climate effects is usually immediate, but often nonlinear: for example, insect populations react to the exceedance of thermal thresholds (Lange et al. 2006; see Chap. 12), wind disturbances only occur above a certain minimum wind speed (Blennow et al. 2010; see Chap. 8), and the danger of avalanches increases in a very limited temperature window with high snow cover (Germain et al. 2009; see Chap. 9).

Fig. 16.5 (a) Bark beetles are increasingly advancing to higher elevations due to climate change in the European Alps. (b) Broadleaved deciduous trees are becoming more competitive in the mountains of Central Europe because of global warming, which reduces the host tree density for bark beetles in the medium term. (Photos: R. Seidl)

16.3.2   Indirect Effects

Indirect climate effects describe climate-induced changes in vegetation structure and composition, which in turn influence the activity of abiotic and biotic disturbances. For example, dry climatic conditions reduce the net primary productivity of vegetation and lead to lower levels of combustible material being available, thus reducing the risk of fire (Pausas and Ribeiro 2017). This indirect effect counteracts the abovementioned direct effect of an increased fire hazard caused by a drier climate. With sufficient precipitation, however, warming and CO_2 fertilization lead to increased productivity (Reyer et al. 2014), especially at higher latitudes and in mountainous regions. In forest ecosystems, this results in taller trees, which are more susceptible to wind disturbances (Blennow et al. 2010). Also, a change in tree species composition because of rising temperatures, for example, from conifer-dominated forests to mixed forests or broadleaved forests in Northern Europe and the Alps, can have a profound effect on the susceptibility to host-specific biotic disturbances such as bark beetles (Temperli et al. 2013), and can thus counteract the direct climate effect of an increase in bark beetle populations (Fig. 16.5b).

Another indirect climate effect on disturbances results from the climate sensitivity of the defence mechanisms of plants against potential disturbing factors. Drought-stressed trees, for example, which are forced to use up their carbohydrate reserves to maintain their physiological functions, have fewer resources at their disposal for the defence against insect infestations and are thus more susceptible toward disturbance (Anderegg et al. 2015). Such effects of drought can predispose ecosystems over large areas and thus contribute to a synchronization of the insect–host system, leading to large waves of mass outbreaks (Seidl et al. 2016b). Other possible indirect climate effects are the reduction of soil frost caused by rising temperatures, resulting in a reduction in the stability of trees against winter storms (Usbeck et al. 2010), as well as the temporal decoupling of insect–host systems as a consequence of temperature-related changes in plant phenology (Schwartzberg et al. 2014).

In contrast to direct climate effects, which have an immediate impact on disturbances, indirect effects sometimes manifest themselves only with a considerable time lag. For example, because of the high inertia in forest dynamics, tree species composition can lag several centuries behind climatic development—the speed with which broadleaved trees extend their ranges to higher elevations has been estimated to be about 0.25 m per year for the northern Alps (Thom et al. 2017). In general, many indirect climate effects have a dampening effect on disturbance regimes and could thus buffer disturbance increases caused by direct climate effects, albeit with a considerable time lag.

16.3.3   Disturbance Interactions

Changes in climate not only have direct and indirect effects on disturbance but also influence disturbance interactions, which are an integral part of disturbance regimes (see Chap. 2; Canelles et al. 2021). The interaction between different disturbance agents can be either positive (increase of "disturbance 1" results in increase of "disturbance 2") or negative (increase of "disturbance 1" results in decrease of "disturbance 2"). Globally, positive interactions are more frequent than negative ones, and especially abiotic disturbances often promote the occurrence of biotic disturbance (Seidl et al. 2017). For example, increased disturbance activity by wind via direct and/or indirect climate effects can lead to an accumulation of combustible biomass on the ground, which in turn can significantly change fire behavior (Hicke et al. 2012). A climate-induced increase in the activity of root pathogens in trees can result in lower stability of trees and thus in higher susceptibility to wind (Whitney et al. 2002). Windthrows, in turn, are important triggers of bark beetle outbreaks (see Chap. 12). Windthrow can lead to a disproportionate increase in bark beetle disturbances under warmer climatic conditions (Fig. 16.6; Seidl and Rammer 2017). A climate-induced increase in bark beetle activity can in turn reduce the resistance of mountain forests to avalanches, which can lead to an intensification of the disturbance regime as a result of the self-reinforcing feedback between avalanches (Zurbriggen et al. 2014).

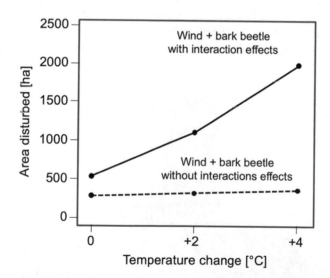

Fig. 16.6 Interaction effects increase the climate sensitivity of a wind–bark beetle disturbance regime. For the 20,900 ha forest landscape of Kalkalpen National Park in Austria, interaction effects were isolated via computer simulation and sensitivities to warmer temperatures were analyzed. The results represent the interaction effect of the natural disturbance regime without human intervention. (Modified from Seidl and Rammer 2017)

Negative disturbance interactions include those in which a disturbance reduces the resource for subsequent disturbances—for example, forest areas that burned with high severity or were recently infested by bark beetles usually have a low probability for further fire or insect disturbances in the short to medium term, as the necessary resources for their formation and spread are lacking (Hart et al. 2015; Seidl et al. 2016a). Climate effects on disturbance interactions usually become effective quickly and do not show long delays (i.e., in contrast to indirect effects of climate change).

16.4   Climate Change and Disturbance Dynamics

Because of the dynamic interplay between climate, vegetation, and disturbances (Fig. 16.7), the future development of disturbances under climate change remains challenging to predict. However, based on the different response times of direct and indirect effects as well as interaction effects, it can be assumed that direct effects and interaction effects will dominate in the short to medium term, whereas indirect effects will gain importance in the long term. Since expected direct and interaction effects often facilitate disturbances, a further increase in disturbances can be expected for the coming decades (Seidl et al. 2017). For Europe, where an increase of the quantitatively most important forest disturbance factors (i.e., wind, fire, and bark beetles) was already observed in the second half of the twentieth century (Seidl et al. 2011), a further increase can be assumed for the future. This is supported by a simulation study by Seidl et al. (2014): Relative to 1971–1980, wind disturbance could increase on average by 229% until 2030, fire disturbance could increase on average by 314%, and bark beetle disturbance could increase by as much as 764%.

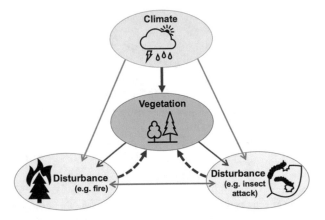

Fig. 16.7 The dynamic interactions between climate, vegetation, and disturbance result in complex responses of disturbance regimes to climate change. The figure shows the different climate effects on disturbances (direct, blue; indirect, violet; interactions, green)

This would mean that the wind disturbance levels that occurred in the past once every 15 years could occur every other year in the future. With regard to bark beetles, the projected mean disturbance level for 2021–2030 even corresponds to a disturbance event that has historically occurred only once every 32 years. It is particularly interesting that none of the short-term trajectories studied for Europe (i.e., a combination of 13 climate scenarios and four management strategies) resulted in a simulated decrease in disturbance until 2030 (Seidl et al. 2014).

In the long term, it can be assumed that indirect climate effects and negative feedback within the disturbance regime will dampen the increase in disturbances. For instance, a simulation study for a Central European forest landscape highlighted the mitigating effect of a changing tree species composition on the susceptibility of forests to bark beetles (Temperli et al. 2013). In resilient ecosystems, disturbed areas eventually recover the structures and functions of the system prior to disturbance (see Chap. 5; Albrich et al. 2021). However, with changing climate and disturbances, regime shifts can occur, for example, a transition from coniferous to broadleaved forests, or from forest to open land (Johnstone et al. 2016). At the same time, a climate-induced tree species shift also changes the susceptibility to disturbance, as illustrated by the abovementioned example (Temperli et al. 2013). Disturbances are thus catalysts for the adaptation of ecosystems to changing environmental conditions (Thom et al. 2017) and can also be metaphorically understood as an "immune reaction" of the system, which reduces the disequilibrium between climate and vegetation.

The extent to which disturbance activity will change under climate change ultimately also depends on human management. Planning interventions (e.g. the promotion of climate-adapted plant species) in combination with measures to avoid disturbances (e.g. fire prevention and containment) can reduce the projected increases in disturbances in the coming decades (see Chap. 17).

References

Aguayo J, Elegbede F, Husson C, Saintonge FX, Marçais B (2014) Modeling climate impact on an emerging disease, the *Phytophthora alni*-induced alder decline. Glob Chang Biol 20:3209–3221

Albrich K, Thom D, Rammer W, Seidl R (2021) The long way back: development of Central European mountain forests towards old-growth conditions after the cessation of management. J Veg Sci 32:e13052

Anderegg WR, Hicke JA, Fisher RA, Allen CD, Aukema J, Bentz B, Hood S, Lichstein JW, Macalady AK, McDowell N (2015) Tree mortality from drought, insects, and their interactions in a changing climate. New Phytol 208:674–683

Auer I, Böhm R, Jurkovic A, Lipa W, Orlik A, Potzmann R, Schöner W, Ungersböck M, Matulla C, Briffa K (2007) HISTALP – historical instrumental climatological surface time series of the Greater Alpine Region. Int J Climatol 27:17–46

Battisti A, Stastny M, Netherer S, Robinet C, Schopf A, Roques A, Larsson S (2005) Expansion of geographic range in the pine processionary moth caused by increased winter temperatures. Ecol Appl 15:2084–2096

Billmire M, French NHF, Loboda T, Owen RC, Tyner M (2014) Santa Ana winds and predictors of wildfire progression in southern California. Int J Wildland Fire 23:1119–1129

Blennow K, Andersson M, Sallnäs O, Olofsson E (2010) Climate change and the probability of wind damage in two Swedish forests. For Ecol Manag 259:818–830

Canelles Q, Aquilué N, James PMA, Lawler J, Brotons L (2021) Global review on interactions between insect pests and other forest disturbances. Landsc Ecol 36:945–972

Dale VH, Joyce LA, McNulty S, Neilson RP, Ayres MP, Flannigan MD, Hanson PJ, Irland LC, Lugo AE, Peterson CJ (2001) Climate change and forest disturbances: climate change can affect forests by altering the frequency, intensity, duration, and timing of fire, drought, introduced species, insect and pathogen outbreaks, hurricanes, windstorms, ice storms, or landslides. Bioscience 51:723–734

Fischer EM, Rajczak J, Schär C (2012) Changes in European summer temperature variability revisited. Geophys Res Lett 39:L19702

Gardiner BA, Blennow K, Jean-Michel Carnus J-M, Fleischer P, Ingemarson F, Landmann G, Lindner M, Marzano M, Nicoll B, Orazio C, Peyron J-L, Reviron M-P, Schelhaas M-J, Schuck A, Spielmann M, Usbeck T (2010) Destructive storms in European forests: past and forthcoming impacts. Final report to DG Environment. European Forest Institute, Atlantic European Regional Office – EFIATLANTIC, 132 p

Germain D, Filion L, Hétu B (2009) Snow avalanche regime and climatic conditions in the Chic-Choc Range, eastern Canada. Clim Chang 92:141–167

Gobiet A, Kotlarski S, Beniston M, Heinrich G, Rajczak J, Stoffel M (2014) 21st century climate change in the European Alps – a review. Sci Total Environ 493:1138–1151

Hart SJ, Veblen TT, Mietkiewicz N, Kulakowski D (2015) Negative feedbacks on bark beetle outbreaks: widespread and severe spruce beetle infestation restricts subsequent infestation. PLoS One 10:e0127975

Hicke JA, Johnson MC, Hayes JL, Preisler HK (2012) Effects of bark beetle-caused tree mortality on wildfire. For Ecol Manag 271:81–90

IPCC (2013) Climate Change 2013: the physical science basis. Working Group I contribution to the IPCC fifth assessment report. Intergovernmental Panel on Climate Change. Cambridge University Press, Cambridge, UK

Jacob D, Petersen J, Eggert B, Alias A, Christensen OB, Bouwer LM, Braun A, Colette A, Déqué M, Georgievski G (2014) EURO-CORDEX: new high-resolution climate change projections for European impact research. Reg Environ Chang 14:563–578

Jentsch A, Kreyling J, Beierkuhnlein C (2007) A new generation of climate-change experiments: events, not trends. Front Ecol Environ 5:365–374

Johnstone JF, Allen CD, Franklin JF, Frelich LE, Harvey BJ, Higuera PE, Mack MC, Meentemeyer RK, Metz MR, Perry GLW, Schoennagel T, Turner MG (2016) Changing disturbance regimes, ecological memory, and forest resilience. Front Ecol Environ 14:369–378

Jönsson AM, Bärring L (2011) Future climate impact on spruce bark beetle life cycle in relation to uncertainties in regional climate model data ensembles. Tellus A 63:158–173

Kjellström E, Nikulin G, Hansson U, Strandberg G, Ullerstig A (2011) 21st century changes in the European climate: uncertainties derived from an ensemble of regional climate model simulations. Tellus A 63:24–40

Knutson TR, McBride JL, Chan J, Emanuel K, Holland G, Landsea C, Held I, Kossin JP, Srivastava AK, Sugi M (2010) Tropical cyclones and climate change. Nat Geosci 3:157

Koštál V, Doležal P, Rozsypal J, Moravcová M, Zahradníčková H, Šimek P (2011) Physiological and biochemical analysis of overwintering and cold tolerance in two Central European populations of the spruce bark beetle, *Ips typographus*. J Insect Physiol 57:1136–1146

Lange H, Økland B, Krokene P (2006) Thresholds in the life cycle of the spruce bark beetle under climate change. Inter J Complex Syst 1648:1–10

Nagel TA, Firm D, Rozenbergar D, Kobal M (2016) Patterns and drivers of ice storm damage in temperate forests of Central Europe. Eur J For Res 135:519–530

Nikulin G, Kjellström E, Hansson U, Strandberg G, Ullerstig A (2011) Evaluation and future projections of temperature, precipitation and wind extremes over Europe in an ensemble of regional climate simulations. Tellus A 63:41–55

Pausas JC, Ribeiro E (2017) Fire and plant diversity at the global scale. Glob Ecol Biogeogr 26:889–897

Pryor SC, Barthelmie RJ, Clausen N-E, Drews M, MacKellar N, Kjellström E (2011) Analyses of possible changes in intense and extreme wind speeds over northern Europe under climate change scenarios. Clim Dyn 38:189–208

Ranson M, Kousky C, Ruth M, Jantarasami L, Crimmins A, Tarquinio L (2014) Tropical and extra-tropical cyclone damages under climate change. Clim Chang 127:227–241

Reyer C, Lasch-Born P, Suckow F, Gutsch M, Murawski A, Pilz T (2014) Projections of regional changes in forest net primary productivity for different tree species in Europe driven by climate change and carbon dioxide. Ann For Sci 71:211–225

Schär C, Vidale PL, Lüthi D, Frei C, Häberli C, Liniger MA, Appenzeller C (2004) The role of increasing temperature variability in European summer heatwaves. Nature 427:332

Schwartzberg EG, Jamieson MA, Raffa KF, Reich PB, Montgomery RA, Lindroth RL (2014) Simulated climate warming alters phenological synchrony between an outbreak insect herbivore and host trees. Oecologia 175:1041–1049

Seidl R, Rammer W (2017) Climate change amplifies the interactions between wind and bark beetle disturbances in forest landscapes. Landsc Ecol 32:1485–1498

Seidl R, Schelhaas M-J, Lexer MJ (2011) Unraveling the drivers of intensifying forest disturbance regimes in Europe. Glob Chang Biol 17:2842–2852

Seidl R, Schelhaas MJ, Rammer W, Verkerk PJ (2014) Increasing forest disturbances in Europe and their impact on carbon storage. Nat Clim Chang 4:806–810

Seidl R, Donato DC, Raffa KF, Turner MG (2016a) Spatial variability in tree regeneration after wildfire delays and dampens future bark beetle outbreaks. Proc Natl Acad Sci USA 113:13075–13080

Seidl R, Müller J, Hothorn T, Bässler C, Heurich M, Kautz M (2016b) Small beetle, large-scale drivers: how regional and landscape factors affect outbreaks of the European spruce bark beetle. J Appl Ecol 53:530–540

Seidl R, Thom D, Kautz M, Martin-Benito D, Peltoniemi M, Vacchiano G, Wild J, Ascoli D, Petr M, Honkaniemi J, Lexer MJ, Trotsiuk V, Mairota P, Svoboda M, Fabrika M, Nagel TA, Reyer CPO (2017) Forest disturbances under climate change. Nat Clim Chang 7:395–402

Seidl R, Honkaniemi J, Akala T, Aleinikov A, Angelstam P, Bouchard M, Boulanger Y, Burton PJ, De Grandpre L, Gauthier S, Hansen WD, Jepsen JU, Jogiste K, Kneeshaw D, Kuuluvainen T, Lisitsyna O, Makoto K, Mori AS, Pureswaran DS, Shorohova E, Shubnitsina E, Tayler AE, Vladimirova N, Vodde F, Senf C (2020) Globally consistent climate sensitivity of natural disturbances across boreal and temperate forest ecosystems. Ecography 43:1–12

Shaw TA, Baldwin M, Barnes EA, Caballero R, Garfinkel CI, Hwang Y-T, Li C, O'Gorman PA, Rivière G, Simpson IR (2016) Storm track processes and the opposing influences of climate change. Nat Geosci 9:656

Solomon S, Plattner G-K, Knutti R, Friedlingstein P (2009) Irreversible climate change due to carbon dioxide emissions. Proc Natl Acad Sci USA 106:1704–1709

Temperli C, Bugmann H, Elkin C (2013) Cross-scale interactions among bark beetles, climate change, and wind disturbances: a landscape modeling approach. Ecol Monogr 83:383–402

Thom D, Rammer W, Seidl R (2017) Disturbances catalyze the adaptation of forest ecosystems to changing climate conditions. Glob Chang Biol 23:269–282

Ulbrich U, Leckebusch GC, Pinto JG (2009) Extra-tropical cyclones in the present and future climate: a review. Theor Appl Climatol 96:117–131

Usbeck T, Wohlgemuth T, Pfister C, Bürgi A, Dobbertin M (2010) Increasing storm damage to forests in Switzerland from 1858 to 2007. Agric For Meteorol 150:47–55

Whitney RD, Fleming RL, Zhou K, Mossa DS (2002) Relationship of root rot to black spruce windfall and mortality following strip clear-cutting. Can J For Res 32:283–294

Williams AP, Abatzoglou JT (2016) Recent advances and remaining uncertainties in resolving past and future climate effects on global fire activity. Curr Clim Chang Rep 2:1–14

Zurbriggen N, Nabel JEMS, Teich M, Bebi P, Lischke H (2014) Explicit avalanche-forest feedback simulations improve the performance of a coupled avalanche-forest model. Ecol Complex 17:56–66

Part VII
Disturbances and Management

Chapter 17
Managing Disturbance Risks

Rupert Seidl (ID), **Sigrid Netherer** (ID), **and Thomas Thaler** (ID)

Abstract Risk describes the impact of uncertainties on objectives. Disturbances are important risk factors for ecosystem management because they can negatively affect the provisioning of ecosystem services and their occurrence and extent cannot be predicted with high accuracy. Risk management is the coordinated activity of managing and controlling risks. The three central elements of risk management are risk identification, risk assessment, and risk treatment. Disturbance risks can be quantified, for example, by estimating predisposition and by scenario analyses. Risk assessment includes both economic and social components (e.g. social vulnerability). The general possibilities to treat risks are to accept them, to reduce them, or to collectivize them.

Keywords Ecosystem management · Predisposition assessment · Risk assessment · Risk identification · Risk management · Risk treatment · Scenario analysis · Social vulnerability · Uncertainties

R. Seidl (✉)
Ecosystem Dynamics and Forest Management Group, School of Life Sciences, Technical University of Munich, Freising, Germany

Berchtesgaden National Park, Berchtesgaden, Germany
e-mail: rupert.seidl@tum.de

S. Netherer
Institute of Forest Entomology, Forest Pathology and Forest Protection, University of Natural Resources and Life Sciences Vienna, Vienna, Austria

T. Thaler
Institute for Mountain Risk Engineering, University of Natural Resources and Life Sciences Vienna, Vienna, Austria

T. Wohlgemuth et al. (eds.), *Disturbance Ecology*, Landscape Series 32,
https://doi.org/10.1007/978-3-030-98756-5_17

17.1   Introduction and Definitions

In general, the term risk refers to the possibility of damage. The etymological origin of the word lies in the Latin term *resecum*, which denotes a rock or cliff and thus a danger to seafaring vessels (Pfeifer 1989). An approach widely used in the natural sciences defines risk R as a function of the probability of occurrence of an event P_S and its damage potential E_S (Eq. 17.1), whereby both variables are usually linked multiplicatively (Eq. 17.2; expected value approach, see Haimes 2004):

$$R = f\left(P_S; E_S\right) \tag{17.1}$$

$$R = P_S \cdot E_S \tag{17.2}$$

While the early, influential work of Knight (1921) distinguished between risk (= quantifiable) and uncertainty (= not quantifiable), the International Organization for Standardization generally defines risk as the impact of uncertainty on objectives (ISO 2009).

In the context of natural resource management, disturbances are sources of uncertainty, as neither their occurrence nor their effects can be predicted with high accuracy. Compared to other risk factors, disturbances often have a special significance in ecosystem management: by definition, they occur abruptly (see Chap. 2) and can change the vegetation structure and composition desired by management within a short period of time. Despite their relatively short duration, the following recovery can take decades and more, and the effects of disturbances can persist in ecosystems for a very long time. After a windthrow in a Central European mountain forest, for example, several decades may pass until the regenerating tree cohort has again developed the appropriate stand structure to protect humans and their infrastructure from natural hazards such as avalanches and rockfall (see Chap. 9). This character of "slow in, fast out" (Körner 2003) makes disturbances a particularly relevant component of risk in the context of natural resource management.

Risk management refers to the coordinated activities of managing and controlling risks with the aim of reducing losses (Purdy 2010). The core questions of risk management are: (1) What can happen? (2) How likely or how frequent is the occurrence? and (3) What are the consequences of the event that has occurred? (Kaplan and Garrick 1981). The central element of risk management is risk assessment, which consists of the identification and analysis of risks (What can happen?) and the evaluation of these risks in the context of concrete goals (What is acceptable?) (BUWAL 1999). From this, possible approaches of reacting to risks are derived (Fig. 17.1). These three core elements of risk management—identifying risk, assessing risk, and dealing with risk—are examined in detail below. In general, risk management is not a one-time activity but must be carried out periodically in order to adequately take into account changing conditions and new goals. Monitoring of measures and their effects is an important tool to continuously improve risk management based on the experience gained.

Fig. 17.1 The risk management process. (Purdy Turner 2011, modified)

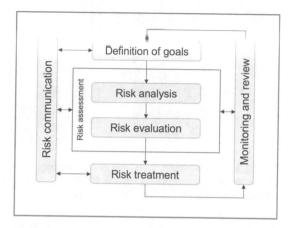

17.2 Recognizing Risks

Identification

The identification of the risks is the first step in the risk assessment process and provides answers to the questions: What can happen? When can it happen? and How does a risk manifest itself? (Purdy 2010). In the context of natural resource management, potentially all disturbances discussed in Chaps. 7, 8, 9, 10, 11, 12 and 13 are risk factors, as they can lead to loss of living biomass and/or mortality of plants. Therefore, a first important aspect in risk assessment is to identify the disturbances occurring in a given area, for which historical sources of past events play an important role (see Chap. 3). However, whether the occurrence of a disturbance constitutes a risk depends on the management objectives. For example, bark beetle infestation significantly affects timber production (Seidl et al. 2008), which is why it is often considered a major risk factor in production forests. If, on the other hand, biodiversity is a priority of management, the same disturbance does not threaten the management objective but may even contribute to its achievement (Thom et al. 2017; Kortmann et al. 2021). Whether or not a disturbance poses a risk cannot be answered generally and depends on the specific objectives of management (Fig. 17.1).

For most disturbance factors, the probability of damage and the extent of damage are negatively correlated; weak/small disturbances occur more frequently (short return intervals), whereas strong/large disturbances occur less frequently (long return intervals). This has been confirmed, for example, for unmanaged mountain forests in the Western Carpathians by means of tree ring analysis (Janda et al. 2017): from 1790 to 1960, disturbances removing up to 20% of canopy cover occurred on average once every 66 years. In contrast, severe disturbances reducing canopy cover by >60% occurred only every 690 years. Since disturbance frequency and severity can have different effects on management objectives, their inverse relationship needs to be considered in risk management. In the context of wood production or carbon storage, for instance, weak disturbances that occur frequently can even have

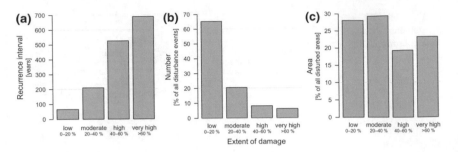

Fig. 17.2 Relationship between disturbance severity and (**a**) disturbance interval, (**b**) the number of disturbance events, and (**c**) the area disturbed in unmanaged mountain forests of the Western Carpathians. (Source: Janda et al. 2017, modified)

positive effects (i.e. thinning effects; Seidl et al. 2008). However, the protective function against natural hazards in mountain forests is strongly reduced by high disturbance frequency (Sebald et al. 2019). It should be noted, however, that in relation to the total disturbed area, small low severity events represent an important part of the disturbance regime because of their high frequency (Fig. 17.2).

Another important aspect in the identification of risks is that disturbances usually do not occur in isolation but can trigger other disturbances. In Central Europe, for example, windthrows are often followed by outbreaks of bark beetles, because low resistance breeding material is available in large quantities allowing for high reproduction success of the bark beetle and eventually leading to a spread into adjacent forest stands (see Chap. 12; Stadelmann et al. 2014). Thus, the occurrence of the two disturbances is spatially and temporally linked (interaction effect, cascading disturbances) and together increases the overall extent of the disturbance. An independent consideration of the two disturbance agents in a risk analysis would therefore underestimate the effect of disturbances (see also Seidl and Rammer 2017). A global overview of the interactions of important disturbance factors has shown that disturbances by wind, fire, and drought are often followed by disturbances by insects and pathogens (Fig. 17.3).

Quantification

Risks can be quantified using various methods. Historical and recent observations can provide a first indication on relevant parameters of the disturbance regime (see Fig. 17.3; Overbeck and Schmidt 2012; Thom et al. 2013). However, such information is often only available at regional to national scales, but not for a specific landscape or management unit. Furthermore, since past conditions often do not adequately represent the current or future situation, other methods are generally preferred for the quantification of risks. An overview of the wide range of available methods is provided by Holthausen et al. (2004), Hanewinkel et al. (2010), and Yousefpour et al. (2012). In the following text, two approaches—predisposition assessment and scenario analysis—are described as examples.

Predisposition assessment is based on the conception of Manion (1981) that tree mortality occurs through an interaction of predisposing, inciting, and contributing

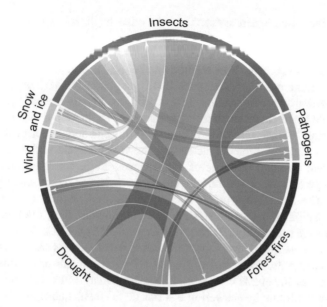

Fig. 17.3 Interaction of disturbance agents. The size of the sectors in the outer circle reflects the distribution of disturbance agents in the considered literature, while the arrows through the center of the circle indicate the relative frequency of interactions between individual agents. The arrows point from the influencing agent to the influenced agent. (Source: Seidl et al. 2017, modified)

factors. In the case of windthrow, for example, stand structure and composition are predisposing factors, while root rot can be a contributing factor, and the gusts occurring during a storm event are factors inciting the disturbance. The predisposition of forests can be affected both positively and negatively by silvicultural interventions such as tree species selection, thinning, and harvesting measures (see Chap. 8). A prerequisite for effective risk management (see below) is to understand the complex interrelations between predisposing stand characteristics and the various abiotic and biotic disturbance agents influencing forest structure and functioning. Therefore, predisposition assessment plays an important role in the risk management of forest stands (Netherer and Nopp-Mayr 2005; Jactel et al. 2009). However, the method is also applied in ecosystems other than forests.

Predisposition assessment quantifies the probability of forest disturbance based on site and stand characteristics. This can be approached in two different ways: (1) based on statistical relationships and (2) with the help of knowledge-based expert systems. Statistical approaches such as classification trees or regression models can be used to derive damage probabilities while considering drivers such as weather, exposition, stand age, or canopy closure (Seidl et al. 2011; Netherer et al. 2019). However, the quantification of risks using statistical methods is limited by the high temporal and spatial variation of disturbances, which result in poor "goodness of fit" of statistical models. Furthermore, statistical models can rarely be generalized, that is, their application in other areas is only possible to a very limited extent (Hanewinkel et al. 2010).

Knowledge-based expert systems, on the other hand, rely on general considerations based on meta-analyses of the literature or on expert knowledge. Different management options can be evaluated in terms of their risk using multi-criteria decision analysis (MCDA). This method can be used to consider a large number of risk factors simultaneously (Jactel et al. 2012). In contrast to MCDA, which is usually geared toward specific management objectives, "bonus–malus" systems reveal when specific stand characteristics either increase risk or help to prevent or reduce risk (Berryman 1986; Speight and Wainhouse 1989). In such predisposition assessment models, site and stand characteristics are weighted and scaled depending on their influence on the occurrence of disturbances. Different forest characteristics (e.g. stand age classes, dominant tree height) are assessed by means of predisposition points. The sum of these points results in a quantitative measure of risk, that is, the predisposition of a specific stand to a certain disturbance, often expressed in terms of damage probability (Führer and Nopp 2001; Fig. 17.4).

In a *scenario analysis*, several possible future developments are explored starting from the current state. Scenario analyses describe the outcome of a sequence of different influences and events in a logically consistent way. Put simply, scenario analyses explore the future by answering a series of "if–then" questions, thus allowing "a journey into possible futures" (UNEP 2002). An advantage of scenario analyses is that possible future developments (e.g. regarding the change of climate and disturbance regimes, see Chap. 16) can be explicitly considered. Scenario analyses can be carried out qualitatively (i.e. descriptive) as well as quantitatively (e.g. using simulation models, see Box 17.1 and Fig. 17.5) (van der Sluijs et al. 2004). In both cases, the value of scenarios lies primarily in a relative comparison of possible developmental pathways under a set of different assumptions, which is why the analysis of a single scenario is usually less informative than the examination of a

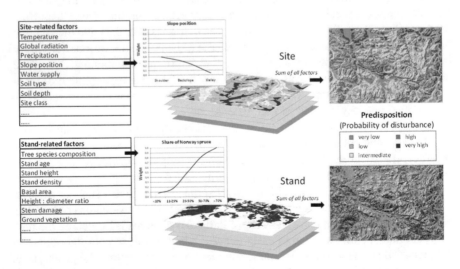

Fig. 17.4 Evaluation of site- and stand-based risk factors in a predisposition assessment system; the predisposition (probability of damage) of a stand results from the sum of all relevant factors

Box 17.1: Disturbance Modelling

Werner Rammer (iD)
Ecosystem Dynamics and Forest Management Group,
School of Life Sciences, Technical University of Munich, Freising, Germany
e-mail: werner.rammer@tum.de

Rupert Seidl (iD)
Ecosystem Dynamics and Forest Management Group,
School of Life Sciences, Technical University of Munich, Freising, Germany
Berchtesgaden National Park, Berchtesgaden, Germany

Why Disturbance Modelling?

A model is generally a simplified representation of reality in mathematical form. In addition to observation and experiment, models are a third, and increasingly important, method of gaining knowledge in science (Winsberg 2010). However, the term "model" itself is not yet clearly defined: "models" can refer to anything from simple statistical correlations (regressions) to very complex software systems. Similarly, models are used for a broad range of applications, for example, to gain a better understanding of empirically observed data and to identify key factors leading to disturbance events. Another example is the mapping of disturbance risk as a function of climatic and/or natural conditions. Furthermore, models are often used to scale local and short-term observations to larger spatial scales and longer time periods (Seidl et al. 2013). Disturbance processes are increasingly integrated into ecosystem models, allowing the simulation of dynamic feedbacks between vegetation and disturbances. Such coupled systems can answer "what-if" questions in the form of scenario analyses, for example, to assess how disturbance regimes respond to changes in the vegetation as a result of climate change or management decisions.

Disturbance Processes and Their Modelling

Five key processes can be distinguished, which are represented in disturbance models (according to Seidl et al. 2011):

- *Vulnerability:* Vulnerability is the susceptibility of an ecosystem to a disturbance factor. For example, the wind susceptibility of a forest stand can be statistically modelled based on stand and site characteristics (Pasztor et al. 2015).
- *Occurrence:* Actual disturbance events are often triggered by external factors, such as the occurrence of wind speeds above a critical wind speed. Such triggering factors can be considered in process-based models (Seidl et al. 2014).

(continued)

Box 17.1 (continued)

- *Effects:* The effects of a disturbance can depend on vegetation properties and particularly on the vulnerability of vegetation. The severity of the disturbance (e.g. the degree of tree mortality on the affected area) is often used as an indicator for disturbance effects. Neighbourhood relationships also play a role in the modelling of impacts (e.g. when previously sheltered trees become exposed after neighbouring trees have been blown down during a storm event) (Seidl et al. 2014).

While these first three processes primarily relate to the modelling of individual disturbance events, the modelling of disturbance regimes includes long-term temporal and spatial interactions between disturbances and vegetation, as well as interactions between disturbance agents.

- *Temporal and Spatial Dynamics:* This relates to the modelling of the spatial patterns of disturbances (e.g. flight patterns and host search behaviour of bark beetles) (Kautz et al. 2014), but also the temporal interactions and feedbacks between vegetation and disturbance (e.g. decreasing bark beetle risk after previous outbreak waves because of a lack of host trees) (Temperli et al. 2013).
- *Disturbance Interactions:* Disturbances often do not occur in isolation but can mutually reinforce or weaken each other. Such interactions (e.g. between wind and bark beetles) need to be considered in simulation models to estimate the dynamic behaviour of disturbance regimes (Seidl & Rammer 2017).

Model Concepts in Disturbance Modelling

Disturbance models include one or more of these key processes, and different modelling techniques can be used (Seidl et al. 2011, simplified):

- *Statistical models* are mathematical approximations of observed data based on statistical methods (e.g. regression analysis, Pasztor et al. 2015). These models can be applied on the level of single disturbance events or on the level of disturbance regimes.
- *Static process-based models* are approaches that model disturbance processes on the basis of mechanistic relationships but without considering feedbacks or changes in the vegetation (e.g. models for the developmental phenology of bark beetles) (Baier et al. 2007).
- *Dynamic process-based models* are also based on mechanistic relationships but explicitly consider the interactions between vegetation and dis-

(continued)

Box 17.1 (continued)

turbances. Such disturbance models are often integrated into complex dynamic ecosystem models. Depending on the research question, the model platforms used can be directed toward the simulation of vegetation processes (Seidl et al. 2007), biogeochemical processes (Wolf et al. 2008), or landscape processes (Sturtevant et al. 2004).

References

Baier P, Pennerstorfer J, Schopf A (2007) PHENIPS—a comprehensive phenology model of *Ips typographus* (L.) (Col., Scolytinae) as a tool for hazard rating of bark beetle infestation. For Ecol Manag 249:171–186

Kautz M, Schopf R, Imron MA (2014) Individual traits as drivers of spatial dispersal and infestation patterns in a host–bark beetle system. Ecol Model 273:264–276

Pasztor F, Matulla C, Zuvela-Aloise M, Rammer W, Lexer MJ (2015) Developing predictive models of wind damage in Austrian forests. Ann For Sci 72:289–301

Seidl R, Rammer W (2017) Climate change amplifies the interactions between wind and bark beetle disturbances in forest landscapes. Landsc Ecol 32:1485–1498

Seidl R, Baier P, Rammer W, Schopf A, Lexer MJ (2007) Modelling tree mortality by bark beetle infestation in Norway spruce forests. Ecol Model 206:383–399

Seidl R, Fernandes PM, Fonseca TF, Gillet F, Jönsson AM, Merganicová K, Netherer S, Arpaci A, Bontemps JD, Bugmann H, González-Olabarria JR, Lasch P, Meredieu C, Moreira F, Schelhaas MJFM (2011) Modelling natural disturbances in forest ecosystems: a review. Ecol Model 222:903–924

Seidl R, Eastaugh CS, Kramer K, Maroschek M, Reyer C, Socha J, Vacchiano G, Zlatanov T, Hasenauer H (2013) Scaling issues in forest ecosystem management and how to address them with models. Eur J For Res 132:653–666

Seidl R, Rammer W, Blennow K (2014) Simulating wind disturbance impacts on forest landscapes: tree-level heterogeneity matters. Environ Model Softw 51:1–11

Sturtevant BR, Gustafson EJ, Li W, He HS (2004) Modeling biological disturbances in LANDIS: a module description and demonstration using spruce budworm. Ecol Model 180:153–174

Temperli C, Bugmann H, Elkin C (2013) Cross-scale interactions among bark beetles, climate change, and wind disturbances: a landscape modeling approach. Ecol Monogr 83:383–402

Winsberg E (2010) Science in the age of computer simulation. University of Chicago Press, Chicago, 152 p

Wolf A, Kozlov MV, Callaghan TV (2008) Impact of non-outbreak insect damage on vegetation in northern Europe will be greater than expected during a changing climate. Clim Change 87:91–106

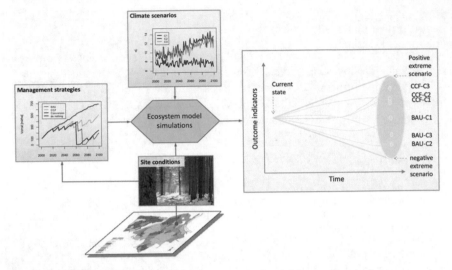

Fig. 17.5 Scheme of model-based scenario analysis for risk assessment in forest ecosystems under climate change. A combination of different management strategies and climate scenarios is analyzed by simulation modelling to quantify possible changes in relevant target variables. The analysis is done on a stand-by-stand basis, that is, different environmental and stand conditions are taken into account. In the scenario envelope to the right the results for two management strategies (business-as-usual BAU, continuous cover forestry CCF) and three climate scenarios (C1–C3) are shown schematically. For a concrete implementation of the schematic analysis shown here, see Seidl et al. (2008)

larger number of contrasting scenarios. In this regard, scenarios often follow "story-lines" to facilitate the interpretation of differences between the individual scenarios. In a typical application, alternative scenarios which describe, for example, a change in climate or management, are compared with a baseline scenario, which assumes continuity in the currently prevailing conditions ("business as usual").

Extreme scenarios are of central importance in risk analysis. These extreme scenarios provide information about the potential effects of a possible chain of negative events. Such "worst case scenarios" can be included in risk management in accordance with the precautionary principle. An important advantage of scenario analysis is that situations and conditions that have not (yet) been observed in the past, but which are possible in the future, can be taken into account. This is of particular importance when one considers that severe disturbance events occur only rarely (Fig. 17.2; Pfister 1999), and it is therefore highly likely that they are not represented in the observations of the past. However, it should be noted that future surprises (i.e. events that we are currently not aware about and have no knowledge about—*unknown unknowns*) are not considered in either qualitative or quantitative scenario analyses. These *unknown unknowns* can also have significant effects on management objectives, for example, if a non-native, invasive pest is unexpectedly introduced and causes major disturbances. Therefore, the risks identified by scenario analyses should be considered as conservative.

17.3   Evaluating Risks

Once risks have been identified and quantified, they must be evaluated with regard to their relevance and priority for objectives. In other words, the question arises of what may and may not happen. An answer to this question must take individual and institutional values and preferences into account. Therefore, both ecological and socioeconomic considerations must be included in this step of the risk management process. Since the focus of this book so far has been largely on the ecological role of disturbances, we focus here on the economic and social aspects of risk assessment.

Economic Risks

Economic valuation methods play an important role in risk assessment (Hanewinkel et al. 2010; Bernetti et al. 2011; Montagné-Huck and Brunette 2018). They are practiced by individual entrepreneurs as well as by government decision makers, for example, in the context of natural disturbances (Knoke et al. 2021). In this context, the damage potential is determined by means of economic indicators (e.g. see Möllmann and Möhring 2017; Knoke et al. 2021). The economic analysis also includes an assessment of risk management strategies, such as silvicultural interventions and their potential to reduce risks. Furthermore, the potential impact of forest disturbances on the ecosystem services, such as the protective effect of forests in the context of disaster risk reduction (Moos et al. 2019) or the provision of other ecosystem services (Lee et al. 2020), can be assessed economically. The analysis is usually quantitative, whereby the risk R for an object is generally calculated as an objectively measurable damage expectation (Hanewinkel et al. 2010; Knoke et al. 2021).

The choice of the economic method is generally based on the valuation of the damage caused by a disturbance. Damage is defined as direct and indirect effects on a system, an object, or an operational process (interruption). Damage can be calculated with monetary or nonmonetary approaches. For monetary approaches, such as the quantification of damage to goods and buildings, market prices are usually used. Knoke et al. (2021), for instance, estimated the economic losses from disturbances in Norway spruce forests of Central Europe at between €2600 and €34,400 per hectare, depending on scenario and valuation approach used. For nonmonetary values, such as loss of species or ecosystem services not traded on markets, the damage is usually measured based on consumer preferences. These include approaches such as the *willingness-to-pay* and the *willingness-to-accept* (Hanley and Spash 1995; Olschewski et al. 2012), allowing the economic valuation of goods that are not provided by the usual market mechanisms and therefore do not have a corresponding market price (Hanley and Spash 1995). In this respect, the first calculation of the value of a bluethroat gained much public attention in German-speaking countries with the publication of "Der Wert eines Vogels: ein Fensterbilderbuch" [The worth of a bird: a picturebook] (Vester 1983).

In the case of direct impacts, the main types of damage taken into account are those that are directly caused by an event, either in full or in part, such as damage to buildings and other infrastructure, goods, dams, or forest stands. Frequently, the damage is valued on the basis of the restoration costs, minus depreciation, or with the loss of assets for the affected party (Hanley and Spash 1995; Knoke et al. 2021). In Switzerland, for example, after Storm Lothar in 1999, the direct loss of assets due to wind breakage and windthrow was estimated at CHF 225 million (Baur et al. 2003). In addition to the loss of assets, a distinction must be made between the loss of income due to the event, that is, the reduction in income compared to income in the absence of the disturbance event. For Storm Lothar, this was calculated at CHF 284 million for the first 6 years after the event, whereby the loss of income here is mainly due to the storm-induced drop in prices on the timber market (Baur et al. 2003; see Chap. 8). In addition to the loss of assets and income, the effects of an interruption in operations can also be assessed in monetary terms. Here, damage is evaluated from an interruption of a business resulting from the inability to fulfil promised services. The economic methods used are computable general equilibrium models and agent-based computational economics (Loomis 1993; Rose 2004).

Indirect effects usually occur as a consequence of an event and must therefore be considered over a longer time frame, such as production declines caused by road closures or changes in the age class distribution of forests caused by disturbances (Senf et al. 2021). This also includes the damage that is caused by increased techno-logical or institutional efforts and which therefore has long-term adverse effects on the overall economy. The assessment of indirect losses can also include losses of property values, landscape values, or carbon sequestration. Indirect impacts are usu-ally determined by questionnaires or by economic modelling, such as input–output modelling or econometric analyses (Holmes et al. 2010; Price et al. 2010).

Societal Evaluation of the Risk

The societal evaluation of risks is fundamentally concerned with the vulnerability of society. The term vulnerability is closely linked to the negative consequences and effects of natural processes such as disturbances, whereby vulnerability is under-stood primarily as potential loss. Vulnerability is analyzed and evaluated differently in different scientific disciplines. In the natural sciences, vulnerability tends to be assessed in terms of monetary losses, that is, using metric data. In contrast, in the social sciences, vulnerability tends to be assessed using ordinal data. In particular, the focus in social sciences is on potential and actual impairments of society in terms of social values and perceptions as well as on the ability of societies to cope and adapt to forest disturbance (Fischer and Frazier 2018). The evaluation of social vulnerability mainly focuses on the questions of how losses caused by natural dis-turbance might influence individuals and how they can respond to these losses. Studies use biophysical and socioeconomic indicators to reach an understanding of how strongly communities/individuals depend on income from forests and how vul-nerable they are to disturbance events. A second research direction assesses how forest disturbances, like forest fires or avalanches, lead to economic losses and increased risks for humans (e.g. O'Neill and Handmer 2012).

Social vulnerability considers three principles: (1) the identification of condi-tions that make people or places susceptible to disturbances (Burton et al. 1993), (2) the assumption that the vulnerability of society is related to social structures (e.g.

adaptive capacity to cope with forest disturbance events) (Keskitalo 2008; Fischer and Frazier 2018), and (3) the evaluation of the hazards of space, such as intensive human use of areas in hotspots of natural disturbance. Social vulnerability therefore refers to the respective position of humans in society, including their social status and relationships. Consequently, social vulnerability is often examined and explained from the perspective of social inequality in society (Watts and Bohle 1993; Blaikie et al. 1994). Fischer and Frazier (2018), for instance, use a wide range of different indicators to assess the social vulnerability, exposure, and adaptive capacity to forest disturbances. Since social relations and social processes are characterized by a high degree of dynamics, social vulnerability can only be understood and evaluated as a dynamic concept (Fischer et al. 2016). Thus, disturbances are usually only implicitly included in the determination of social vulnerability—in contrast to a more explicit consideration in the natural sciences (Fischer et al. 2016).

The literature distinguishes between the taxonomic and the situational approach for measuring social vulnerability (Cutter et al. 2003; Kuhlicke et al. 2011). The taxonomic approach calculates social vulnerability on the basis of individual indicators, which are often presented spatially using maps (Fischer and Frazier 2018). In particular, statistical data on social inequality are used as input variables (Table 17.1). The basic assumption is that the socioeconomic status of a person correlates

Table 17.1 Examples of socioeconomic variables measuring social vulnerability to disturbance

Criteria and indicator (example)	Explanation	Relationship with social vulnerability
Education: proportion of population with only basic school education	Higher education usually leads to higher pay as well as to a better understanding of complex ecological dynamics, such as disturbances (adaptive behaviour)	Higher education reduces vulnerability
Socioeconomic status: missing security because of a lack of political and economic influence (e.g. unemployment)	People with higher income recover faster economic setbacks (including those arising from ecosystem disturbances) because of higher resources, insurance, and well-established social networks	Higher status and higher income reduce vulnerability
Economic dependency on income from ecosystem management (e.g. forestry)	People with diverse income streams recover faster from disturbance-induced losses in the management of ecosystems	High dependence on income from ecosystem management increases vulnerability
Connections in the local community: for example, membership of associations, family members in the vicinity, trust in the local community	People with a strong social network are better informed and can rely on support from the local community to cope with damage related to disturbances	A strong social network reduces vulnerability
Family structure: for example, family size	Large families often have lower resources available per capita, and thus reduced ability to compensate losses related to disturbances	A large family increases social vulnerability

negatively with social vulnerability, that is, poor groups are more affected by disturbance and disaster than rich ones (Cutter et al. 2003; Fischer and Frazier 2018). The
situational approach, in contrast, is based on the actual experiences of people rather
than statistical data. Here, social vulnerability is determined based on an individual's interpretation. In a combination of these two approaches, Parkins and
MacKendrick (2007), for example, showed that the vulnerability of society to a
large-scale bark beetle outbreak can vary greatly between affected communities and
that certain population groups are more affected by an outbreak than others.

17.4   Treating Risks

There are three general options for treating risks: acceptance, reduction, and
collectivization.

Accepting Risks
If the risk assessment results in the insight that an identified risk is small enough to
be accepted, the risk treatment is complete. Note that both acceptance and ignorance of a risk lead to inaction, that is, no risk management measures are taken.
However, the acceptance of a risk as the result of a well-founded risk analysis
(Fig. 17.1) is preferable to the general ignorance of risks. The risk assessment process highlights remaining uncertainties and facilitates the communication of risks
and uncertainties. In other words, risks only become a problem when ignored (Ebert
2013). Accepting a disturbance risk is usually possible where both the probability of
occurrence of a disturbance and the potential for damage, that is, the negative consequences on objectives, are small (Fig. 17.6). At the other end of the spectrum, that

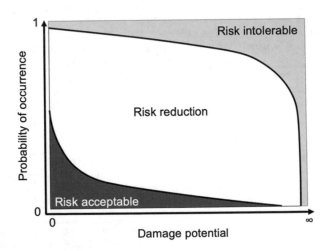

Fig. 17.6 Schematic relationship between risk treatment, probability of occurrence, and damage
potential

is, where the potential for damage is very high or the probability of occurrence is very high, the possibilities of risk reduction are usually strongly limited, which is why it is unlikely that the set objectives can be achieved. As a result, a re-evaluation and modification of objectives are necessary, in order to align them with the prevailing disturbance regime.

Reducing Risks

The reduction of risks is usually most effective when probability of occurrence and damage potential are intermediate (Fig. 17.6). Risk reduction is based on understanding the processes underlying the disturbance (see Chaps. 7, 8, 9, 10, 11, 12 and 13) and the factors influencing the disturbance; this understanding can be achieved by undertaking the risk assessment process described in Sects. 17.2 and 17.3. A distinction can be made between cause-related and effect-related risk reduction measures. The former aims to reduce the occurrence of a disturbance, and the latter aims to reduce their negative effects. When considering risk reduction measures for a certain disturbance, care must be taken to ensure that they do not increase other risks. This is particularly important in long-lived ecosystems such as forests, where the short-term effects of management measures can differ considerably from the long-term effects. For example, the risk management of forest fires in the western United States throughout most of the twentieth century consisted of strong measures to prevent and contain forest fires. This resulted in denser and more structured forests and the accumulation of fuel on the ground. Today, such stands are very susceptible to fire (see Chap. 7) since fuel availability is high and the presence of "ladder fuels" allows ground fires to develop into crown fires. Thus, past risk reduction measures have contributed to a currently increased fire risk (Stephens et al. 2013).

Measures for targeted risk reduction are advisable if the predictability of the risk factor is relatively high and sufficient knowledge about risk reduction is available (Fig. 17.7; Wildavsky 1988; Seidl 2014). If, on the other hand, a risk factor occurs nearly randomly in space and time (high stochasticity) and is therefore not predictable, risk reduction measures cannot be targeted, and the costs of risk reduction often exceed the benefits. Moreover, measures to reduce a risk need to be effective, which requires a good understanding of the risk factor and a thorough evaluation of the measures before they can be broadly applied. Information on both of these aspects can limit risk treatment, for example, in the case of newly introduced invasive diseases and pests (see Chaps. 10, 11, 12 and 13). In such a situation the focus must be shifted from risk reduction to fostering the resilience of the system to a risk factor (see Chap. 5; Seidl 2014).

Collectivizing Risks

In addition to accepting and reducing them, risks can also be transferred to a collective as part of a risk transfer system. This is particularly common for risks with a high potential for damage and a low probability of occurrence (Steinrücken 2008; see also Fig. 17.6). Risk transfer approaches can be divided into insurance

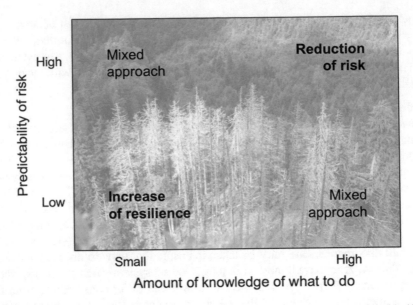

Fig. 17.7 Risk management under uncertainty. Risk reduction measures are advisable if the uncertainties regarding the occurrence and treatment of a risk factor are relatively small. Since uncertainties can never be eliminated fully (*unknown unknowns*), resilience should be the primary focus in a combined approach, which simultaneously aims to reduce risk and increase resilience. (Source: Seidl 2014, modified)

solutions and compensation payments, for example, in the form of government subsidies. In Austria, for instance, there is a disaster relief fund for this purpose (BMI 1996). In the United States, the Federal Emergency Management Agency (FEMA) coordinates responses to disasters (including extreme climatic events). FEMA aims to build a culture of preparedness, but for events that exceed a state's capacity to respond the governor of the state can declare a "state of emergency," and resources and federal funds can be made available for the response. In the context of insurance, a distinction is made between: (1) voluntary private insurance, (2) compulsory insurance, and (3) a state monopoly insurance (Steinrücken 2008). Furthermore, insurance solutions can be based on mixed public and private financial schemes (Ebert 2013)—for example, the Austrian hail insurance [German: *Hagelversicherung*], which is financed by private payments but also receives public subsidies. Both public and private sector approaches have different advantages and disadvantages. Overall, the design of efficient risk transfer systems is usually based on the optimal economic level of precaution. This raises the questions of how the negative effects of natural hazards can be sufficiently buffered while economic losses from such risks are increasing and how forest owners can be motivated to take appropriate precautionary measures (Steinrücken 2008; Fischer 2019).

17.5 Outlook

Because of the growing demand for ecosystem services and a possible increase in disturbance activity as a consequence of climate change, considerations of disturbance risk management will become increasingly important in the future. One focus will need to be on an improved quantification of disturbance risks as well as the economic consequences of disturbance. A second focus should be on the social vulnerability to disturbances and the capacity of societies to cope with disturbances. Furthermore, a stronger integration of risk management considerations into operational management can increase the stability of ecosystem service supply. However, it should also be considered that a complete reduction of uncertainties and risks in ecosystem management is neither possible nor desirable (see Holling & Meffe 1996). A central element in dealing with risks and uncertainties is therefore the promotion of biodiversity, which contributes to risk spreading.

References

Baur P, Bernath K, Holthausen N, Roschewitz A (2003) LOTHAR Ökonomische Auswirkungen des Sturms Lothar im Schweizer Wald, Teil I. Einkommens- und Vermögenswirkungen für die Waldwirtschaft und gesamtwirtschaftliche Beurteilung des Sturms. BUWAL Umwelt-Materialien 157:1–190

Bernetti I, Ciampi C, Fagarazzi C, Sachelli S (2011) The evaluation of forest crop damages due to climate change. An application of Dempster–Shafer method. J For Econ 17:285–297

Berryman AA (1986) Forest insects. Plenum Press, New York, London, Principles and practice of population management, 279 p

Blaikie P, Cannon T, Davis I, Wisner B (1994) At risk: natural hazards, people's vulnerability, and disasters. Routledge, London, 284 p

BMI (1996) Bundesgesetz über Maßnahmen zur Vorbeugung und Beseitigung von Katastrophenschäden. Katastrophenfondsgesetz 1996—KatFG 1996, Austria

Burton I, Kates RW, White GF (1993) The environment as hazard (2nd edn). Guildford Press, London, 290 p

BUWAL (1999) Risikoanalyse bei gravitativen Naturgefahren. BUWAL Umwelt-Materialien 107(1):1–115

Cutter SL, Boruff BJ, Shirley WL (2003) Social vulnerability to environmental hazards. Soc Sci Q 84:242–261

Ebert C (2013) Risikomanagement kompakt. Risiken und Unsicherheiten bewerten und beherrschen (2nd edn). Springer, Berlin, 206 p

Fischer AP (2019) Characterizing behavioral adaptation to climate change in temperate forests. Landsc Urban Plan 188:72–79

Fischer AP, Frazier TG (2018) Social vulnerability to climate change in temperate forest areas: new measures of exposure, sensitivity, and adaptive capacity. Ann Am Assoc Geogr 108:658–678

Fischer AP, Spies TA, Steelman TA, Moseley C, Johnson BR, Bailey JD, Ager AA, Bourgeron P, Charnley S, Collins BM, Kline JD, Leahy JE, Littell JS, Millington DA, Nielsen-Pincus M, Olsen CS, Paveglio TB, Roos CI, Steen-Adams MM, Stevens FR, Vukomanovic J, White EM, Bowman DMJS (2016) Wildfire risk as a socioecological pathology. Front Ecol Environ 14:276–284

Führer E, Nopp U (2001) Ursachen. Facultas-University Verlag, Wien, Vorbeugung und Sanierung von Waldschäden, 514 p

Haimes YY (2004) Risk modeling, assessment, and management (2nd edn). Wiley, Hoboken, 837 p

Hanewinkel M, Hummel S, Albrecht A (2010) Assessing natural hazards in forestry for risk management: a review. Eur J For Res 130:329–351

Hanley N, Spash CL (1995) Cost-benefit analysis and the environment. Edward Elgar Publishing, Cheltenham, 278 p

Holling CS, Meffe GK (1996) Command and control and the pathology of natural resource management. Conserv Biol 10:328–337

Holmes TP, Liebhold AM, Kovacs KF, von Holle B (2010) A spatial-dynamic value transfer model of economic losses from a biological invasion. Ecol Econ 70:86–95

Holthausen N, Hanewinkel M, Holecy J (2004) Risikomanagement in der Forstwirtschaft am Beispiel des Sturmrisikos, vol forstarchiv 75, pp 149–157

ISO (2009) Risk management—principles and guidelines, ISO 31000. International Organisation for Standardization (ed), Geneva

Jactel H, Nicoll BC, Branco M, Gonzalez-Olabarria JR, Grodzki W, Långström B, Moreira F, Netherer S, Orazio C, Piou D, Santos H, Schelhaas MJ, Tojic K, Vodde F (2009) The influences of forest stand management on biotic and abiotic risks of damage. Ann For Sci 66:701–701

Jactel H, Branco M, Duncker P, Gardiner B, Grodzki W, Långström B, Moreira F, Netherer S, Nicoll B, Orazio C, Piou D, Schelhaas MJ, Tojic K (2012) A multicriteria risk analysis to evaluate impacts of forest management alternatives on forest health in Europe. Ecol Soc 17:52

Janda P, Trotsiuk V, Mikoláš M, Bače R, Nagel TA, Seidl R, Seedre M, Morrissey RC, Kucbel S, Jaloviar P, Jasík M, Vysoký J, Šamonil P, Čada V, Mrhalová H, Lábusová J, Nováková MH, Rydval M, Matějů L, Svoboda M (2017) The historical disturbance regime of mountain Norway spruce forests in the Western Carpathians and its influence on current forest structure and composition. For Ecol Manag 388:67–78

Kaplan S, Garrick BJ (1981) On the quantitative definition of risk. Risk Anal 1:11–27

Keskitalo ECH (2008) Vulnerability and adaptive capacity in forestry in northern Europe: a Swedish case study. Clim Change 87:219–234

Knight FH (1921) Risk, uncertainty, and profit. Houghton Mifflin Company, Boston. 381 p

Knoke T, Gosling L, Thom D, Chreptun C, Rammig A, Seidl R (2021) Economic losses from natural disturbances in Norway spruce forests—a quantification using Monte-Carlo simulations. Ecol Econ 185:107046

Körner C (2003) Slow in, rapid out—carbon flux studies and Kyoto targets. Science 300:1242–1243

Kortmann M, Müller JC, Baier R, Bässler C, Buse J, Cholewinska O, Förschler MI, Georgiev KB, Hilszczanski J, Jaroszewicz B, Jaworski T, Kaufmann S, Kuijper D, Lorz J, Lotz A, Lubek A, Mayer M, Mayerhofer S, Meyer S, Moriniere J, Popa F, Reith H, Roth N, Seibold S, Seidl R, Stengel E, Wolski GJ, Thorn S (2021) Ecology versus society: impacts of bark beetle infestations on biodiversity and restorativeness in protected areas of Central Europe. Biol Conserv 254:108931

Kuhlicke C, Scolobig A, Tapsell S, Steinführer A, De Marchi B (2011) Contextualizing social vulnerability: findings from case studies across Europe. Nat. Hazards 58:789–810

Lee J, Kim H, Song C, Kim GS, Lee WK, Son Y (2020) Determining economically viable forest management option with consideration of ecosystem services in Korea: a strategy after successful national forestation. Ecosyst Serv 41:101053

Loomis JB (1993) An investigation into the reliability of intended visitation behavior. Environ Resour Econ 3:183–191

Manion PD (1981) Tree disease concepts. Prentice-Hall, Cornell University, 399 p

Möllmann TB, Möhring B (2017) A practical way to integrate risk in forest management decisions. Ann For Sci 74:75

Montagné-Huck C, Brunette M (2018) Economic analysis of natural forest disturbances: a century of research. J For Econ 32:42–71

Moos C, Thomas M, Pauli B, Bergkamp G, Stoffel M, Dorren L (2019) Economic valuation of ecosystem-based rockfall risk reduction considering disturbances and comparison to structural ⚊⚊⚊⚊⚊⚊ ⚊⚊ ⚊⚊⚊⚊ ⚊⚊⚊⚊⚊⚊ ⚊⚊⚊, ⚊⚊ ⚊⚊⚊⚊

Netherer S, Nopp-Mayr U (2005) Predisposition assessment systems (PAS) as supportive tools in forest management—rating of site and stand-related hazards of bark beetle infestation in the High Tatra Mountains as an example for system application and verification. For Ecol Manag 207:99–107

Netherer S, Panassiti B, Pennerstorfer J, Matthews B (2019) Acute drought is an important driver of bark beetle infestation in Austrian Norway spruce stands. Front For Glob Change 2:Article 39:1–21

Olschewski R, Bebi P, Teich M, Wissen Hayek U, Gret-Regamy A (2012) Avalanche protection by forests—a choice experiment in the Swiss Alps. For Policy Econ 17:19–24

O'Neill SJ, Handmer J (2012) Responding to bushfire risk: the need for transformative adaptation. Environ Res Lett 7:014018

Overbeck M, Schmidt M (2012) Modelling infestation risk of Norway spruce by *Ips typographus* (L.) in the Lower Saxon Harz Mountains (Germany). For Ecol Manag 266:115–125

Parkins JR, MacKendrick NA (2007) Assessing community vulnerability: a study of the mountain pine beetle outbreak in British Columbia, Canada. Glob Environ Chang 17:460–471

Pfeifer W (1989) Etymologisches Wörterbuch des Deutschen. Q-Z. Akad.-Verlag, Berlin

Pfister C (1999) Wetternachhersage—500 Jahre Klimavariationen und Naturkatastrophen (1496–1995). Paul Haupt, Bern, 304 p

Price JI, McCollum DW, Berrens RP (2010) Insect infestation and residential property values: a hedonic analysis of the mountain pine beetle epidemic. For Policy Econ 12:415–422

Purdy G (2010) ISO 31000: 2009—setting a new standard for risk management. Risk Anal 30:881–886

Rose A (2004) Economic principles, issues, and research priorities in hazard loss estimation. In: Okuyama Y, Chang SE (eds) Modeling spatial and economic impacts of disasters. Springer, Berlin, pp 13–36

Sebald J, Senf C, Heiser M, Scheidl C, Pflugmacher D, Seidl R (2019) The effects of forest cover and disturbance on torrential hazards: large-scale evidence from the Eastern Alps. Environ Res Lett 14:114032

Seidl R (2014) The shape of ecosystem management to come: anticipating risks and fostering resilience. BioScience 64:1159–1169

Seidl R, Rammer W (2017) Climate change amplifies the interactions between wind and bark beetle disturbances in forest landscapes. Landsc Ecol 32:1485–1498

Seidl R, Rammer W, Jäger D, Lexer MJ (2008) Impact of bark beetle (*Ips typographus* L.) disturbance on timber production and carbon sequestration in different management strategies under climate change. For Ecol Manag 256:209–220

Seidl R, Fernandes PM, Fonseca TF, Gillet F, Jönsson AM, Merganicová K, Netherer S, Arpaci A, Bontemps JD, Bugmann H, González-Olabarria JR, Lasch P, Meredieu C, Moreira F, Schelhaas MJ, Frits M (2011) Modelling natural disturbances in forest ecosystems: a review. Ecol Model 222:903–924

Seidl R, Thom D, Kautz M, Martin-Benito D, Peltoniemi M, Vacchiano G, Wild J, Ascoli D, Petr M, Honkaniemi J, Lexer MJ, Trotsiuk V, Mairota P, Svoboda M, Fabrika M, Nagel TA, Reyer CPO (2017) Forest disturbances under climate change. Nat Clim Change 7:395–402

Senf C, Sebald J, Seidl R (2021) Increasing canopy mortality affects the future demographic structure of Europe's forests. One Earth 4:1–7

Speight M, Wainhouse D (1989) Ecology and management of forest insects. Clarendon Press, Oxford, 374 p

Stadelmann G, Bugmann H, Wermelinger B, Bigler C (2014) Spatial interactions between storm damage and subsequent infestations by the European spruce bark beetle. For Ecol Manag 318:167–174

Steinrücken T (2008) Wirtschaftspolitische Grundsätze der Gestaltung von Risikotransfersystemen für Naturgefahren. Vierteljahrsh Wirtsch forsch 77:80–97

Stephens SL, Agee JK, Fulé PZ, North MP, Romme WH, Swetnam TW, Turner MG (2013) Managing forests and fire in changing climates. Science 342:41–42

Thom D, Seidl R, Steyrer G, Krehan H, Formayer H (2013) Slow and fast drivers of the natural disturbance regime in Central European forest ecosystems. For Ecol Manag 307:293–302

Thom D, Rammer W, Dirnböck T, Müller J, Kobler J, Katzensteiner K, Helm N, Seidl R (2017) The impacts of climate change and disturbance on spatio-temporal trajectories of biodiversity in a temperate forest landscape. J Appl Ecol 54:28–38

UNEP (2002) Global Environmental Outlook 3: past, present and future perspectives. United Nations Environmental Program, Earthscan

van der Sluijs JP, Janssen PHM, Petersen AC, Kloprogge P, Risbey JS, Tuinstra W, Ravetz JR (2004) Tool catalogue for uncertainty assessment. RIVM/MNP guidance for uncertainty assessment and communication, vol 4. Copernicus Institute, Utrecht, 60 p

Vester F (1983) Der Wert eines Vogels: ein Fensterbilderbuch. Kösel-Verlag, München, 17 p

Watts MJ, Bohle HG (1993) The space of vulnerability: the causal structure of hunger and famine. Prog Hum Geogr 17:43–67

Wildavsky A (1988) Searching for safety. Transaction Publishers, New Brunswick, 253 p

Yousefpour R, Jacobsen JB, Thorsen BJ, Meilby H, Hanewinkel M, Oehler K (2012) A review of decision-making approaches to handle uncertainty and risk in adaptive forest management under climate change. Ann For Sci 69:1–15

Chapter 18
Disturbances and Ecosystem Services

Dominik Thom (iD), Anke Jentsch (iD), and Rupert Seidl (iD)

Abstract Ecosystem services are the benefits people obtain from ecosystems. Disturbances can have multiple, often negative, effects on ecosystem services. Primary production is temporarily reduced by disturbances, while water and nutrient cycles are stimulated by disturbances. Consequently, the production of plant biomass (wood, animal fodder) may be temporarily decreased. In the context of climate regulation, disturbances reduce carbon storage (warming effect) but simultaneously increase albedo (cooling effect). Furthermore, disturbances reduce the protection function of forests against natural hazards. The way disturbances affect cultural services, such as the recreational function of ecosystems, depends on the subjective perception of people.

Keywords Fire · Wind · Bark beetles · Drought · Supporting services · Provisioning services · Regulating services · Cultural services

D. Thom (✉)
Ecosystem Dynamics and Forest Management Group, School of Life Sciences, Technical University of Munich, Freising, Germany

Gund Institute for Environment, University of Vermont, Burlington, VT, USA
e-mail: dominik.thom@tum.de

A. Jentsch
Bayreuth Center for Ecology and Environmental Sciences (BayCEER), University of Bayreuth, Bayreuth, Germany

R. Seidl
Ecosystem Dynamics and Forest Management Group, School of Life Sciences, Technical University of Munich, Freising, Germany

Berchtesgaden National Park, Berchtesgaden, Germany

© The Author(s), under exclusive license to Springer Nature
Switzerland AG 2022
T. Wohlgemuth et al. (eds.), *Disturbance Ecology*, Landscape Series 32,
https://doi.org/10.1007/978-3-030-98756-5_18

413

18.1   Introduction

Ecosystem services embrace all properties of an ecosystem that have a beneficial effect on human well-being. The concept of ecosystem services originally developed from environmental services in the United States. The Millennium Ecosystem Assessment (MEA) addressed the influence of humans on ecosystems globally and promoted a concept of ecosystem services that found broad acceptance. Since then, ecosystem services have been categorized as supporting, providing, regulating, and cultural services (Hassan et al. 2005; Table 18.1).

Terrestrial ecosystems generate a variety of different ecosystem services. Because of the high spatial heterogeneity of ecosystems and the temporal variability of ecosystem processes, the amount of services varies depending on season, elevation, and geographical location. Individual ecosystem services can be distributed very differently across a landscape (Lamy et al. 2016). Landscape heterogeneity can thus significantly increase the range and stability of services (Turner et al. 2013). The value of ecosystem services is determined by local societies and global developments and therefore differs across time and space. Prior to the widespread use of fossil energy sources, severe timber shortages were common in many places, which is why almost all other ecosystem services provided by forests were subordinate to timber production. Today, in times of high population densities with diverse needs, the importance of other ecosystem services is increasing. Consequently, wood is only one of many ecosystem services demanded by society. In densely populated regions of the European Alps, where tourism is often the main income source, the

Table 18.1 Ecosystem services distinguished into four categories as defined by the Millennium Ecosystem Assessment (Hassan et al. 2005)

Categories	Services
Supporting services	Nutrient cycling
Necessary for the production of all other ecosystem services	Soil formation
	Primary production
	Oxygen production
Provisioning services	Food and feed
Products people obtain from ecosystems	Medicine
	Fuel
	Wood and fiber
Regulating services	Water purification
Benefits people obtain from the regulation of ecosystem processes	Climate regulation
	Protection from natural hazards
	Disease regulation
Cultural services	Aesthetic
Nonmaterial benefits people obtain from ecosystems	Spiritual
	Educational
	Recreational

protection of human infrastructure against natural hazards has become a priority (Fig. 18.1). Permanent grasslands have played a central role in producing fodder for livestock for many centuries, but its economic profitability has decreased significantly in recent years (see also Chap. 15). At the same time, these ecosystems have often been shaped by humans over generations, and they have obtained a cultural as well as aesthetic value. They also constitute an important habitat for a number of specialized species, which is why, for instance, alpine and wood pastures (Stuber and Bürgi 2001) are still grazed despite their limited economic value (Fig. 18.2). Thus, societies define both the benefits of ecosystems and the importance of disturbance-induced changes in services in a context-specific way.

Disturbances can have a negative, neutral, or positive impact on ecosystem services. The magnitude of disturbance impacts on ecosystem services depends not only on disturbance type but also on the spatial and temporal context (Vanderwel et al. 2013), disturbance severity and frequency (Wang et al. 2001), site characteristics (Huang et al. 2013), and climate scenarios (Seidl et al. 2008). Ecosystem services are influenced by disturbances in different ways. For example, severe flooding

Fig. 18.1 Forest ecosystems protect houses and human infrastructure from natural hazards, as shown here in the example of Hallstatt, Austria. (Photo: R. Seidl)

Fig. 18.2 Alpine meadows are not only of aesthetic value, but they also constitute an important habitat for specialized species. (Photo: A. Jentsch)

can lead to negative impacts (e.g. erosion and local damage to buildings), but across larger spatial scales, such as in floodplain areas, it can also have positive impacts (e.g. transport of organic sediments that increase soil fertility). Windthrow can reduce timber production, but it can also foster berry bushes or edible herbs as light and temperature conditions change. Since most ecosystems provide several ecosystem services simultaneously, a comprehensive understanding of disturbance effects is needed to define priorities for sustainable management. This chapter summarizes the impacts of major disturbance agents (e.g. fire, wind, insects, and drought) on ecosystem services provided by forests and grasslands.

18.2 Supporting Services

Supporting ecosystem services refer to basic properties of ecosystems on which all other services are based. They include material cycles and the production of biomass.

18.2.1 Net Primary Production

The term net primary production (NPP) refers to the dry matter bound by photo- or chemoautotrophic organisms. That is the fixation of carbon in above- and below-ground parts of plants minus the carbon consumed by plant respiration. Disturbances cause a more or less long-lasting reduction of NPP (Peters et al. 2013). How fast NPP recovers after a disturbance (one characteristic of resilience; see Chap. 5) depends on disturbance severity as well as on ecosystem development before the

disturbance. NPP of forest ecosystems typically peaks in intermediate developmen-
tal stages, a period of relatively high productivity and high accumulation of biomass (Chen et al. 2006). If
a disturbance occurs in forests of intermediate development, it may thus have a
stronger impact on NPP than if the disturbance occurred in forests in earlier or later
development stages. While the NPP of a forest ecosystem usually returns to the
level before disturbance within a few years after a low-severity disturbance (e.g.
after windthrow of individual trees), the recovery process may take decades after a
high-severity event (e.g. after a large-scale forest fire) (Peters et al. 2013). In very
frequently disturbed grasslands, the typical disturbance regime of grazing and mowing as well as drought and fire reduces the aboveground biomass in the short term
(see Chap. 15); however, the NPP of temperate grasslands is considered to be very
resilient to disturbances (Fig. 18.1; Jentsch et al. 2011). Under otherwise comparable site conditions, the productivity and performance of many ecosystems increase
with increasing plant species diversity (e.g. promoted by disturbances, see Chap. 4)
in the long term (Hector et al. 1999; Silva Pedro et al. 2016).

18.2.2   Water and Nutrient Cycles

Natural vegetation reflects the prevailing environmental conditions. With ongoing
succession, ecosystems develop specific material cycles. Management alters the
natural disturbance regime and the vegetation in a regionally characteristic way. As
a result, material cycles often change significantly. This can be illustrated by comparing arable and forest ecosystems. Water and nutrient runoff are problematic in
open fields. The shallow rooting vegetation cannot absorb and utilize all precipitation and nutrients that are often added in the form of fertilizers. In addition, arable
lands tend to dry out quickly during a drought, which can cause mortality of cultivated crops. In contrast, forests are effective systems storing large amounts of water
and nutrients (Schume et al. 2004; Berger et al. 2006). The roots of trees penetrate
deeply into the ground. Moreover, evapotranspiration regulates the local climatic
conditions (microclimate) efficiently (De Frenne et al. 2019). Thus, forests are self-
maintaining, and trees can withstand droughts better than crops (Stuart-Haëntjens
et al. 2018). For permanent grasslands, it has been demonstrated that species-rich
plant communities have more efficient biogeochemical cycles than species-poor
plant communities. For example, less nitrate from fertilization and atmospheric
inputs runs off with seepage water, contaminating the groundwater and, thus, reducing water quality (Scherer-Lorenzen et al. 2003). Disturbances can strongly influence water and nutrient cycles. Since disturbances produce large amounts of dead
biomass, nutrients released by disturbances cannot always be absorbed entirely by
the remaining vegetation (e.g. Beudert et al. 2015). Depending on soil type, this
leads to nutrient leaching and thus to a more or less long-lasting impoverishment of
the ecosystem. In grassland systems, changes in the soil water balance, for example,
driven by droughts or heavy rainfall events, can greatly reduce the density and functionality of the soil microflora and microfauna for several months to years (Sheik

et al. 2011). Direct effects (nutrient transport, redox conditions, water supply) are of similar importance as indirect effects (i.e. reduced photosynthesis rates of plants during longer dry periods and the associated reduced release of carbohydrates into the soil) (Wall et al. 2013). On the other hand, minor disturbance events in ecosystems can also have long-term positive effects on the water and nutrient cycle if they contribute to the revitalization of a system by returning organic material to the nutrient cycle. For example, after a disturbance-induced reduction in canopy cover in coniferous forests, more light reaches the forest floor, warms it, and accelerates the decomposition of the humus layer that has formed during stand development (Mladenoff 1987). The newly developing stand may benefit from the additional release of nutrients, and provides improved conditions for the establishment of multiple plant species.

18.3 Provisioning Services

This category of ecosystem services includes products that humans obtain from ecosystems and which can usually be quantified in monetary terms. Disturbances often negatively affect provisioning services. Important examples of provisioning services are wood from forests and fodder for livestock from arable fields (e.g. corn), meadows, and pastures.

18.3.1 Biomass

With the disturbance-induced decrease of live biomass, reductions in forest productivity can last for several years because the necessary structures for optimal photosynthesis performance as well as nutrient and water uptake need time to recover after disturbances. Also, an early reset of forest development before the rotation age is reached may cause growth losses reducing wood production. The value of windthrown timber is often reduced by mechanical damages (e.g. wind breakage) or fungal infestation (e.g. blue stain fungi). Furthermore, harvesting is more difficult in wind-disturbed stands, and consequently economic revenues are reduced. After large-scale storm events (see Chap. 8) and bark beetle calamities (see Chap. 12), an oversupply of timber causes price reductions that may last a few years (Prestemon and Holmes 2004). Timber growth of the remaining forest may also be affected by disturbances. Growth can be reduced for several years after a disturbance event as a result of root, trunk, or crown damage (Seidl and Blennow 2012). The influence of disturbances on timber production is thus usually negative in the first years after a disturbance event.

However, long-term impacts of disturbances on timber production remain uncertain. After the culmination of timber growth in the optimal phase, productivity often decreases with increasing stand age but may increase again as forests develop more complex structures as they age (Thom et al. 2019). Disturbances may reset forest development cycles and promote complex structures which may enhance productivity due to niche complementarity. In addition, disturbed areas with fast-growing, early successional species have comparatively high growth rates (Kaohian et al. 2005). Juvenile trees benefit from the release of nutrients resulting from an accelerated mineralization as temperatures are elevated in stand openings. They may also benefit from the decomposition of deadwood in the longer term, provided that the deadwood is not salvaged. In this context, the variability of natural disturbance regimes should also be noted. In ecosystems characterized by wildfire, disturbances can be essential for the regeneration of certain tree species and their long-term productivity (e.g. *Pinus palustris* Mill. in North America; Gilliam and Platt 1999) as fires consume part of the accumulating litter layer which prevents seed germination and stores nutrients (e.g. in boreal ecosystems; Pollock and Payette 2010).

In permanent grasslands primarily used for fodder production, disturbances also initially reduce biomass. However, the resilience of this system is significantly higher than the resilience of forests (Stuart-Haëntjens et al. 2018). In grassland, returning to the original state usually takes a few years, whereas forest recovery may take decades or centuries. The discussion about the potential for biomass production in permanent grasslands to contribute to the generation of renewable energy has gained traction. However, the impact of disturbances on biomass production in permanent grassland is difficult to evaluate as neither the risks nor the benefits have been sufficiently investigated (Eggers et al. 2009; Hellmann and Verburg 2010).

18.3.2 Food and Medicinal Products

While food obtained from forests through hunting animals and collecting wild fruits or plants plays a subordinated role in industrialized countries, in other regions of the world, these natural resources represent a significant share of the population's food supply. Especially people living in tropical and subtropical regions meet their food demands to a large degree from forests or small-scale agriculture, such as in home gardens or other agroforestry systems. The survival of these people can be endangered by large-scale disturbances, in particular, by wildfire. Fires do not only combust edible plants; they can also reduce the populations of hunted animals (Zamora et al. 2010). This negative disturbance effect lasts at least as long as these animals cannot find fodder on the disturbed area. However, once a rich herb layer has been established, an improvement in habitat conditions for wildlife or edible plants is also possible, so that in the mid-term after a fire, there might be an increase in the populations of hunted wildlife species and the amounts of fruits and vegetables.

For the agricultural sector, disturbances are regarded as threat. Warming, as well as increases in frequency and severity of extreme weather events, leads to reductions in the food production. Pest infestations, diseases, hailstorms, flooding, and drought negatively affect crop plants and livestock (Tubiello et al. 2007). In addition, drought and heavy rainfall alter the production of leaf substances that can reduce the fodder quality, the rate of litter decomposition, and the activity of soil microbes (Joos et al. 2010; Walter et al. 2012).

In the context of pharmaceutical products, a disturbance-driven increase in species diversity might be positive. Some plants, such as deadly nightshade (*Atropa belladonna* L.), whose ingredients have a fever-reducing and pain-relieving effect, need canopy openings for their growth. Other plants, such as the antispasmodic hedge woundwort (*Stachys sylvatica* L.), prefer undisturbed forests. Mixed-severity disturbances provide different environmental conditions and may thus increase the beta-diversity of plants. In addition to well-known medicinal plants and their ingredients for the production of drugs for humans, plants contribute to animal health. Phytotherapeutics are particularly important with respect to diseases that are resistant to conventional pharmaceutical products—for instance, plant-based products are promising as anthelmintics for treatment of infections by drug-resistant parasitic helminth worms (of both livestock and humans). Phytomedicine may gain importance in the future as the use of antibiotics and other drugs is increasingly being critically questioned in human and veterinary medicine.

18.4   Regulating Services

Regulating services are essentially ecosystem services that protect people and infrastructure against harmful environmental influences and pollution as well as natural hazards. Disturbance impacts on regulating functions are often quantifiable, for instance, via changes in the vegetation such as canopy cover (as an indicator of avalanche protection) or via carbon fluxes (as an indicator of the climate-regulating function of forests). However, the net effect of disturbances on regulating ecosystem services is usually complex and requires consideration of the temporal and spatial dynamics of the disturbance regime. For example, the protection against rockfall and avalanches is not impaired after a low-severity disturbance event and especially if deadwood remains in the forest. In the long term, such a disturbance may even lead to an increase of the protection function (e.g. due to an increasing stand density). However, the loss of biomass after a large-scale disturbance of high severity usually leads to a strong reduction of the protection function against natural hazards (Cordonnier et al. 2008). The regulation of natural hazards in mountainous areas is becoming more and more important because of the increase in population density (e.g. in the Alps, where tourism has increased considerably). The same holds true for water and air purification in polluted regions (e.g. in metropolitan areas like Beijing). Furthermore, the climate-regulating function of ecosystems is of increasing importance as greenhouse gas emissions continue to increase.

18.4.1 Climate-Regulating Services

The sequestration and storage of carbon is by far the most frequently discussed climate-regulating service of ecosystems. Compared to other terrestrial ecosystems, forests are of particular importance for the uptake and long-term storage of atmospheric carbon. It is estimated that forests currently absorb about 30% of anthropogenic CO_2 emissions (Pan et al. 2011). The vast majority of studies reveal a negative effect of disturbances on carbon storage, mainly driven by a reduction of living biomass in ecosystems and increasing soil respiration after disturbances (Thom and Seidl 2016). Prominent examples are the recent outbreaks of the mountain pine bark beetle (*Dendroctonus ponderosae* Hopkins) in British Columbia, Canada (see also Box 12.1 in Chap. 12) as well as of the European spruce bark beetle (*Ips typographus* L.) outbreaks after windthrow and extreme droughts in Europe (Kurz et al. 2008; Lindroth et al. 2009; Hlásny et al. 2021). In boreal and (sub)tropical forest ecosystems, fire is the most important disturbance agent altering the carbon balance (see Conard et al. 2002). In terms of their climate impact, forests are referred to as "slow in, fast out" systems (Körner 2003), because the carbon stored over decades can be released quickly into the atmosphere through disturbances. However, there are also examples of long-term positive disturbance impacts. At a study site in Ontario, carbon in aboveground live biomass was not highest in old forests, but peaked 92 years after fire (Seedre and Chen 2010). At the same site, the highest amounts of soil carbon were found between 29 and 149 years after fire (Chen and Shrestha 2012). That means the maximum aboveground carbon storage in this ecosystem is already reached within decades after a disturbance and does not level off or continue to increase as forests age, as suggested by other studies (Keeton et al. 2011; Thom et al. 2019). A disturbance-driven transformation of forests into a different ecosystem type can also have a positive effect on climate regulation. For instance, a study from Alaska suggested an accelerated expansion of a bog after forest fire ultimately storing more carbon in the long term (Myers-Smith et al. 2008).

In comparison to forest ecosystems, grasslands are a weaker CO_2 sink (see also Chap. 15). The strength of the sink increases with cultivation intensity; however, this comes at the expense of higher emissions of the greenhouse gas nitrous oxide (N_2O) (Soussana et al. 2007; Chang et al. 2015). In particular, extensively managed meadows of Central European low mountain ranges (meadows with annual or bi-annual mowing cycles) are carbon sinks (Foken and Lüers 2015). Management intensification increases CO_2 exchange rates of grasslands (photosynthesis or respiration) but reduces CO_2 exchange rates on unfertilized fallow land (Schmitt et al. 2010).

Despite intensive research, some disturbance effects on the ecosystems' carbon balance remain poorly understood. While aboveground processes are relatively well understood, belowground processes are more challenging to investigate. Hence, there is considerable uncertainty in estimating disturbance impacts on belowground carbon fluxes (Thom and Seidl 2016) (Box 18.1).

Box 18.1: Ecological Novelty: A Challenge for Nature Conservation and Society

Anke Jentsch (iD)
Bayreuth Center for Ecology and Environmental Sciences (BayCEER),
University of Bayreuth, Bayreuth, Germany
e-mail: anke.jentsch@uni-bayreuth.de

Ecological novelty is a term used to describe newly formed ecosystems that arise from rapid anthropogenic environmental changes and differ from the original ecosystems in composition, function, and character (Hobbs et al. 2006; Kueffer 2015). Typical examples include: massive land-use changes, such as the introduction of offshore wind power plants along North European coasts, progressive fragmentation (e.g. as a result of oil exploration drilling in boreal primary forests), the spread of non-native species such as nitrogen-fixing legumes after their planting for slope stabilization, atmospheric nitrogen inputs near urban agglomerations, various aspects of climate change (e.g. heat waves with new maximum temperatures as well as rapidly changing temperatures), and new disturbance regimes (e.g. heat pulses in running water because of the discharge of cooling water). New ecosystems can thus emerge in a variety of ways (Box Fig. 1)—for example, by degradation of original and seminatural ecosystems, by creation of previously nonexistent structures (Ayanu et al. 2015), or by abandonment of intensively used agricultural land under changed environmental conditions. A global example of novel landscapes are nocturnal light spaces created by street lights, which disrupt the behavior of nocturnal animals, including innumerable insects that orient themselves using polarized sun and moon light, but lose this ability to orient themselves in the diffuse light patterns of street lamps (e.g. the incandescent

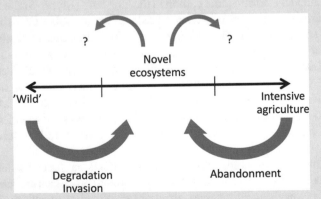

Box Fig. 1 Schematic representation of the development of novel ecosystems through degradation of near-natural ecosystems or after the abandonment of the use of intensively farmed agricultural land. (From Hobbs et al. 2006, repainted)

(continued)

Box 18.1 (continued)

lights commonly used in the past, or LED lights that are becoming increasingly common as street lights).

The introduction of alien species (*novel organisms*), including plants, animals, pathogens, and genetically modified organisms, and the disappearance of native or long-cultivated species as a result of land abandonment or degradation lead to *novel species communities* (Box Fig. 2) and to *novel species interactions* between species and finally to *novel communities* and *novel ecosystems*. Under the changed environmental conditions, so-called hybrid systems (Hobbs et al. 2006) are created, in which both original and new species are mixed. Further environmental changes lead to completely new ecosystems with different species, interactions, and functions than in the original ecosystem. The introduction of alien species can also change the disturbance regime. For example, fire frequencies increase after immigration of invasive species in Mediterranean grasslands or after planting of eucalypts (*Eucalyptus* spp.) in Mediterranean forest areas. Another example is the heat pulse downstream from power plants that occurs as the result of the drained cooling water

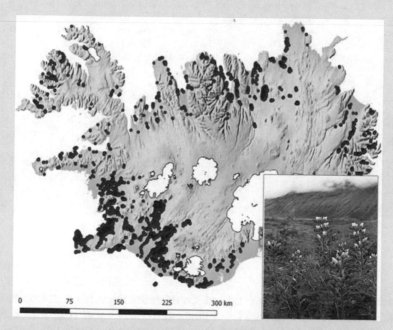

Box Fig. 2 Distribution of the invasive Nootka lupin (*Lupinus nootkatensis* Donn ex Sims) in Iceland, which was introduced several decades ago to stabilize slopes and improve soil quality. Today, the lupin has become a problem there because it colonizes almost all of the island's lowland ecosystems and reduces the local biodiversity. (Vetter et al. 2018)

(continued)

Box 18.1 (continued)

entering rivers; the resulting temperatures are beyond the ecological niche of the original organisms.

Ecological novelty is a growing challenge for nature conservation research and protected area management worldwide (Bridgewater et al. 2011). In ecosystems that have already been subject to severe environmental changes over a long period of time, linking historical environmental conditions and natural disturbance regimes is often difficult and makes little sense. Therefore, management and protection of species, processes, as well as resources/land focus on ecosystems that fulfil particular functions. In doing so, the traditional, site-specific focus on the existing species assemblages and historical conservation targets has been abandoned (Hobbs et al. 2009; Larson and Kueffer 2013).

References

Ayanu Y, Jentsch A, Müller-Mahn D, Rettberg S, Romankiewicz C, Koellner T (2015) Ecosystem engineer unleashed: *Prosopis juliflora* threatening ecosystem services? Reg Environ Change 15:155–167

Bridgewater P, Higgs ES, Hobbs RJ, Jackson ST (2011) Engaging with novel ecosystems. Front Ecol Environ 9:423–423

Hobbs RJ, Arico S, Aronson J, Baron JS, Bridgewater P, Cramer VA, Epstein PR, Ewel JJ, Klink CA, Lugo AE (2006) Novel ecosystems: theoretical and management aspects of the new ecological world order. Glob Ecol Biogeogr 15:1–7

Hobbs RJ, Higgs E, Harris JA (2009) Novel ecosystems: implications for conservation and restoration. Trends Ecol Evol 24:599–605

Kueffer C (2015) Ecological novelty: towards an interdisciplinary understanding of ecological change in the Anthropocene. In: Greschke HM, Tischler J (eds) Grounding global climate change: contributions from the social and cultural sciences. Wiley-Blackwell, pp 19–38

Larson BM, Kueffer C (2013) Managing invasive species amidst high uncertainty and novelty. Trends Ecol Evol 28:255–256

Vetter V, Tjaden N, Jaeschke A, Buhk C, Wahl V, Wasowicz P, Jentsch A (2018) Invasion of a legume ecosystem engineer in a cold biome alters plant biodiversity. Front Plant Sci 9:Article 715

Numerous studies have investigated disturbances impacts on carbon storage (e.g. Seidl et al. 2014). There are fewer studies on other climate regulating ecosystem services (e.g. Thom et al. 2017). One of these services is albedo, that is, the reflection of solar radiation from the Earth's surface back to the atmosphere. A high albedo means less absorption of radiation and thus less warming of the

Earth's surface. In general, the reflectivity of snow is highest, whereas, for instance, a dark coniferous forest has a low albedo value. Disturbances create gaps with potentially greater snow depth and snow cover duration in winter. Furthermore, disturbances in coniferous forests may increase the proportion of broadleaved trees, ultimately resulting in a long-term increase in albedo (Euskirchen et al. 2009; Dore et al. 2012). In grasslands, mowing and grazing, fire, and drought all reduce the leaf area index. This leads to a reduction of the radiation absorbed by the vegetation and, depending on soil color and moisture, to an increasing albedo (see Chap. 15). Disturbances may thus have a positive influence on the radiation balance of ecosystems. Hence, the positive effect of disturbances on albedo may (partially) offset the negative effect of disturbances on the carbon balance of ecosystems.

Another climate-regulating service that is affected by disturbances and that has, as yet, received little attention is the latent heat flux. Latent heat is energy absorbed or released by a phase change of water. When water evaporates, energy is transferred from the Earth's surface into the atmosphere and induces a local cooling effect. However, water vapor is a greenhouse gas itself and can thus contribute to warming on a larger scale. Such complex secondary effects, which also include cloud formation, remain highly uncertain to date. Therefore, in the following, we only discuss the direct disturbance effect on the climate-regulating function of ecosystems driven by latent heat fluxes. Disturbance-induced alterations of the vegetation can have different effects on the latent heat flux of ecosystems over time. For example, a short-term reduction of biomass results in less evapotranspiration (i.e. less cooling), whereas changes in vegetation characteristics may have the opposite effect in the long term. In particular, changes in tree species composition may have a significant influence on the amount of water released into the atmosphere. A simulation study of forest ecosystems in a mountainous landscape in Austria indicates that a disturbance-induced increase in the dominance of broadleaved trees under climate change elevates evapotranspiration rates. Thus, from this point of view, disturbances have a positive effect on the climate-regulating function of the forest (Thom et al. 2017). A disturbance and climate-related change in evapotranspiration can also be assumed in grasslands, although the effect is likely much smaller compared to forest ecosystems.

18.4.2 Water Pollution Control

Soil and vegetation filter and recycle water from deeper soil layers back to the surface and into the atmosphere. In open landscapes, mowing, grazing, fire, and drought interfere in many ways with the water balance and its components with consequences for transpiration and evaporation, precipitation infiltration, surface runoff, and soil water content (see Chap. 15). Thus, the reduction of evapotranspiration during grazing and mowing leads to a higher soil water content compared to undisturbed grasslands. A disturbance of the ecosystem also leads to a faster decomposition of biomass and increased leaching of the topsoil, so that more substances enter the groundwater (Mikkelson et al. 2013). The nitrogen concentration is an important characteristic of drinking water quality. Moreover, heavy metals such as mercury can also be released (Emelko et al. 2011). In disturbance impact assessments, the point in time when a disturbance occurred and the rate of leaching into the groundwater must be considered. For example, in open landscapes, fertilization greatly increases nitrogen inputs which may lead to nitrogen leaching into the groundwater as well as increased N_2O emissions (see Chap. 15). Since fires cause an abrupt and rapid transformation of biomass in grasslands and forests, the nitrogen levels in groundwater are particularly high in the first years after a fire and can affect drinking water quality (Kučerová et al. 2008; Emelko et al. 2011).

In contrast, bark beetle infestation of trees is a slower process. It takes several months or even years for trees to die. Moreover, the decomposition of deadwood may take decades. In addition, as the overstorey canopy opens, the herbaceous and shrubby vegetation of the understorey can establish, which requires nitrogen and water and therefore reduces precipitation runoff peaks (although to a lesser extent than trees). The release of harmful substances into the groundwater is thus buffered, and that means the nitrogen content in the water increases in the first years after a bark beetle disturbance, but in many cases remains below a critical threshold (Huber 2005; Beudert et al. 2015). Disturbances caused by wind result in a faster decomposition rate of organic material than bark beetle disturbances, because most windthrown material is in contact with the forest floor. In turn, the growth of a buffering herbaceous and shrubby vegetation can be delayed by downed deadwood. Hence, it can be assumed that the effects of wind disturbances on drinking water are in-between those of fire and bark beetle outbreaks and have, at least in the mid-term, only a moderate influence on water quality (Scharenbroch and Bockheim 2008).

18.4.3 Protection Against Natural Hazards and Erosion Control

A continuous vegetation cover provides protection in many ways. In particular, forests provide effective and cost-efficient protection against gravitational natural hazards (i.e. rockfalls, avalanches, mudflows, floods) for humans and infrastructure (see Chap. 9). The protection against natural hazards is of greatest importance in densely populated mountain areas and especially in the European Alps. The demands on the protection forest vary with the locally relevant natural hazards. However, in general, the aim should be to maintain a stable permanent stocking with relatively high stand density. To ensure long-term protection against natural hazards in mountainous regions, a mixture of different tree species, sufficient natural regeneration, and an appropriate forest structure is recommended (Dorren et al. 2004). Disturbances open forest canopies and reduce stand densities. Thus, disturbances are considered a threat to the overall protective function of forests (Zurbriggen et al. 2014). Because of the long recovery process of mountain forests, it is often very difficult to quickly restore sufficient protection (Brang et al. 2006; Brang et al. 2015). However, a few studies have revealed that the high structural diversity in the first years after a disturbance (e.g. windthrow) can lead to increased protection against avalanches and rockfall (Wohlgemuth et al. 2017). This can be explained by the high terrain roughness caused by root plates and downed deadwood after a storm event. However, advancing decomposition and the associated destabilization of structural elements reduces the protection function over time, while a rapid establishment of the new tree regeneration is necessary to ensure long-term protection against natural hazards (Schwitter et al. 2015). For erosion protection, any kind of vegetation cover is better than bare ground. Comparing different forms of land use, forests provide the highest level of protection against erosion, especially to protect steep slopes (El Kateb et al. 2013). Larger gaps created by natural or anthropogenic disturbances should thus be avoided in areas with high erosion risk. In grasslands, grazing and (machine-supported) mowing often lead to soil compaction which reduces the infiltration of precipitation into the soil and consequently increases surface runoff and soil erosion (see also Chap. 15). Thus, with increasing rainfall intensity in agricultural landscapes, permanent grasslands play an important role in erosion control. Surface runoff is reduced in grasslands during heavy rainfall events as they have a higher surface roughness and seepage than bare ground. Therefore, soil erosion is not an issue, even in species-poor grasslands.

18.5 Cultural Services

In comparison to other ecosystem services, cultural services are more challenging to assess quantitatively, since they are often difficult to measure, nonmaterial, and intrinsic ecosystem services, and in addition, they are valued very differently depending on culture and individual perception. Landscapes offer nature experience, recreation, inspiration, and cultural identity and are sometimes even perceived as sacred sites. For example, the mighty kauri (*Agathis australis* (D.Don) Lindl.) trees are sacred to the Maori in New Zealand, and lime (*Tilia* spp.) trees were once sacred to the Germanic peoples in Europe. A diversity of structural elements, the abundance of flowers, and a variety of animals, such as butterflies or birds, improve the positive perception about a landscape (Clergeau et al. 2001). There is comparatively little literature available regarding the disturbance impacts on this group of ecosystem services. The influence of disturbances on the recreational effects of ecosystems has been investigated the most and will be explained in more detail in this section.

18.5.1 Recreation

The evaluation of disturbance impacts on the recreational value of forests and other ecosystems depends on the individual and cultural perception. If a major disturbance event occurs, the local population and people seeking recreation often react negatively at first. However, the high diversity of plant and animal species on disturbed areas can subsequently improve this perception significantly.

A good example of a change in the attitude toward disturbances is the Bavarian Forest National Park, where several thousand hectares of forest have been disturbed by bark beetles (especially *Ips typographus*, see Chap. 12) and wind in recent decades. While the local population and tourists were concerned and skeptical about the development of the national park after the beginning of the bark beetle mass outbreak in the early 1990s, the attitude toward this disturbance, which is unique in Germany, is overall neutral now and varies with age and level of education: younger and more educated people tend to see the disturbance more positively than older and less educated people (Müller and Job 2009). A similar study has been conducted in two national parks in British Columbia (Canada), where a bark beetle mass outbreak has been ongoing since the 1990s (see Box 12.1 in Chap. 12). Interviewed tourists described their emotions toward the bark beetle disturbance significantly rather negatively (McFarlane and Watson 2008). Furthermore, a survey has been conducted in Canada on the aesthetic value of forests disturbed by forest fire. In this case, people indicated an overall negative perception toward the disturbance (Hunt and Haider 2004).

Fig. 18.3 Soil disturbances with (**a**) topsoil removal and (**b**) sheep grazing as specific nature conservation measures for the restoration of sand dune habitats. (Photos: A. Jentsch)

Another example concerning the disturbance impacts on recreational services is the restoration of nature reserves created on military training zones. In certain cases, the designated nature reserves have been abandoned by the military. As a result, these areas often lack the soil disturbance regimes that had previously been important for many protected and endangered species (Jentsch 2007; Jentsch et al. 2009). Soil disturbances, such as tank movement (in the case of the military training zones), or tractor movement, topsoil erosion, nomadic sheep herding, or horse grazing, drive the vegetation dynamics in inland sand dune habitats of Central Europe. These disturbances need to be incorporated into nature conservation measures in order to maintain the presence of certain species (Fig. 18.3); however, the measures are often perceived negatively by the public, and there is a need for information campaigns to explain why such measures are needed.

18.6 Summary

Disturbance effects on ecosystem services depend on a variety of factors. A comprehensive analysis of temperate and boreal forests based on 478 scientific publications on disturbance effects on ecosystem services (Thom and Seidl 2016) revealed that natural disturbances have a predominantly negative effect on forest ecosystem services. Only albedo—a climate-regulating service—is clearly positively affected by disturbances (Fig. 18.4). The review study further indicated that all ecosystem service categories are similarly affected by disturbances (Fig. 18.5).

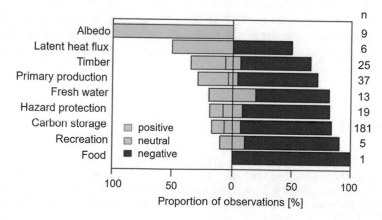

Fig. 18.4 Disturbance impacts on individual forest ecosystem services (n = number of observations). (Source: Thom and Seidl 2016)

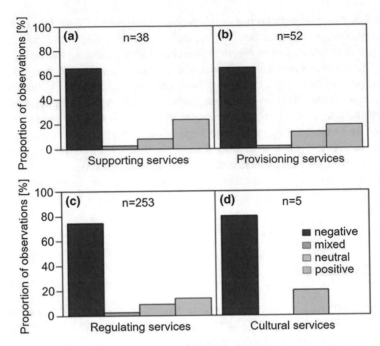

Fig. 18.5 Effects of disturbances on ecosystem service categories—according to the Millennium Ecosystem Assessment (Hassan et al. 2005): (**a**) supporting, (**b**) provisioning, (**c**) regulating, and (**d**) cultural ecosystem services (*n* = number of observations). (Source: Thom and Seidl 2016)

Supporting and provisioning services of Central European grasslands, such as biomass production and fodder quality, are highly resistant and resilient to disturbances. Resistance and resilience increase with species diversity. Experimental studies from different countries confirm a continued biomass supply even under severe summer drought or changes in the mowing regime. However, other supporting, provisioning, and regulating services, such as nutrient and water cycling, are more strongly influenced by extreme weather events and land-use changes than biomass production.

References

Berger TW, Swoboda S, Prohaska T, Glatzel G (2006) The role of calcium uptake from deep soils for spruce *(Picea abies)* and beech *(Fagus sylvatica)*. For Ecol Manag 229:234–246

Beudert B, Bässler C, Thorn S, Noss R, Schröder B, Dieffenbach-Fries H, Foullois N, Müller J (2015) Bark beetles increase biodiversity while maintaining drinking water quality. Conserv Lett 8:272–281

Brang P, Schönenberger W, Frehner M, Schwitter R, Thormann J-J, Wasser B (2006) Management of protection forests in the European Alps: an overview. For Snow Landsc Res 80:23–44

Brang P, Hilfiker S, Wasem U, Schwyzer A, Wohlgemuth T (2015) Langzeitforschung auf Sturmflächen zeigt Potenzial und Grenzen der Naturverjüngung. Schweiz Z Forstwes 100,147–158

Chang J, Ciais P, Viovy N, Vuichard N, Sultan B, Soussana J-F (2015) The greenhouse gas balance of European grasslands. Glob Chang Biol 21:3748–3761

Chen HYH, Shrestha BM (2012) Stand age, fire and clearcutting affect soil organic carbon and aggregation of mineral soils in boreal forests. Soil Biol Biochem 50:149–157

Clergeau P, Mennechez G, Sauvage A, Lemoine A (2001) Human perception and appreciation of birds: a motivation for wildlife conservation in urban environments of France. In: Marzluff JM, Bowman R, Donnelly R (eds) Avian ecology and conservation in an urbanizing world. Kluwer Academic Publishers, Boston, pp 68–88

Conard SG, Sukhinin AI, Stocks BJ, Cahoon DR, Davidenko EP, Ivanova GA (2002) Determining effects of area burned and fire severity on carbon cycling and emissions in Siberia. Clim Chang 55:197–211

Cordonnier T, Courbaud B, Berger F, Franc A (2008) Permanence of resilience and protection efficiency in mountain Norway spruce forest stands: a simulation study. For Ecol Manag 256:347–354

Coursolle C, Margolis HA, Barr AG, Black TA, Amiro BD, McCaughey JH, Flanagan LB, Lafleur PM, Roulet NT, Bourque CP-A, Arain MA, Wofsy SC, Dunn A, Morgenstern K, Orchansky AL, Bernier PY, Chen JM, Kidston J, Saigusa N, Hedstrom N (2006) Late-summer carbon fluxes from Canadian forests and peatlands along an east west continental transect. Can J For Res 36:783–800

De Frenne P, Zellweger F, Rodríguez-Sánchez F, Scheffers BR, Hylander K, Luoto M, Vellend M, Verheyen K, Lenoir J (2019) Global buffering of temperatures under forest canopies. Nat Ecol Evol 3:744–749

Dore S, Montes-Helu M, Hart SC, Hungate BA, Koch GW, Moon JB, Finkral AJ, Kolb TE (2012) Recovery of ponderosa pine ecosystem carbon and water fluxes from thinning and stand-replacing fire. Glob Chang Biol 18:3171–3185

Dorren LKA, Berger F, Imeson AC, Maier B, Rey F (2004) Integrity, stability and management of protection forests in the European Alps. For Ecol Manag 195:165–176

Eggers J, Tröltzsch K, Falcucci A, Maiorano L, Verburg PH, Framstad E, Louette G, Maes D, Nagy S, Ozinga W, Delbaere B (2009) Is biofuel policy harming biodiversity in Europe? GCB Bioenergy 1:18–34

El Kateb H, Zhang H, Zhang P, Mosandl R (2013) Soil erosion and surface runoff on different vegetation covers and slope gradients: a field experiment in Southern Shaanxi Province, China. Catena 105:1–10

Emelko MB, Silins U, Bladon KD, Stone M (2011) Implications of land disturbance on drinking water treatability in a changing climate: demonstrating the need for «source water supply and protection» strategies. Water Res 45:461–472

Euskirchen ES, McGuire AD, Rupp TS, Chapin FS, Walsh JE (2009) Projected changes in atmospheric heating due to changes in fire disturbance and the snow season in the western Arctic, 2003–2100. J Geophys Res-Biogeo 114:G04022

Foken T, Lüers J (2015) Regionale Ausprägung des Klimawandels in Oberfranken. In: Obermaier G, Samimi C (eds) Folgen des Klimawandels. Bayreuther Kontaktstudium Geographie, Band 8. Universität Bayreuth, pp 33–42

Gilliam FS, Platt WJ (1999) Effects of long-term fire exclusion on tree species composition and stand structure in an old-growth Pinus palustris (longleaf pine) forest. Plant Ecol 140:15–26

Hassan R, Scholes R, Ash N (2005) Millennium Ecosystem Assessment. In: Ecosystems and human well-being: current state and trends, vol 1. Island Press, Washington, DC, 917 p

Hector A, Schmid B, Beierkuhnlein C, Caldeira MC, Diemer M, Dimitrakopoulos PG, Finn J, Freitas H, Giller PS, Good J, Harris R, Högberg P, Huss-Danell K, Joshi J, Jumpponen A, Körner C, Leadley PW, Loreau M, Minns A, Mulder CPH, O'Donovan G, Otway SJ, Pereira JS, Prinz A, Read DJ, Scherer-Lorenzen M, Schulze ED, Siamantziouras A-SD, Spehn EM, Terry AC, Troumbis AY, Woodward FI, Yachi SL, Lawton JH (1999) Plant diversity and productivity experiments in European grasslands. Science 286:1123–1127

Hellmann F, Verburg PH (2010) Impact assessment of the European biofuel directive on land use and biodiversity. J Environ Manag 91:1389–1396

Hlásny T, Zimová S, Merganičová K, Štěpánek P, Modlinger R, Turčáni M (2021) Devastating outbreak of bark beetles in the Czech Republic: drivers, impacts, and management implications. For Ecol Manag 490:119075.

Huang S, Liu H, Dahal D, Jin S, Welp LR, Liu J, Liu S (2013) Modeling spatially explicit fire impact on gross primary production in interior Alaska using satellite images coupled with eddy covariance. Remote Sens Environ 135:178–188

Huber C (2005) Long lasting nitrate leaching after bark beetle attack in the highlands of the Bavarian Forest National Park. J Environ Qual 34:1772–1779

Hunt LM, Haider W (2004) Aesthetic impacts of disturbances on selected boreal forested shorelines. For Sci 50:729–738

Jentsch A (2007) The challenge to restore processes in face of nonlinear dynamics – on the crucial role of disturbance regimes. Restor Ecol 15:334–339

Jentsch A, Friedrich S, Steinlein T, Beyschlag W, Nezadal W (2009) Assessing conservation action for substitution of missing dynamics on former military training areas in Central Europe. Restor Ecol 17:107–116

Jentsch A, Kreyling J, Elmer M, Gellesch E, Glaser B, Grant K, Hein R, Lara M, Mirzae H, Nadler SE, Nagy L, Otieno D, Pritsch K, Rascher U, Schädler M, Schloter M, Singh BK, Stadler J, Walter J, Wellstein C, Wöllecke J, Beierkuhnlein C (2011) Climate extremes initiate ecosystem-regulating functions while maintaining productivity. J Ecol 99:689–702

Joos O, Hagedorn F, Heim A, Gilgen AK, Schmidt MWI, Siegwolf RTW, Buchmann N (2010) Summer drought reduces total and litter-derived soil CO_2 effluxes in temperate grassland – clues from a 13C litter addition experiment. Biogeosciences 7:1031–1041

Kashian DM, Turner MG, Romme WH (2005) Variability in leaf area and stemwood increment along a 300-year lodgepole pine chronosequence. Ecosystems 8:48–61

Keeton WS, Whitman AA, McGee GC, Goodale CL (2011) Late-successional biomass development in northern hardwood-conifer forests of the northeastern United States. For Sci 57:489–505

Körner C (2003) Slow in, rapid out – carbon flux studies and Kyoto targets. Science 300:1242–1243

Kučerová A, Rektoris L, Štechová T, Bastl M (2008) Disturbances on a wooded raised bog – How windthrow, bark beetle and fire affect vegetation and soil water quality? Folia Geobot 43:49–67

Kurz WA, Dymond CC, Stinson G, Rampley GJ, Neilson ET, Carroll AL, Ebata T, Safranyik L (2008) Mountain pine beetle and forest carbon feedback to climate change. Nature 452:987–990

Lamy T, Liss KN, Gonzalez A, Bennett EM (2016) Landscape structure affects the provision of multiple ecosystem services. Environ Res Lett 11:124017

Lindroth A, Lagergren F, Grelle A, Klemedtsson L, Langvall O, Weslien P, Tuulik J (2009) Storms can cause Europe-wide reduction in forest carbon sink. Glob Chang Biol 15:346–355

McFarlane BL, Watson DOT (2008) Perceptions of ecological risk associated with mountain pine beetle (Dendroctonus ponderosae) infestations in Banff and Kootenay National Parks of Canada. Risk Anal 28:203–212

Mikkelson KM, Dickenson ERV, Maxwell RM, McCray JE, Sharp JO (2013) Water-quality impacts from climate-induced forest die-off. Nat Clim Chang 3:218–222

Mladenoff DJ (1987) Dynamics of nitrogen mineralization and nitrification in hemlock and hardwood treefall gaps. Ecology 68:1171–1180

Müller M, Job H (2009) Managing natural disturbance in protected areas: tourists' attitude towards the bark beetle in a German national park. Biol Conserv 142:375–383

Myers-Smith IH, Harden JW, Wilmking M, Fuller CC, McGuire AD, Chapin FS III (2008) Wetland succession in a permafrost collapse: interactions between fire and thermokarst. Biogeosciences 5:1273–1286

Pan Y, Birdsey RA, Fang J, Houghton R, Kauppi PE, Kurz WA, Phillips OL, Shvidenko A, Lewis SL, Canadell JG, Ciais P, Jackson RB, Pacala SW, McGuire AD, Piao S, Rautiainen A, Sitch S, Hayes D (2011) A large and persistent carbon sink in the world's forests. Science 333:988–993

Peters EB, Wythers KR, Bradford JB, Reich PB (2013) Influence of disturbance on temperate forest productivity. Ecosystems 16:95–110

Pollock JL, Favette J (2010) Stability in the patterns of long-term development and growth of the Canadian spruce-moss forest. J Biogeogr 37:1684–1697

Prestemon JP, Holmes TP (2004) Market dynamics and optimal timber salvage after a natural catastrophe. For Sci 50:495–511

Scharenbroch BC, Bockheim JG (2008) The effects of gap disturbance on nitrogen cycling and retention in late-successional northern hardwood-hemlock forests. Biogeochemistry 87:231–245

Scherer-Lorenzen M, Palmborg C, Prinz A, Schulze E-D (2003) The role of plant diversity and composition for nitrate leaching in grasslands. Ecology 84:1539–1552

Schmitt M, Bahn M, Wohlfahrt G, Tappeiner U, Cernusca A (2010) Land use affects the net ecosystem CO_2 exchange and its components in mountain grasslands. Biogeosciences 7:2297–2309

Schume H, Jost G, Hager H (2004) Soil water depletion and recharge patterns in mixed and pure forest stands of European beech and Norway spruce. J Hydrol 289:258–274

Schwitter R, Sandri A, Bebi P, Wohlgemuth T, Brang P (2015) Lehren aus Vivian für den Gebirgswald – im Hinblick auf den nächsten Sturm. Schweiz Z Forstwes 166:159–167

Seedre M, Chen HYH (2010) Carbon dynamics of aboveground live vegetation of boreal mixedwoods after wildfire and clear-cutting. Can J For Res 40:1862–1869

Seidl R, Blennow K (2012) Pervasive growth reduction in Norway spruce forests following wind disturbance. PLoS One 7:e33301

Seidl R, Rammer W, Jäger D, Lexer MJ (2008) Impact of bark beetle (*Ips typographus* L.) disturbance on timber production and carbon sequestration in different management strategies under climate change. For Ecol Manag 256:209–220

Seidl R, Schelhaas MJ, Rammer W, Verkerk PJ (2014) Increasing forest disturbances in Europe and their impact on carbon storage. Nat Clim Change 4:806–810

Sheik CS, Beasley WH, Elshahed MS, Zhou X, Luo Y, Krumholz LR (2011) Effect of warming and drought on grassland microbial communities. ISME J 5:1692

Silva Pedro M, Rammer W, Seidl R (2016) A disturbance-induced increase in tree species diversity facilitates forest productivity. Landsc Ecol 31:989–1004

Soussana J-F, Allard V, Pilegaard K, Ambus P, Amman C, Campbell C, Ceschia E, Clifton-Brown J, Czobel S, Domingues R, Flechard C, Fuhrer J, Hensen A, Horvath L, Jones M, Kasper G, Martin C, Nagy Z, Neftel A, Raschi A, Baronti S, Rees RM, Skiba UM, Stefani P, Manca G, Sutton M, Tuba Z, Valentini R (2007) Full accounting of the greenhouse gas (CO_2, N_2O, CH_4) budget of nine European grassland sites. Agric Ecosyst Environ 121:121–134

Stuart-Haëntjens E, De Boeck HJ, Lemoine NP, Mänd P, Kröel-Dulay G, Schmidt IK, Jentsch A, Stampfli A, Anderegg WRL, Bahn M, Kreyling J, Wohlgemuth T, Lloret F, Classen AT, Gough CM, Smith MD (2018) Mean annual precipitation predicts primary production resistance and resilience to extreme drought. Sci Total Environ 636:360–366

Stuber M, Bürgi M (2001) Agrarische Waldnutzungen in der Schweiz 1800–1950. Waldweide, Waldheu, Nadel-und Laubfutter I Agricultural use of forest in Switzerland 1800–1950. Wood pasture, wood hay collection, and the use of leaves and needles for fodder. Schweiz Z Forstwes 152:490–508

Thom D, Seidl R (2016) Natural disturbance impacts on ecosystem services and biodiversity in temperate and boreal forests. Biol Rev 91:760–781

Thom D, Rammer W, Seidl R (2017) The impact of future forest dynamics on climate: Interactive effects of changing vegetation and disturbance regimes. Ecol Monogr 87:665–684

Thom D, Golivets M, Edling L, Meigs GW, Gourevitch JD, Sonter L, Galford GL, Keeton WS (2019) The climate sensitivity of carbon, timber, and species richness covaries with forest age in boreal–temperate North America. Glob Chang Biol 25:2446–2458

Tubiello FN, Soussana J-F, Howden SM (2007) Crop and pasture response to climate change. Proc Natl Acad Sci USA 104:19686–19690

Turner MG, Donato DC, Romme WH (2013) Consequences of spatial heterogeneity for eco-system services in changing forest landscapes: priorities for future research. Landsc Ecol 28:1081–1097

Vanderwel MC, Coomes DA, Purves DW (2013) Quantifying variation in forest disturbance, and its effects on aboveground biomass dynamics, across the eastern United States. Glob Chang Biol 19:1504–1517

Wall DH, Bardgett RD, Behan-Pelletier V, Herrick JE, Jones H, Ritz K, Six J, Strong DR, van der Putten WH (2013) Soil ecology and ecosystem services. Oxford University Press, Oxford, p 424

Walter J, Grant K, Beierkuhnlein C, Kreyling J, Weber M, Jentsch A (2012) Increased rainfall variability reduces biomass and forage quality of temperate grassland largely independent of mowing frequency. Agric Ecosyst Environ 148:1–10

Wang C, Gower ST, Wang Y, Zhao H, Yan P, Bond-Lamberty BP (2001) The influence of fire on carbon distribution and net primary production of boreal *Larix gmelinii* forests in north-eastern China. Glob Chang Biol 7:719–730

Wohlgemuth T, Schwitter R, Bebi P, Sutter F, Brang P (2017) Post-windthrow management in protection forests of the Swiss Alps. Eur J For Res 136:1029–1040

Zamora R, Molina-Martínez JR, Herrera MA, Rodríguez Y, Silva F (2010) A model for wildfire prevention planning in game resources. Ecol Model 221:19–26

Zurbriggen N, Nabel JEMS, Teich M, Bebi P, Lischke H (2014) Explicit avalanche-forest feed-back simulations improve the performance of a coupled avalanche-forest model. Ecol Complex 17:56–66

Index

© The Author(s), under exclusive license to Springer Nature
Switzerland AG 2022
T. Wohlgemuth et al. (eds.), *Disturbance Ecology*, Landscape Series 32,
https://doi.org/10.1007/978-3-030-98756-5

Printed in the United States
by Baker & Taylor Publisher Services